Phthalonitrile Resins and Composites

PLASTICS DESIGN LIBRARY (PDL) PDL HANDBOOK SERIES

Series Editor: Sina Ebnesajjad, PhD (sina@FluoroConsultants.com)
President, FluoroConsultants Group, LLC Chadds Ford, PA, USA
www.FluoroConsultants.com

The **PDL Handbook Series** is aimed at a wide range of engineers and other professionals working in the plastics industry, and related sectors using plastics and adhesives.

PDL is a series of data books, reference works and practical guides covering plastics engineering, applications, processing, and manufacturing, and applied aspects of polymer science, elastomers and adhesives.

Recent titles in the series

The Effect of Sterilization Methods on Plastics and Elastomers, McKeen (9780128145111)
Polymeric Foams Structure-Property-Performance, Obi (ISBN: 9781455777556)
Technology and Applications of Polymers Derived from Biomass, Ashter (ISBN: 9780323511155)
Fluoropolymer Applications in the Chemical Processing Industries, 2e, Ebnesajjad & Khaladkar (ISBN: 9780323447164)
Reactive Polymers, 3e, Fink (ISBN: 9780128145098)
Service Life Prediction of Polymers and Plastics Exposed to Outdoor Weathering, White, White & Pickett, (ISBN: 9780323497763)
Polylactide Foams, Nofar & Park (ISBN: 9780128139912)
Designing Successful Products with Plastics, Maclean-Blevins (ISBN: 9780323445016)
Waste Management of Marine Plastics Debris, Niaounakis, (ISBN: 9780323443548)
Film Properties of Plastics and Elastomers, 4e, McKeen, (ISBN: 9780128132920)
Anticorrosive Rubber Lining, Chandrasekaran (ISBN: 9780323443715)
Shape-Memory Polymer Device Design Safranski & Griffis, (ISBN: 9780323377973)
A Guide to the Manufacture, Performance, and Potential of Plastics in Agriculture, Orzolek, (ISBN: 9780081021705)
Plastics in Medical Devices for Cardiovascular Applications, Padsalgikar, (ISBN: 9780323358859)
Industrial Applications of Renewable Plastics, Biron (ISBN: 9780323480659)
Permeability Properties of Plastics and Elastomers, 4e, McKeen, (ISBN: 9780323508599)
Expanded PTFE Applications Handbook, Ebnesajjad (ISBN: 9781437778557)
Applied Plastics Engineering Handbook, 2e, Kutz (ISBN: 9780323390408)
Modification of Polymer Properties, Jasso-Gastinel & Kenny (ISBN: 9780323443531)
The Science and Technology of Flexible Packaging, Morris (ISBN: 9780323242738)
Stretch Blow Molding, 3e, Brandau (ISBN: 9780323461771)
Chemical Resistance of Engineering Thermoplastics, Baur, Ruhrberg & Woishnis (ISBN: 9780323473576)
Chemical Resistance of Commodity Thermoplastics, Baur, Ruhrberg & Woishnis (ISBN: 9780323473583)
Color Trends and Selection for Product Design, Becker, (ISBN: 9780323393959)
Fluoroelastomers Handbook, 2e, Drobny (ISBN: 9780323394802)
Introduction to Bioplastics Engineering, Ashter (ISBN: 9780323393966)
Multilayer Flexible Packaging, 2e, Wagner, Jr. (ISBN: 9780323371001)
Fatigue and Tribological Properties of Plastics and Elastomers, 3e, McKeen (ISBN: 9780323442015)
Emerging Trends in Medical Plastic Engineering and Manufacturing, Schönberger & Hoffstetter (ISBN: 9780323370233)
Manufacturing and Novel Applications of Multilayer Polymer Films, Langhe & Ponting (ISBN: 9780323371254)

To submit a new book proposal for the series, or place an order, please contact Edward Payne, Acquisitions Editor at edward.payne@elsevier.com

Phthalonitrile Resins and Composites

Properties and Applications

Mehdi Derradji
Harbin Engineering University, Harbin, China
Military Polytechnic School, Algiers, Algeria

Jun Wang
Harbin Engineering University, Harbin, China

Wenbin Liu
Harbin Engineering University, Harbin, China

ELSEVIER
William Andrew
Applied Science Publishers

William Andrew is an imprint of Elsevier
The Boulevard, Langford Lane, Kidlington, Oxford, OX5 1GB, United Kingdom
50 Hampshire Street, 5th Floor, Cambridge, MA 02139, United States

British Library Cataloguing-in-Publication Data
A catalogue record for this book is available from the British Library

Library of Congress Cataloging-in-Publication Data
A catalog record for this book is available from the Library of Congress

ISBN: 978-0-12-812966-1

For Information on all William Andrew publications
visit our website at https://www.elsevier.com/books-and-journals

Publisher: Matthew Deans
Acquisition Editor: Edward Payne
Editorial Project Manager: Katie Chan
Production Project Manager: Anitha Sivaraj
Cover Designer: Christian J. Bilbow

Typeset by MPS Limited, Chennai, India

Contents

About the Authors ix
Preface xi

1 Introduction to the Phthalonitrile Resins and Their Curing Behavior 1
- 1.1 Overview and Historical Development of the Phthalonitrile Resins 1
- 1.2 Curing Behavior and Mechanism 5
  - 1.2.1 Curing Behavior/Monomer Structure Relation 5
  - 1.2.2 Curing Agents/Cure Behavior Relationship 7
  - 1.2.3 Self-catalyzed Phthalonitrile Monomers and Oligomers 29
- 1.3 Conclusions 50
- References 50

2 Phthalonitrile Resins' Properties 55
- 2.1 Introduction 55
- 2.2 Early Phthalonitrile Resins 56
- 2.3 First-Generation Phthalonitrile Resins 59
  - 2.3.1 Thermal Resistance Properties 60
  - 2.3.2 Mechanical and Thermomechanical Properties 71
  - 2.3.3 Water Absorption Properties 78
- 2.4 Second-Generation Phthalonitrile Resins 80
  - 2.4.1 Thermal Resistance Properties 82
  - 2.4.2 Mechanical and Thermomechanical Properties 93
  - 2.4.3 Water Absorption Properties 99
- 2.5 Actual and Potential Applications of the Phthalonitrile Resins 100
- 2.6 Conclusions 104
- References 104

3 X-Functional Phthalonitrile Monomers and Polymers 107
- 3.1 Introduction 107
- 3.2 X-Functional Phthalonitrile Monomers and Polymers 109
  - 3.2.1 Alkene-Functional Phthalonitrile 109

3.2.2 Alkyne-Functional Phthalonitrile 112
3.2.3 Imide-Functional Phthalonitrile 118
3.2.4 Bismaleimide-Functional Phthalonitrile 135
3.2.5 Novolac-Functional Phthalonitrile 136
3.2.6 Benzoxazine-Functional Phthalonitrile 140
3.2.7 Benzimidazole-Functional Phthalonitrile 151
3.2.8 Cyanate Ester-Functional Phthalonitrile 155
3.2.9 Naphthyl-Functional Phthalonitrile 158
3.2.10 Pyridine-Functional Phthalonitrile 161
3.2.11 Poly(Ether Ether Ketone)-Functional Phthalonitrile 164
3.2.12 Poly(Arylene Ether Nitrile)-Functional Phthalonitrile 166
3.3 Conclusions 170
References 170

4 Phthalonitrile-Based Blends and Copolymers 175
4.1 Introduction 175
4.2 Phthalonitrile Blends and Copolymers 176
4.2.1 Phthalonitrile/Phthalonitrile 176
4.2.2 Phthalonitrile/Epoxy 182
4.2.3 Phthalonitrile/Benzoxazine 197
4.2.4 Phthalonitrile/Novolac 210
4.2.5 Phthalonitrile/Bismaleimide 217
4.2.6 Phthalonitrile/Propargyl 222
4.2.7 Phthalonitrile/Poly(Arylene Ether Nitrile) 226
4.2.8 Phthalonitrile/Poly(Ether Ether Ketone) 231
4.2.9 Phthalonitrile/Carboxyl-Terminated Butadiene Acrylonitrile Rubber 233
4.3 Conclusions 235
References 235

5 Fiber-Reinforced Phthalonitrile Composites 241
5.1 Introduction 241
5.2 Carbon Fibers/Phthalonitrile Composites 243
5.2.1 Properties of Carbon Fibers 243
5.2.2 Carbon Fibers Reinforced Phthalonitrile Composites 247
5.3 Glass Fibers/Phthalonitrile Composites 270
5.3.1 Properties of Glass Fibers 270
5.3.2 Glass Fibers Reinforced Phthalonitrile Composites 274

5.4 Conclusions 291
References 291

6 Ceramic-Based Phthalonitrile Micro- and Nanocomposites 295
6.1 Introduction 295
6.2 Properties of the Studied Ceramic Fillers 299
6.3 Preparation Methods of the Ceramic-Based Phthalonitrile Hybrid Materials 301
6.3.1 Sol–Gel Processing 301
6.3.2 In Situ Polymerization 303
6.3.3 Solution Blending Technique 304
6.4 Surface Modification of Ceramic Particles 305
6.4.1 Chemical Treatments 305
6.4.2 Grafting of Synthetic Polymers 308
6.5 Properties of Ceramic-Based, Phthalonitrile Micro- and Nanocomposites 310
6.5.1 Properties of Phthalonitrile/SiC Microcomposites 310
6.5.2 Properties of Phthalonitrile/Si_3N_4 Nanocomposites 322
6.5.3 Properties of Phthalonitrile/Al_2O_3 Nanocomposites 328
6.5.4 Properties of Phthalonitrile/TiO_2 Nanocomposites 341
6.5.5 Properties of Phthalonitrile/ZnO Nanocomposites 348
6.5.6 Properties of Phthalonitrile/BN Nanocomposites 351
6.5.7 Properties of Phthalonitrile/ZrO_2 Nanocomposites 356
6.5.8 Properties of Phthalonitrile/MAX Phase Nanocomposites 362
6.6 Conclusions 369
References 369

Index 377

About the Authors

Dr. Mehdi Derradji is a research fellow at the Harbin Engineering University, Harbin, China and holds a teaching position at the Military Polytechnic School of Algiers, Algeria. Within the field of materials science and chemical engineering, his research interest includes high-performance polymers and composites, nanomaterials, computational chemistry and nuclear shielding.

Prof. Wenbin Liu gained his PhD in 2008 from Harbin Engineering University, Harbin, China, majoring in Materials' Science. He is a Professor at the College of Materials Science and Chemical Engineering, Harbin Engineering University, China. His research interest includes the synthesis, polymerization mechanism, and structure-property relationship of high-performance polymers and polymer matrix composites.

Prof. Jun Wang received her PhD in 2010 from Harbin Engineering University, Harbin, China, majoring in Materials' Science. She is a Professor at the College of Materials Science and Chemical Engineering, Harbin Engineering University, China. Her research interest includes the synthesis, curing kinetics, and properties of high-performance resins and related composites.

Preface

After decades of extensive research on phthalonitrile resins, it is a pleasure to see the publication of *Phthalonitrile Resins and Composites*, as the first book exclusively dedicated to this class of high-performance, polymeric materials. The main reason for writing this book was the perception that there is an exponential increase in the number of published papers and filed patents, causing any new researchers and those already in the field to fall behind or become overwhelmed. The book intends to gather, classify, and review all the works related to the phthalonitrile resins with emphasis on the major findings and the areas that need further investigation.

Demand for lightweight polymeric materials with metal equivalent performance continues to be increasingly important in many industrial sectors. Phthalonitrile polymers belong to the restricted family of high-performance thermosets satisfying the stringent requirements of exigent applications. The ease of their synthesis with huge molecular design flexibility allows tailor-made properties for a broad range of applications. The unique characteristics of phthalonitrile resins include a very low, A-stage viscosity, outstanding thermal resistance with starting decomposition temperatures over 450°C for typical phthalonitrile polymers, superior flame resistance, absence of a viscoelastic transition before thermal decomposition, low water uptake, advanced UV shielding, and excellent corrosion protective properties. Indeed, the phthalonitrile-based materials are particularly appealing for use in extreme environmental conditions.

The book contains six chapters and historical facts that led to the appearance of this class of materials are presented in Chapter 1, Introduction to the Phthalonitrile Resins and Their Curing Behavior. Since the development of phthalonitrile resins has been closely related to their curing behavior, this chapter scrutinizes the relation between curing behavior and other parameters, such as the monomers' structures and the type and amount of curing agent used. This chapter also aims to provide a broad perspective of the field and to demonstrate the importance of this new class of polymers. Chapter 2, Phthalonitrile Resins' Properties, discusses in greater detail the structure and properties of typical phthalonitrile polymers. The emergence of phthalonitrile resins as one of the leading, high-performance polymers with many exceptional properties, led many researchers to investigate their combination with other polymers as a part of a one-component system. The as-designed monomers, oligomers, and polymers, known as X-functional phthalonitrile materials, are discussed in Chapter 3, X-Functional Phthalonitrile Monomers and Polymers. Considering the great potential of phthalonitrile resins, many researchers investigated their blends and copolymers with various other polymeric systems. Thus, Chapter 4, Phthalonitrile-Based Blends and Copolymers, scrutinizes all the parameters influencing the performance of these materials and adds the most recent data to the subject. Chapter 5, Fiber-Reinforced Phthalonitrile Composites, reviews all of the studied, fiber-reinforced, phthalonitrile composites and highlights their exceptional properties. Finally, the recently developed, ceramic-based, micro- and nanocomposites are described in Chapter 6, Ceramic-Based Phthalonitrile Micro- and Nanocomposites.

Considering the many interesting features of phthalonitrile-based materials and the recent exponential increase in the number of researchers in the field, the chapters of this book were arranged to accommodate new topics as they should arise. Our hope is that this book will act as a catalyst for the future development and scientific progress of phthalonitrile science and technology by allowing an easy and effective access to the latest data and research.

1 Introduction to the Phthalonitrile Resins and Their Curing Behavior

1.1 Overview and Historical Development of the Phthalonitrile Resins

With the recent advances in the field of polymers, different high-performance polymeric materials have been proposed to meet the harsh and exigent applications in aerospace, aircraft, and naval industries. The replacement of metallic parts by polymer-based materials is cost effective because important weight reductions as well as energy saving are easily achieved. Various high-performance polymers are currently in use, such as epoxies, cyanate esters, and polyimides. However, these materials still present many shortcomings. For instance, the epoxy-based composites have a poor thermal resistance, weak toughness, and readily absorb water. Although cyanate ester resins are less hydrophilic than epoxies, the problems related to the hydrolytic stability can greatly weaken their strengths and glass transition temperatures (T_g). Bismaleimide resins have poor processability and their long-term oxidative stability is also limited. Materials derived from polyimides, such as the polyimides prepared by the polymerization of monomeric reactants (PMRs), are high temperature resistant but they also readily absorb water and have some engineering limitations. On the other hand, phthalonitrile (PN) resins, known for their combination of excellent properties, can bridge the gap between the currently used, high-performance resins and the metals/ceramics materials. The PN resins are advanced, high-performance polymeric materials characterized by a large panel of interesting physical and chemical properties. PN resins are recognized for their outstanding thermal resistance with initial decomposition temperatures over 450°C for the typical PN polymers. The good thermal stability of the PN resins is accompanied by an ameliorated processability, superior flame resistance, absence of a T_g before the thermal decompositions, low water uptake, advanced UV shielding, and excellent corrosion protective properties. Such a combination of properties promotes the PN resins over the traditional high-performance thermosets, such as epoxies, cyanate ester, bismaleimide, and polyimides in many applications.

Phthalonitrile Resins and Composites.
DOI: https://doi.org/10.1016/B978-0-12-812966-1.00001-9

The term PN, first introduced by Johannes Pinnow in 1896, refers to any organic structure derivative of benzene bearing two adjacent nitrile groups. In 1934, Dent and Linstead reported that the PN can easily be used as a precursor for the preparation of phthalocyanines, whose typical structure is presented in Fig. 1.1 [1–4]. The PN compounds have been readily combined with metallic copper at high temperature, offering copper phthalocyanines good with yields and no apparent reaction byproducts. The authors also reported that the copper phthalocyanines were exceptionally heat-resistant and can be sublimed at 580°C under nitrogen or carbon dioxide atmospheres.

Later in 1934, Linstead and Lowe, during the preparation of magnesium phthalocyanines, successfully isolated an acetic acid soluble byproduct analyzed to be a polymer of PN [5]. Molecular weight analyses confirmed that the compound was a trimer. The trimer of PN was the first-ever reported polymerization of a PN structure and it was the starting point for further investigations. Sixteen years later in 1950, Ross and Fineman showed an interest in the trimer and performed infrared and X-ray spectroscopies of the compound [6]. A full preparation procedure of the trimer and some of its properties were also reported. In 1958, Marvel and Martin attempted to prepare high molecular cross-linked polymeric phthalocyanines by reacting a tetracyanodiphenyle ether and PN with copper-bronze [7]. The co-condensation reaction only resulted in the formation of dimeric or trimeric compounds and no high molecular phthalocyanines were achieved.

In 1975, Griffith et al. at the naval research laboratory (NRL) were the first to introduce the term "Phthalonitrile resin" after an effective

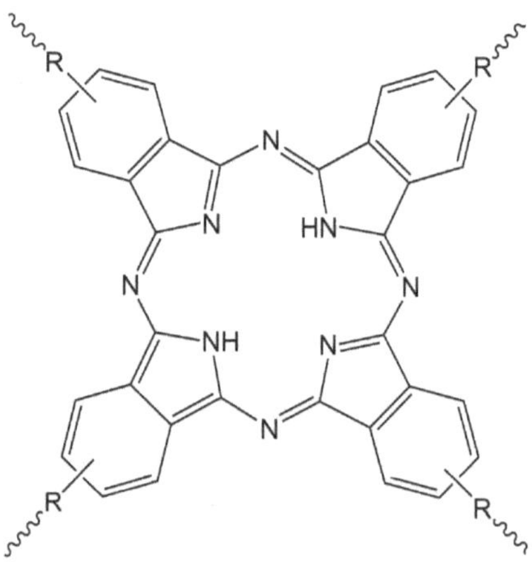

Figure 1.1 Phthalocyanine structure.

preparation of a high molecular weight phthalocyanine [8]. The general structure of the PN monomers is given in Fig. 1.2. The researchers estimated that the weak polymerization process reported by Marvel and Martin was essentially due to the sterical hindrance of the tetranitrile structure. They proposed the inclusion of flexible moieties between the PN ends in order to increase the freedom of movements of the chains. This explanation allowed for an orientation that included an end into the phthalocyanine structure. The authors also investigated the mechanical properties of the polymerized products and stated that the prepared PN polymers had better performances than certain epoxy resins. Encouraged by the obtained good results, Griffith et al. synthesized and investigated the properties of many bifunctional PN monomers obtained by varying the radical between the two PN ends [9–13]. The PN polymers were achieved after curing the monomers alone for a long time or 20%–25% faster with the inclusion of metals or metallic salts. The afforded PN polymers' properties differed according to the type of the radical, curing procedure, amount of the metal or metallic salt and the postcure conditions. Globally, all the polymers exhibited interesting results and further ameliorations were expected.

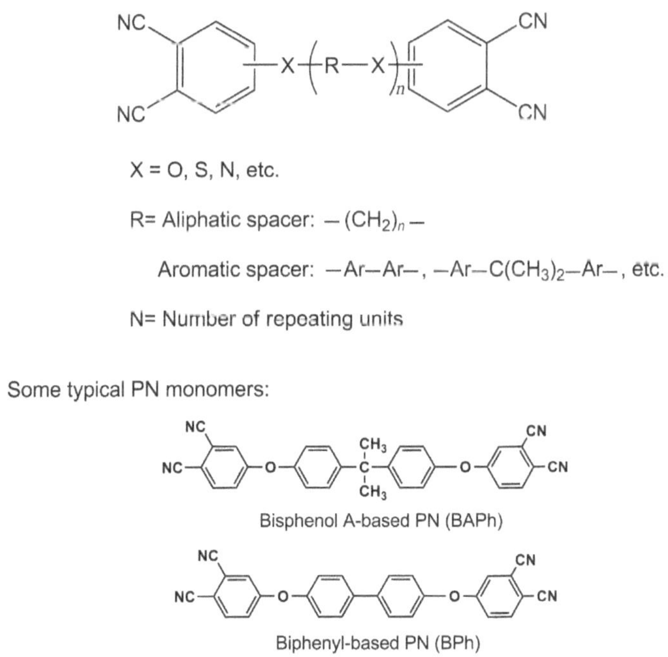

Figure 1.2 General structure of the PN monomers. *PN*, phthalonitrile.

In 1980, Keller introduced a new method for the synthesis of the PN monomers by a simple nucleophilic displacement of a nitro substituent activated by the cyano functions [14]. The new method allowed a cost effective preparation of multiple PN monomers containing different moieties, such as alkoxy or phenoxy linkages [15–20]. The new synthetic route also allowed the preparation of highly aromatized PN monomers. Meanwhile, their cured polymers were found to exhibit a long-term operational stability (in excess of 250°C) and a water absorption less than 1.5%. However, the curing of highly aromatized monomers was difficult to achieve, even with the addition of metals or metallic salts. The author attributed the curing difficulties to the rigidity of the aromatic spacers, determining that they greatly reduced the mobility and reactivity of the cyano groups. To overcome the sluggish curing of the PN monomers, Keller et al. first proposed the addition of bisphenol compounds to initiate the cyano addition reaction [21]. Although the used bisphenols effectively catalyzed the addition reaction, catastrophic reductions in the thermomechanical properties were noted. Further research at the NRL indicated that the addition of small amounts of aromatic amines could also effectively catalyze the cyano addition reaction without any changes in the mechanical and thermal properties [22]. Various types and amounts of amines were then applied and promising results were found. It is important to point out that the work performed by Keller and coworkers, especially the introduction of the amines curing agents, constituted the basis for all the future researches. After 1982, the number of patents and scientific articles exponentially increased and the PN resins started to emerge as a new class of high-performance polymers.

PN monomers can be classified by their functionality into monofunctional PN monomers, bifunctional PN monomers, and multifunctional PN monomers. The PN monomers can also be categorized according to their structures into either the first-generation resins or the oligomeric, second-generation PN resins. Another classification groups the PN monomers according to their ability to be cured with or without a curing agent; monomers cured without any catalyst are known as self-catalyzed PN monomers. A multitude of PN structures were designed to tailor the PN resins properties according to specific needs. The PN-based materials reached an advanced level of maturity and they are now used in certain applications. For instance, some PN resins were designated as the only candidate to satisfy the flame standards of the United States Navy (MIL-STD-2031) for use as the matrix resins in polymeric composites aboard naval submarines [23,24]. Moreover, some recently developed PN resins were measured to have excellent dielectric permittivity for possible use in radomes and other applications requiring radiofrequency transparency. A recent NASA technical report also indicated that the PN-based materials could be a good alternative to the currently used phenolic resins in aerospace applications [25]. Nowadays, some PN monomers and resins are commercialized by "Chromak Research" and "AKRON Polymer System," both companies can provide adapted solutions to different clients requirements.

This chapter will be exclusively concerned with the chronological developments in the curing behavior and processability of the PN monomers. The relationship between the curing behavior and different parameters, such as the monomers structures and the type and amount of the used curing agent, will be scrutinized. Moreover, the newly developed self-catalyzed PN monomers are introduced with examples of their curing behavior and processability.

1.2 Curing Behavior and Mechanism

1.2.1 Curing Behavior/Monomer Structure Relation

As stated in the introduction, the limited molecular weight obtained by Marvel and Martin was essentially due to the sterical hindrance of the tetranitrile used as a precursor for the phthalocyanine polymers. Therefore, Griffith et al. synthesized series of PN monomers bearing aliphatic moieties with different chain lengths varying between 4 and 18 carbon atoms [8]. The preferred synthetic route was the direct reaction between 4-aminophthalonitrile and the aliphatic diacide cholorides (see Scheme 1.1). The relationship between the aliphatic spacer length and the melting point is given in Fig. 1.3. Clearly, as the number of carbon atoms increases the melting point reduces. After C12, a plateau is reached and the melting point stabilizes around 170°C. The prepared PN monomers were fully transformed into phthalocyanine polymers with and without the use of a curing agent.

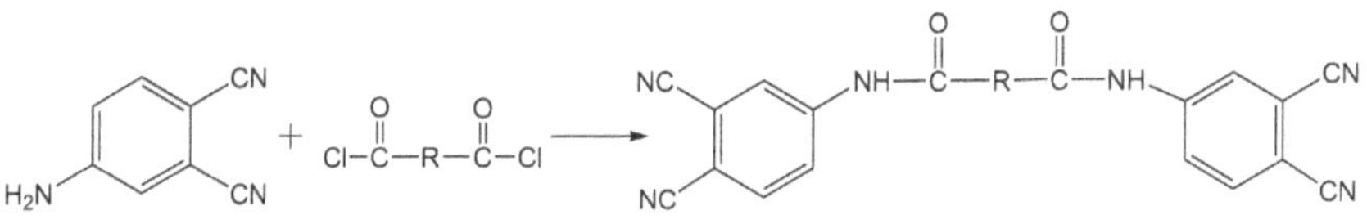

Scheme 1.1 Synthetic route of the PN monomers with aliphatic spacer. *PN*, phthalonitrile. Reprinted from Griffith JR, O'Rear JG, Walton TR. Phthalonitrile resins. Adv chem; Am Chem Soc 1975;142:458–64.

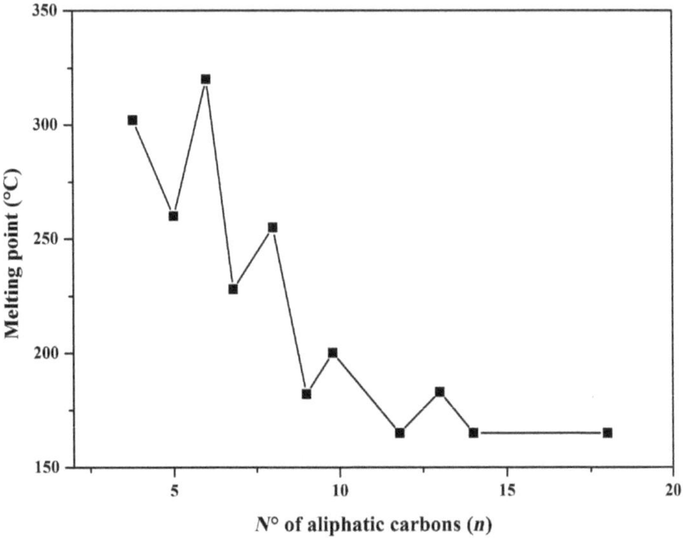

Figure 1.3 The relation between the aliphatic spacer length and the melting point. Reprinted from Griffith JR, O'Rear JG, Walton TR. Phthalonitrile resins. Adv chem; Am Chem Soc 1975;142:458–64.

Keller and Griffith also compared the reactivity of PN monomers containing aliphatic and aromatic spacers [18]. The monomers were prepared by the nucleophilic displacement of a nitro group activated by the cyano functions. An example route of preparation is presented in Scheme 1.2. The curing behavior of the aliphatic-based PN monomers depended upon the length of the chain. At a chosen temperature, longer spaced chains cured relatively quickly. Regardless of the temperature, the curing needed to be continued until solidification. Although the PN monomers containing aromatic linkages were prepared as easily as the aliphatic ones, the aromatic linkages' reactivity was lower. The curing was very sluggish and required elevated temperatures and increased time to afford fully

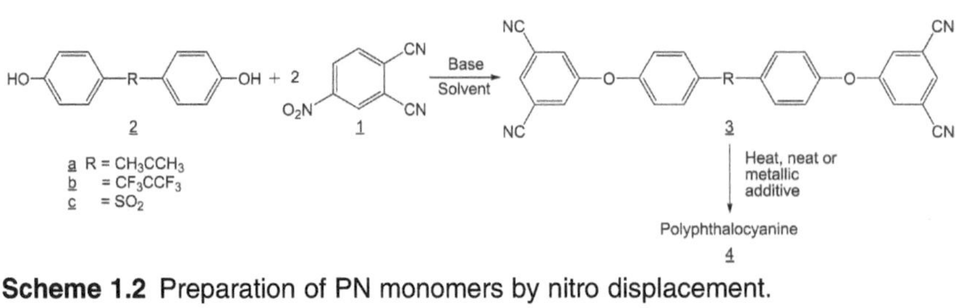

Scheme 1.2 Preparation of PN monomers by nitro displacement. *PN*, phthalonitrile.

cured polymers. For instance, the monomers were cured at temperatures higher than 260°C for several days with a nonstop mixing before an increase in the viscosity was noticed. In this section, we compare the effect of the monomers structures on the curing behavior. The different existing solutions to cure the highly aromatics PN monomers will be discussed later. Another general interesting feature of the PN monomers is their ability to be polymerized stepwise to different stages. Stopping the heating of any PN monomers at early viscosity changes is called a B-stage, the prepolymer can be cooled, grinded into fine powder, and indefinitely stored under ambient temperature without any changes in its stability or properties. This stable B-stage is also appreciated in the industry, since no constraints rise from the storage of the PN prepolymers. The prepolymers can be remelted and heated at any time to achieve the final solidification or C-stage. The curing of any PN monomers, without any catalyst, depends upon its melting point. The B-stage can be generally reached at a temperature from about 15°C to 20°C above the melting point. On the other hand, the C-stage curing temperatures are limited by the decomposition temperature of the monomers.

1.2.2 Curing Agents/Cure Behavior Relationship

1.2.2.1 Metals and Metallic Salts Curing Agents

Considering the curing difficulties of the first developed PN monomers, early researches attempted to use metals and their salts to accelerate and promote the polymerization reaction. Thus, the metals and metallic salts are the first known curing additives for the PN monomers. Many metals can be used, the most preferred chosen for the impact on the final properties of the polymer. These include copper, iron, zinc, and nickel. Other metals can be envisaged, such as chromium, molybdenum, vanadium, beryllium, silver, mercury, tin, lead, antimony, calcium, barium, manganese, magnesium, cobalt, palladium, and platinum. For the metallic salts, the stannous chloride is preferred over other possible metallic salts such as cuprous bromide, cuprous cyanide, cuprous ferri-cyanide, zinc chloride, zinc bromide, zinc iodide, zinc cyanide, zinc ferro-cyanide, zinc acetate, zinc sulfide, silver chloride, ferrous chloride, ferric chloride, ferrous ferri-cyanide, ferrous chloro-platinate, ferrous flouride, ferrous sulfate, cobaltous chloride, cobaltic sulfate, cobaltous cyanide, nickel chloride, nickel cyanide, nickel sulfate, nickel carbonate, stannic chloride, stannous chloride hydrate, and any mixture of the previous compounds

[16–20]. The stannous chloride catalyzes the polymerization reaction better than metals and the resulting polymers contain fewer voids. Moreover, the dispersion of this metallic salt was found to be better than any other metallic-based curing agent [13].

In this kind of activated polymerization, the amount of the curing agent plays a crucial role in the reaction's degree of conversion and the properties of the final product. Generally, the metals or their salts are added in a stoichiometric amount to a powdered or melted PN monomers. If the amount of the curing agent is less than the stoichiometric amount, the reaction will not be well coordinated with the metal or salt. If an amount more than necessary is added, the remaining metal and salt can attack the benzene ring in temperatures exceeding 280°C. Therefore, the reaction requires a high degree of purity and needs to be performed with a great care.

Globally, both metals and metallic salts reduce the curing temperature from 10°C to 20°C, without any changes in the curing rate or curing reaction speed at any given temperature. Moreover, the stannous chloride reduces the curing temperatures by 40°C. However, this curing agent needs to be first introduced as stannous chloride dehydrate as a melt or powder. If the PN monomers are solid powder, heating to the melting point or 20°C above it is required before the addition of the salt. In all instances, the dispersion problems along with the reduced thermo-oxidative stability obtained by the metals and metallic salts greatly limit their development. In 1984, an attempt was made by Snow et al. to improve the compatibility between the metals and the PN monomers by introducing an organic reducing agent such hydroquinone; however, large amounts of it did not get incorporated into the final structure [26]. Nowadays, these kinds of catalysts are considered as obsolete because more efficient curing agents were later found to afford better compatibility and more ameliorated PN polymers properties.

1.2.2.2 Phenols Curing Agents

The limited catalyzing effect of the metals and metallic salts brought researchers to look for other, more effective curing additives such as the phenolic compounds. The hydroxyl groups of the phenols can initiate the polymerization by attacking the nitrile groups of the PN monomers to produce an isoindoline structure. This structure can react with other nitrile groups to propagate the reaction. An example of this mechanism is presented in Scheme 1.3.

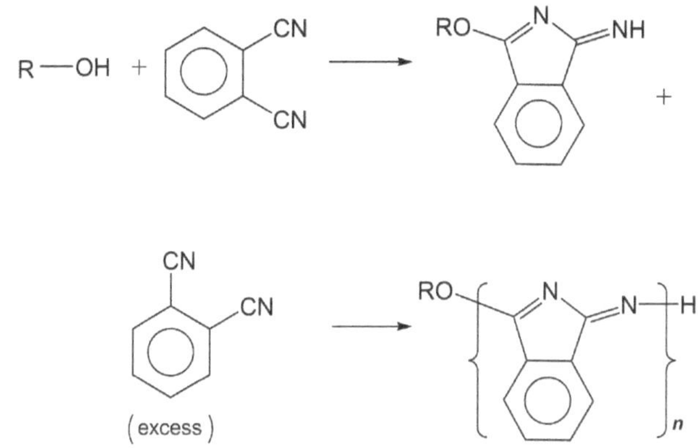

Scheme 1.3 Hydroxyl activated cyano addition reaction.

Aliphatic alcohols can also be considered, but the reduced reactivity of their hydroxyl groups compared to those of aromatic rings results in a lower catalyzing effect. Moreover, the aliphatic alcohols reduced boiling points and decomposition temperatures may cause voids within the cured resin that cannot be tolerated, especially if high mechanical properties are pursued. For some restricted applications where voids can be tolerated, any aliphatic alcohol or phenol can be used if enough of these compounds are added or reintroduced back into the reaction in a manner that compensates for the loss caused by the volatilization. However, the decomposition temperatures of the aliphatic alcohols or phenols need to be at least higher than the polymerization temperature of the PN monomers. If void free PN polymers are required, the boiling point and decomposition temperature of the phenols needs to be between 50°C and 75°C higher than the processing temperature of the PN monomers [27]. Examples of some phenols meeting the minimum temperature requirement are presented in Fig 1.4.

The curing of the PN monomers with phenols curing agents can be achieved either by solvent polymerization or heating polymerization. Solvent polymerization implies the following steps: dissolving the PN monomer in a specific solvent, adding the curing agent to form a suspension, and heating the suspension until B-stage is reached. After that, the solvent is removed by any suitable method prior to final curing to the C-stage. The solvent polymerization allows lower curing temperatures; however, the final product is expected to contain voids. The heat polymerization method consists of adding the phenol at temperatures between the melting point of the monomer and the decomposition temperature of the polymer. The mixture can be stopped and indefinitely stored at the B-Stage or directly completed to the C-stage [27].

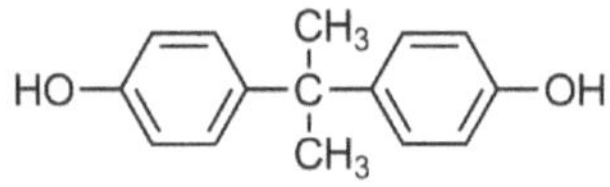

Bisphenol A

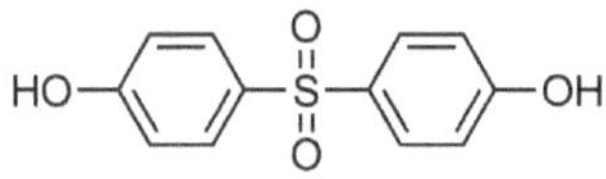

Bisphenol S

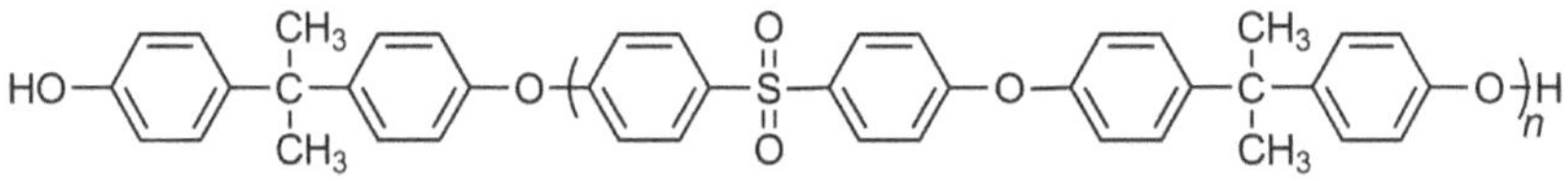

Bisphenol A polysulfone

Figure 1.4 Chemical structures of some phenolic curing agents.

Ting et al. investigated the effect of different amounts of bisphenol A (BPA) on the curing behavior and properties of a BPA-based PN resin (BAPh) (Fig. 1.2) [21]. The free hydroxyl groups of the BPA were found to first attack the nitrile groups forming isoindoline structures, which further react with other nitriles to form a cross-linked polymer structure. Differential scanning calorimetry (DSC) analyses revealed that the mixture of 90:10 molar ratio of the PN monomers and curing agent was the most suitable. The mixture investigated under DSC presented a melting point at 137°C indicating the presence of BPA. At the same time, the melting point of the PN monomers remained unchanged. Staging the mixture at 277°C for half an hour followed by a 15 min DSC isothermal scan resulted in the appearance of an exothermic peak, indicating the initiation of the polymerization process. The PN monomers and curing agent were cured at 210°C for 48 h and postcured at 250°C for 46 h. Thermogravimetric analysis (TGA) indicated that the postcured samples had a better thermal stability than that of the neat PN resins cured at 280°C for 6 days. An Fourier-transform infrared spectroscopy (FTIR) was used to assess the advancement of the polymerization by measuring the time required for 50% and 75% loss of cyano groups. The results indicated that the neat PN resin cured at 300°C had a 50% and 75% loss of cyano groups after 27 and 43 h, respectively. On the other hand, the 90:10 mixture sample cured at 210°C only exhibited the same losses at 7 and 12 h, respectively. Therefore, the curing agent's impact on reducing the time and temperature of curing is evident.

Chen et al. also prepared series of BPA-based PN prepolymers and polymers using BPA as a curing agent [28,29]. The polymerization reaction as well as the curing behavior were scrutinized in terms of the proportion of the curing agent and curing time. The PN monomers were dissolved in *N*-methylpyrrolidinone (NMP) solvent and heated for about 30 min. Then, different molecular ratios of BPA were added into the solution. The PN prepolymers were obtained after drying the mixtures at 150°C for 12 h in an air circulating oven. The PN polymers were obtained by heating the dried mixtures according the following cure schedule; 200°C for 2 h, 220°C for 2 h, 250°C for 4 h, and 300°C for 4 h. The resulting PN polymers, investigated under FTIR, revealed that the cross-linked structure was composed of phthalocyanine, triazine, and isoindoline in different proportions. Since the phthalocyanine was the only known possible structure, the authors performed a full investigation under UV-Vis, FTIR, and DSC analyses to better understand to reaction process and to determine the factors influencing the final structure type. Hence, the UV-Vis analysis was performed for the PN polymers at various amounts of BPA. All the spectra showed an absorption band between 370 and 380 nm related to the isoindole ring. The huge absorption at around 620–750 nm (Q band) was attributed to the phthalocyanine structure. It was also noticed that the prepolymers corresponding to the 2:1 and 2:1.5 molar ratios of BAPh to BPA possessed similar absorptions intensities, and the 2:0.5 and 2:2 prepolymers had lower Q band. Thus, in all the tested prepolymers, the isoindole ring and phthalocyanine structure were present.

By focusing upon the UV-Vis spectra of the 2:2 PN prepolymers at various polymerization reaction hours, it was noted that the absorption bands at around 370–380 nm and those between 620 and 750 nm exhibited the same trend. Therefore, the isoindoline and phthalocyanine structures that were depicted prior to the polymerization were also present after the starting of the cure reaction. Further investigation under FTIR for the PN prepolymers at different amounts of BPA revealed the presence of some characteristic bands. For instance, the bands at 1010, 1520, and 1660 cm^{-1} are related to the NH vibrations in the phthalocyanine structure, stretching vibration of the triazine ring, and the CN characteristic bond in isoindoline structure, respectively. The previous observations indicated that the polymerization reaction occurred between the PN monomers and the phenolic curing agent.

By analyzing the data from the FTIR spectra of the PN prepolymers after 4, 8, and 12 h of curing reactions, it seemed that increasing the curing time led to changes in the FTIR band's intensities. Precisely, the band at 2243 cm^{-1}, characteristic of the cyano group absorption, decreased while increasing the curing time. In fact, with the polymerization reaction going forward, the cyano groups are reduced due to the reaction propagation. Therefore, the intensity of this band gives important information about the degree of curing. Further FTIR investigations of the PN polymers at various amounts of BPA and at different curing times were also performed. The polymers containing fewer amount of curing agent did not display bands at around 1630–1670 cm^{-1}, which means that the isoindoline structure was not formed. The PN polymers of 2:1.5 and 2:2 clearly exhibited the bands related to isoindoline. Thus, it is understood that the isoindoline structure is closely related to the amount of the phenolic curing agent. On the other hand, all the polymers clearly showed the bands related to the phthalocyanine and triazine structures. In light of these facts, the authors proposed the polymerization mechanism presented in Scheme 1.4.

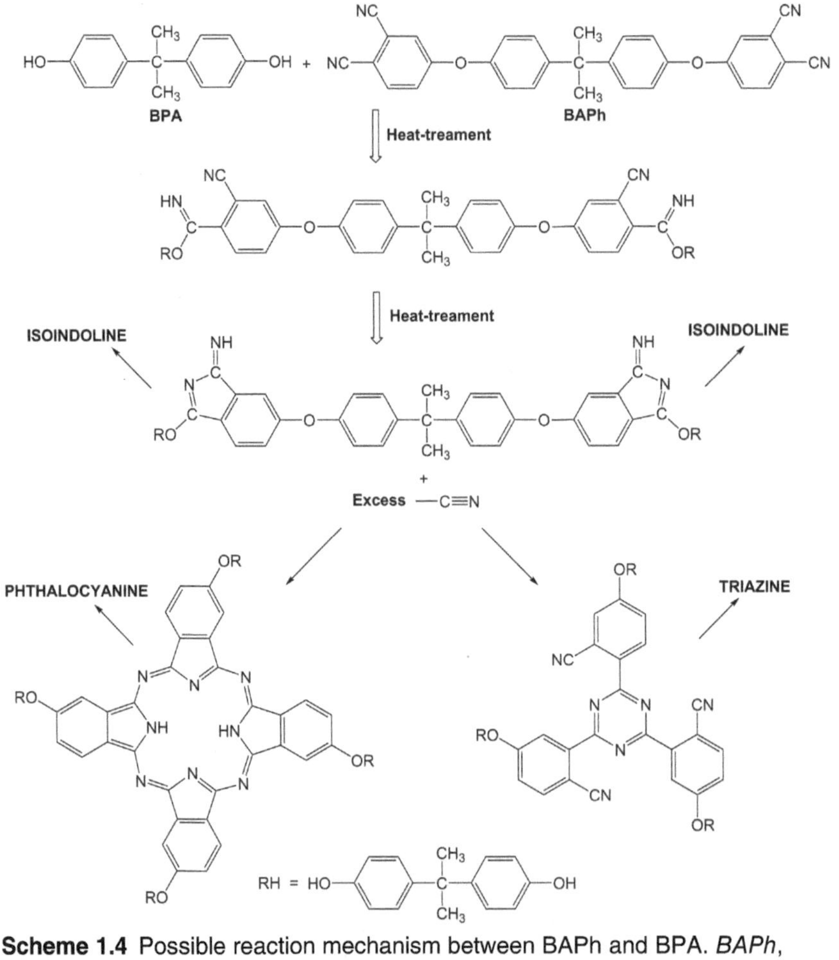

Scheme 1.4 Possible reaction mechanism between BAPh and BPA. *BAPh*, BPA-based PN resin; *BPA*, bisphenol A. Reprinted from Chen Z, Guo H, Tang H, Yang X, Xu M, Liu X. Preparation and properties of bisphenol A-based bis-phthalonitrile composite laminates. J Appl Polym Sci 2013;129:2621–8.

In addition to the spectroscopic test, DSC was employed to assess the curing behavior of the studied materials. The DSC curves of all the samples presented some similarities. For instance, all the samples showed two endothermic peaks and one small exothermic peak. The endothermic peaks were attributed to the melting points of the BPA and the PN monomers, respectively. The exothermic peak was due to the BPA activated polymerization. The heat of polymerization has been found to increase with the contents of the BPA. The processing window, which is the difference between the curing and the melting point temperatures, was also enlarged by more amount of curing agent. A large processing window is favorable for the industrial preparation of composite materials. The PN prepolymer of 2:1.5 was found to exhibit the lowest melting and curing temperatures and could be chosen as the preferred mixture for its favorable curing behavior.

The impact of the BPA on the processability of the PN monomers was investigated by rheometric analyses. The evolution of the complex viscosity of the PN prepolymers at different BPA amounts is presented in Fig. 1.5. The initial viscosities for the PN prepolymers were between 0.1 and 1 Pa.s. All the compositions showed an abrupt change in the complex viscosity at 215°C, 225°C, 250°C, and 290°C for the 2:2,

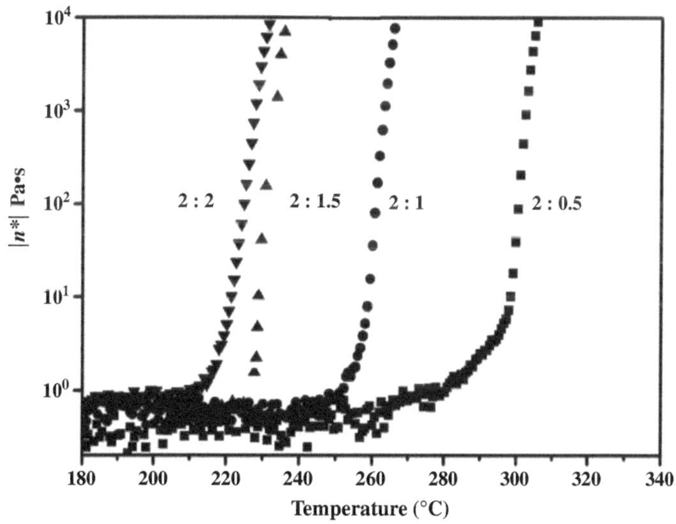

Figure 1.5 Complex viscosity variations of the PN prepolymers at different BPA amounts. *BPA*, bisphenol A; *PN*, phthalonitrile. Reprinted from Chen Z, Guo H, Tang H, Yang X, Xu M, Liu X. Preparation and properties of bisphenol A-based bis-phthalonitrile composite laminates. J Appl Polym Sci 2013;129:2621–8.

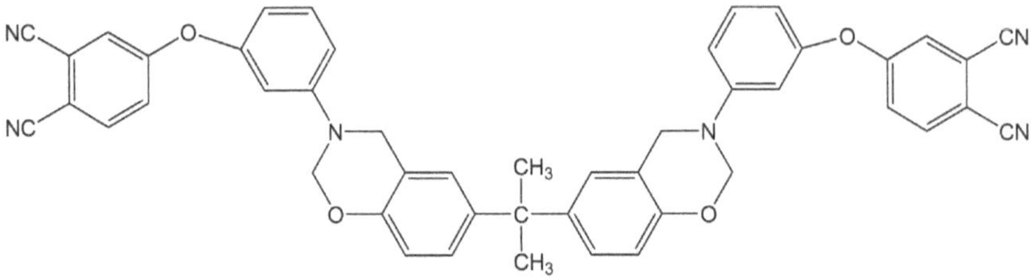

Figure 1.6 Chemical structure of the PN monomer bearing benzoxazine moiety. *PN*, phthalonitrile. Reprinted from Xu M, Jia K, Liu X. Effect of bisphenol-A on the structures and properties of phthalonitrile-based resin containing benzoxazine. Express Polym Lett 2015;9.

2:1.5, 2:1, and 2:0.5, respectively. It also appeared that the 2:2 mixture complex viscosities increased at lower temperatures compared to the other PN prepolymers. Thus, increasing the amount of the curing agent led to relatively lower complex viscosities. In summary, all the prepolymers exhibited ameliorated processability at moderate temperatures.

Xu et al. also reported the effect of BPA on the curing behavior and processability of an interesting PN monomer bearing benzoxazine moiety [30]. The chemical structure of the studied benzoxazine-functional PN (BZPN) monomer is presented in Fig. 1.6. The processability and curing behavior of these PN monomers was found to be greatly enhanced by adding more BPA contents. For more clarifications, the investigations of the curing behavior under DSC are presented in Fig. 1.7. Fig. 1.7A presents the DSC curves of both BZPN monomers and BPA. The BZPN monomers exhibited a small endothermic transition at around 75°C and two exothermic peaks at 230°C and 263°C. The first exothermic peak was related to the ring opening of the oxazine ring of the benzoxazine moieties and the second one was attributed to the cyano addition of the nitrile groups. The impacts of different amounts of BPA on the curing behavior are shown in Fig. 1.7B. The addition, the phenolic curing agent shifted the exothermic peaks to lower temperatures. Lower exothermic peaks mean great opportunities to cure the PN monomers at reduced temperatures.

Following the same path, Luo et al. scrutinized the effect of an *ortho*-diallyl BPA (DBPA) on the curing behavior and processability of the same PN monomer described in Fig. 1.6 [31]. The DSC investigations revealed better results than those obtained with the BPA. As a matter of fact, the introduction of the DBPA lowered the melting temperature of the mixture by 45°C. At the same time, the exothermic transition positions were also markedly lowered. However, when an

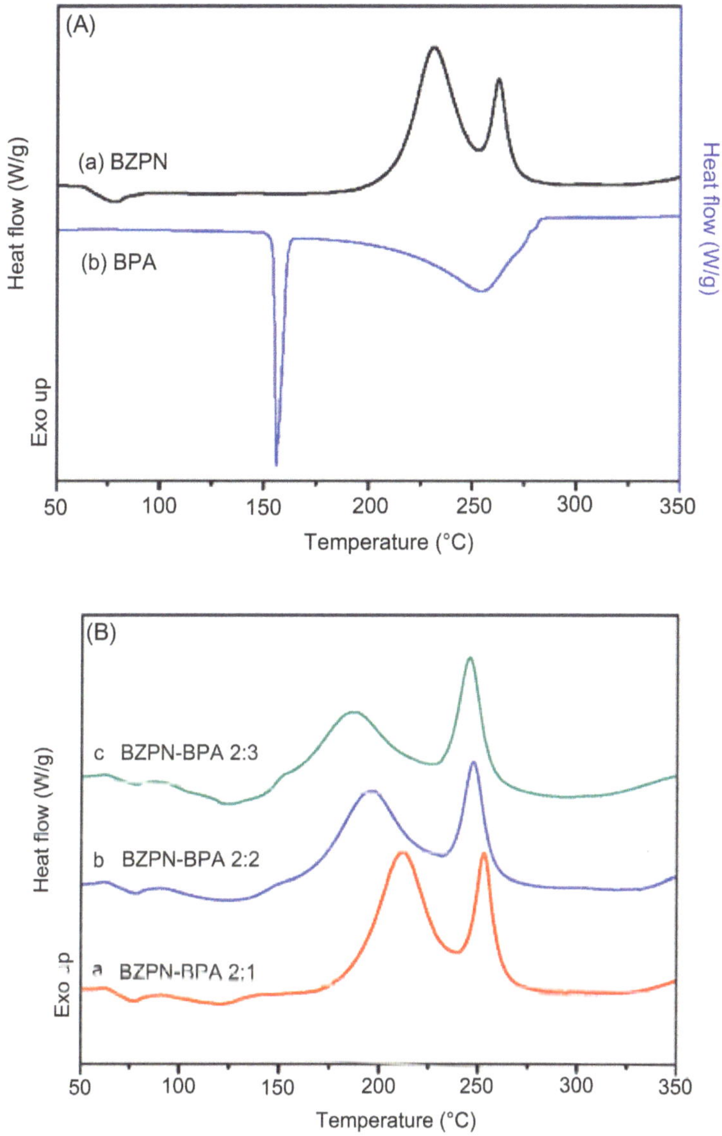

Figure 1.7 (A) DSC curves of both BZPN and BPA; (B) impacts of different amounts of BPA on the BZPN curing behavior. *DSC*, differential scanning calorimetry; *BPA*, bisphenol A; *BZPN*, benzoxazine-functional PN. Reprinted from Xu M, Jia K, Liu X. Effect of bisphenol-A on the structures and properties of phthalonitrile-based resin containing benzoxazine. Express Polym Lett 2015;9.

excessive amount of the curing agent was added, the exothermic peaks were found to increase. This can be attributed to an excessive amount of the curing agent diluting the PN monomers and delaying the polymerization reaction.

1.2.2.3 Organic Amine-Curing Agents

The use of organic amines to cure the PN monomers is the idea behind the development of the PN-based materials. The extremely stubborn curing behavior of the PN monomers, especially those containing a substantial proportion of aromatic rings, can be easily overcome by adding some organic amine additives [22]. Primary amines greatly increase the polymerization reaction of the highly aromatic PN monomers. The amines first attack the nitrile groups to produce *N*-substituted-3-iminoisoindolenine, then the reaction propagates to the other nitrile groups [32]. The reaction mechanism is presented in Scheme 1.5. Secondary and tertiary amines are less suitable because the steric hindrances of these amines produce a weaker catalyzing effect. The choice of the suitable primary amine is only limited by its boiling point or its decomposition temperature. If void can be tolerated in the cured polymers, a great number of amines can be used. For void free applications, the decomposition temperature of the amine needs to be at least 50°C higher than the polymerization temperature. The amines that cannot be used as curing agents are those with decomposition temperatures lower than the curing temperature of the polymer. If the amine starts to volatilize during the polymerization reaction, an equivalent quantity needs to be reintroduced into the reaction to assure fully polymerized materials. Some of the amines meeting these minimum requirements are meta and para alkylene dianiline, where the alkylene group has from 1 to 10 carbon atoms, alkylamine, where the alkyl group has from 15 to 25 carbon atoms, and aminophenylsulfone. Some amines from this group are preferred, such as meta and para phenylenediamine, meta and para methylene dianiline, 4-aminophenyl ether, 4-aminophthalonitrile, 1,4-cyclohexanebis(methylamine), and 4-aminophenylsulfone [32–36]. For void free applications and/or for curing highly aromatic PN monomers, 4-aminophthalonitrile and 4-aminophenylsulfone are privileged over the other amines due to their high boiling points and decomposition temperatures, low vapor pressure, and low degree of reactivity at reduced temperatures which allow their incorporation to the PN monomers at large amounts, up to 20 wt.% [32].

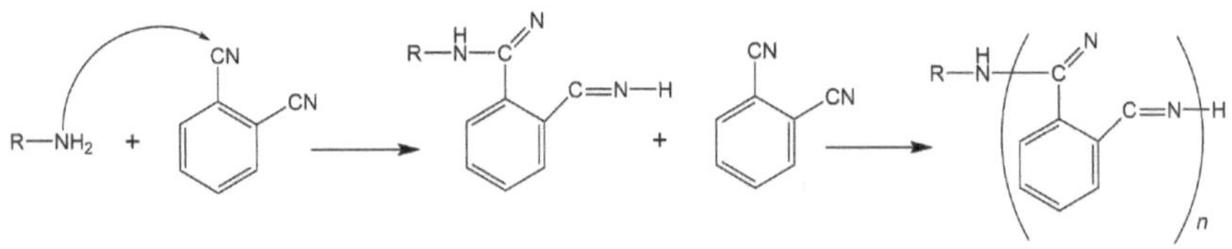

Scheme 1.5 Amine activated cyano addition reaction.

The curing of the PN monomers with amine-curing agents also can be achieved by solvent polymerization or heating polymerization. Solvent polymerization implies the following steps: dissolving the PN monomer in a specific solvent, adding the curing agent to form a suspension, and heating the suspension until B-stage is reached. After that, the solvent is removed by any suitable method prior to final curing at the C-stage. The solvent polymerization allows lower curing temperatures; however, the final product is expected to contain voids. The heat polymerization method consists of adding the phenol at temperatures between the melting point of the monomer and the decomposition temperature of the polymer. The mixture can be stopped and indefinitely stored at the B-Stage or fully completed at the C-stage.

The required amount of amines to be added to the PN monomers depends on the type of the amine and the amount of voids tolerated in the application. It is important to point out that the amine reactivity will determine the ideal amount to be blended with the PN monomers. If void free materials are needed, the possible amines to be used are categorized in three groups [32]. The first group contains 4-aminophthalonitrile and 4-aminophenylsulfone to be added from 1 to 20 wt.% and preferably from 5 to 15 wt.%. The second group includes cyclohexane bis(methylamine) and hexadecylamine generally added from 1 to 15 wt.% and more precisely from 3 to 10 wt.%. The third group contains the most reactive amines to be added from 1 to 15 wt.% and preferably from 3 to 10 wt.%. Such reactive amines include meta and para methylene dianiline, meta and para phenylenediamine and amino phenyl ether, and 4,4′-(*p*-phenylenedioxy)dianiline. Beside the amount of amine, the processing temperatures play a key role in obtaining, or not, void free materials. If the voids can be allowed in the polymer, the amount to be added can reach up to 40 wt.%. The maximum quantity is limited by the stoichiometric amount. If more than necessary amount is added, the unreacted amine can affect the physical properties of the cured PN resin. If the amine amount is less than 1 wt.%, the reaction may not be effectively catalyzed and the improvement in the curing rate and processability would be minimal.

Although the previous organic amines can afford acceptable results, further researches indicated that void problems still occur for amounts greater than 5 wt.%. Therefore, most of the efforts were focused on developing new kind of organic amines. Thus, in 1991, Keller reported on the preparation of an advanced amine that can cure some PN monomers at temperatures lower than their melting points [37]. The new organic amine is known as 1,3-bis(3-aminophenoxy)benzene (*m*-APB) (see Fig. 1.8). This compound makes it possible to cure the PN

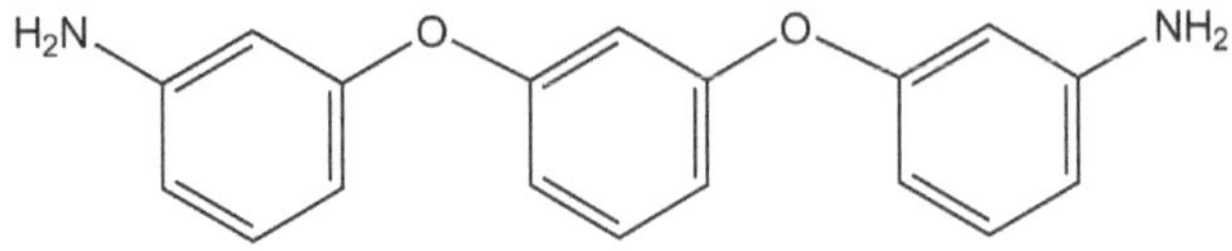

1,3-bis(3-aminophenoxy)benzene (*m*-APB)

Figure 1.8 Structure of 1,3-bis(3-aminophenoxy)benzene (*m*-APB).

monomers at reduced temperatures and time. Moreover, this amine-curing agent markedly decreased the formation of voids during the polymerization process and the resulting PN polymers exhibited ameliorated physical properties. Indeed, the APB curing agent remained the most preferred amine-curing agent for several years [38–42].

Burchill investigated the effect of *m*-APB of the final cross-linked structure of 4,4′-bis(3,4-dicyanophenoxy)biphenyl-based PN (BPh) polymer [43]. The FTIR results showed that during the curing process new absorptions at 1520 and 1357 cm^{-1} were noticed along with a decrease in the nitrile absorptions at 2230 cm^{-1}. The author attributed these changes to the formation of the triazine structure. Further curing the PN monomers with *m*-APB at 240°C for 24 h followed by a post-cure at 315°C for 24 h seemed to slightly affect the nitrile intensity which remained at about 30% of its original value. On the other hand, the effect of the postcure was an increase in the bands related to the triazine. Evidences of the presence of triazine in the early stage of polymerization were also found by comparing the FTIR spectra before, during, and after the polymerization. Thus, the triazine structure was confirmed to be the main structure forming this PN polymer.

Sastri et al. during the preparation of carbon fiber/PN composites also studied the effect of *m*-APB on the curing behavior and processability of the BPh resin [23]. First, a DSC scan of the PN monomers revealed a clear and well-defined endothermic transition resulting from the BPh monomers melting. Investigations on the BPh prepolymers under UV-Vis indicated the presence of several phthalocyanine related absorptions such as those at 609.7, 639.3, 667.9, and 701.5 nm. The T_g of the prepolymers were found to be somewhere between 70°C and 85°C, depending on the amount of the *m*-APB. The DSC curve of the BPh prepolymers showed a small exotherm at 250°C related to the initial reaction between the amine and the PN monomers and the polymerization reaction performed at this temperature. This progressed at a very slow rate.

The curing reaction was also monitored by FTIR spectroscopy. The FTIR spectra of the BPh monomers and prepolymers were compared with those of the BPh polymers obtained after 6 and 24 h of curing at 250°C to get some insight about the curing mechanism. The bands of triazine were depicted at 1520 and 1360 cm^{-1} and their intensities depended on the curing time. The small shift in the nitrile absorption position was also noticed, such phenomena are characteristic of the triazine structure formation. In addition, the bands at 3290 and 1010 cm^{-1} attributed to the phthalocyanine were also depicted from the early stages of the polymerization reaction. These observations correlate with those found by other researchers who state that the organic amines lead to a final polymer structure principally formed by phthalocyanine and triazine.

Rheometric studies were performed to investigate the BPh prepolymers processability at different times and temperatures during the polymerization reaction. The melt viscosity changes were monitored as a function of the curing agent amount. The initial melt viscosities for all the mixtures ranged from 0.05 to 0.3 Pa.s. Interestingly, lower initial melt viscosity is pursued in the industry, especially for some processing techniques such as resin transfer molding (RTM), vacuum assisted resin transfer molding (VARTM), and filament winding. At low amine amounts (1.5–1.7 wt.%), the viscosity gradually increased from 0.05 to 0.25 Pa.s over a period of 2 h. When a high amount of *m*-APB was added (up to 2.3 wt.%), the viscosity was also greatly increased. For instance, after heating the prepolymers containing 2.1 and 2.3 wt.% *m*-APB for 1 h, viscosity increased up to forty fold. Further addition of the curing agent resulted in glassy thermosets after 5 min of reaction time at 250°C.

The state of purity of the PN monomers can also affect the processability of the PN prepolymers. As previously discussed, the preparation of the typical PN monomers involved the use of a bisphenol to execute the nucleophilic displacement of a nitro substituent from 4-nitrophthalonitrile. The unreacted bisphenol in form of impurities may initiate the polymerization reaction even without adding the amine-curing agent. The effect of the phenolic impurities of the processability was also studied by rheometric analyses. Results indicated that the increase in the melt viscosity for the pure PN prepolymers was slow. However, if the PN prepolymers were prepared from monomers containing phenolic impurities, the increase in the viscosity was faster. The reason is that the phenolic compounds are also a class of curing agents for the PN monomers and their presence even in form of impurities can add a complementary catalyzing effect resulting in changes in the melt viscosity. In the lights of these facts, it becomes important to

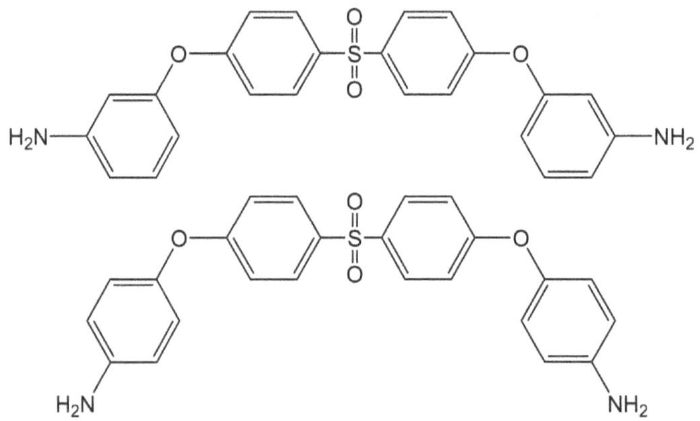

Figure 1.9 Chemical structure of two sulfone-containing diamines curing agents (*m*-BAPS and *p*-BAPS).

eliminate all form of impurities to control the polymerization process. Some methods to afford PN monomers with a high degree of purity are accomplished by precipitating the PN products into a dilute basic solution or by performing multiple recrystallizations from acetonitrile.

Although the APB curing agent offers good results, there are some concerns regarding the loss of the amine from volatility impact on the polymerization process. This problem is more accentuated when thick sections are cured in open molds. Therefore, there was a need to develop amines with higher boiling points and decomposition temperatures. Moreover, improving the processability of the PN prepolymers by reducing the initial melt viscosities and enlarging the processing windows are important challenges to be addressed because not all high-performance polymeric materials can be amenable to processing by RTM, VARTM, or filament winding. In an attempt to overcome these limitations, Sastri et al. suggested the use of two sulfone-containing diamines (*m*-BAPS and *p*-BAPS) as an ameliorated amine-curing agents (see Fig. 1.9) [44].

In contrast with APB, these two compounds have lower volatilities and higher molecular weights. The thermal stability of the aromatic diamines was investigated by TGA and the results are plotted in Fig. 1.10. The *m*-APB extensively used as curing agent is thermally stable at around 250°C and it loses 10–12 wt.% at about 300°C. Since the PN polymerization reaction is accomplished at temperatures in excess of 250°C, the problem of volatility will certainly affect the cure reaction. On the other hand, both *m*- and *p*-BAPS exhibited enhanced thermal stability with initial decomposition temperatures above 350°C. Therefore, the volatility problems with these two compounds are not in question.

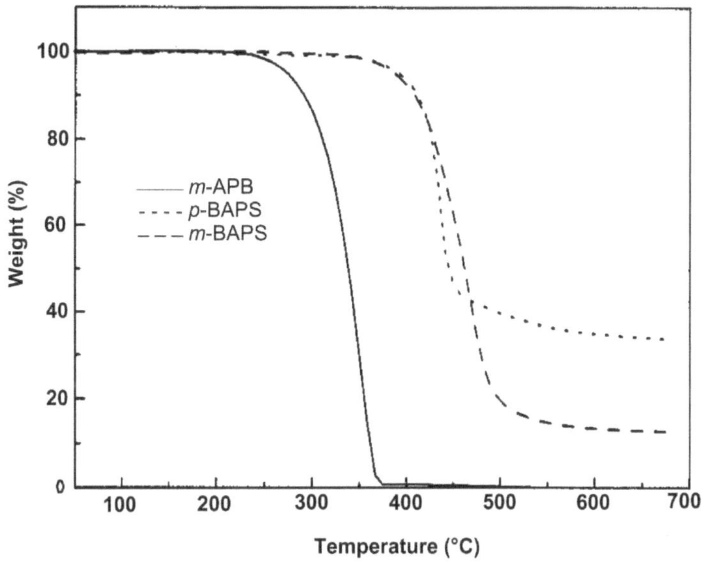

Figure 1.10 Thermal stability of the aromatic diamines investigated by TGA. *TGA*, thermogravimetric analysis. Reprinted from Sastri SB, Keller TM. Phthalonitrile cure reaction with aromatic diamines. J Polym Sci: Part A: Polym Chem 1998;36:1885–90.

The DSC curve of the BPh monomers with 4 wt.% of *p*-BAPS is presented in Fig. 1.11. Two endothermic transitions were depicted at 186 and at around 236°C related to the melting of the amine and PN monomers, respectively. A small exothermic transition at 253°C was attributed to the initial reaction exotherm between the amine and the PN monomers. Previous DSC investigations of the APB with the same PN monomers showed similar exothermic transition positions. The curing reaction was also found to progress at a very slow rate, since the original baseline following the initial exotherm was not attained.

In terms of processability, the melt viscosity of the PN prepolymers containing APB was proven to increase at a faster rate than that of the PN prepolymers prepared from an equivalent amount of BAPS. The evolution of the melt viscosity with time at 260°C for the BPh prepolymers prepared with the same amount of different amines is shown in Fig. 1.12. When the diamines were compared in terms of their isomeric conformations, it appeared that *p*-APB and *m*-BAPS affected the cure at a faster rate. A plausible explanation for this difference in reactivity may be correlated to the dissimilarity in the basicity of the amines. The chemical structure of the BAPS indicated a lower basicity due to the presence of the electron-withdrawing sulfone group compared to the APB which has an electron-donating phenoxy group. Moreover the

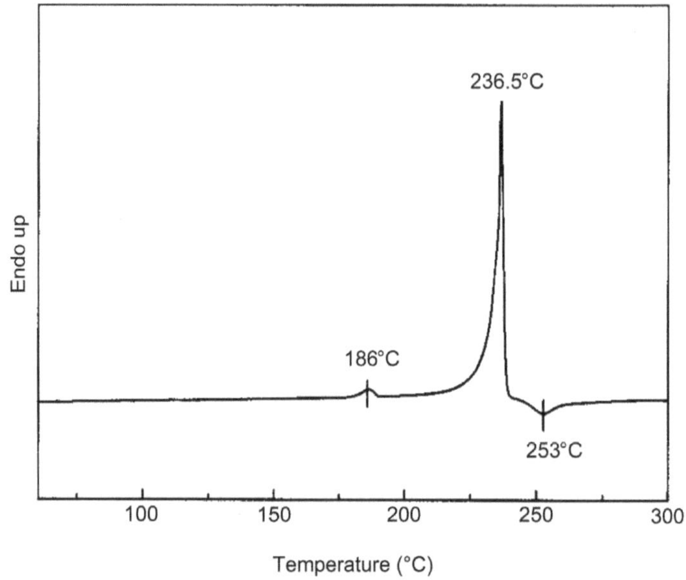

Figure 1.11 DSC curve of the BPh monomers with 4 wt.% of *p*-BAPS. *DSC*, differential scanning calorimetry; *BPh*, 4,40-bis(3,4-dicyanophenoxy)biphenyl-based PN. Reprinted from Sastri SB, Keller TM. Phthalonitrile cure reaction with aromatic diamines. J Polym Sci: Part A: Polym Chem 1998;36:1885–90.

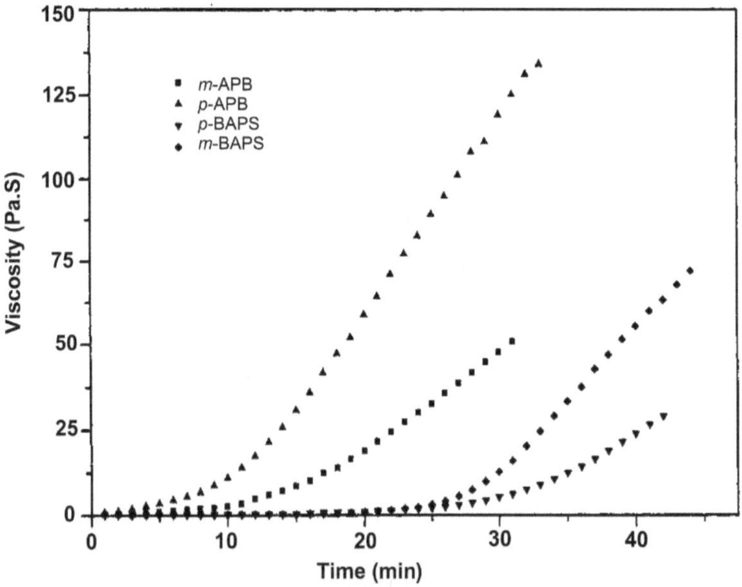

Figure 1.12 Evolution of the melt viscosity over time at 260°C for PN prepolymers prepared with the same amount of different amines. *PN*, phthalonitrile. Reprinted from Sastri SB, Keller TM. Phthalonitrile cure reaction with aromatic diamines. J Polym Sci: Part A: Polym Chem 1998;36:1885–90.

decrease in the basicity is more accentuated in the para position than in the meta position for both APB and BAPS. Based on the previous explanations, the *p*-BAPS was chosen as the preferred amine-curing agent to slow down the curing and enlarge the processing window.

Keller et al. also compared the effect of *m*-APB and *p*-BAPS on the curing behavior of resorcinol-based PN (RPh) polymers [45]. The preparation route and chemical structure of the RPh monomer are presented in Scheme 1.6. DSC analyses confirmed the presence of three transitions for each case; two endothermic transitions related to the melting of the amine and PN monomers and one exothermic transition attributed to the curing reaction. In both cases, adding more contents of the amine shifted the exothermic transition to lower temperatures. In addition, large processing windows, between 70°C and 100°C, were also noticed for both curing agents with a small advantage for the *p*-BAPS over *m*-APB.

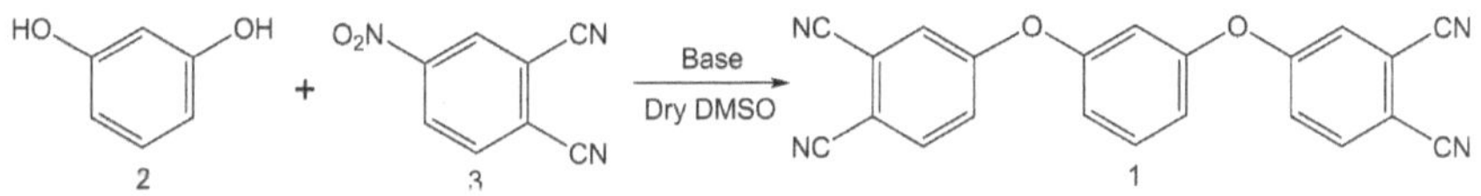

Scheme 1.6 The preparation route and chemical structure of the resorcinol-based PN (RPh) monomer. Reprinted from Keller TM, Dominguez DD. High temperature resorcinol-based phthalonitrile polymer. Polymer 2005;46:4614–18.

Seeking to overcome the volatility issues of the APB curing agents, Liu et al. proposed to promote the curing behavior and processability of a PN oligomer containing biphenyl ethernitrile using three new amine-curing agents: 4-aminophenoxyphthalonitrile (4-APN), 2,6-bis (4-diaminobenzoxy) benzonitrile (BDB), and 4,4′-diaminediphenyl sulfone (DDS) [46]. The chemical structures of the PN oligomer and the amines are presented in Fig. 1.13.

The impact of the amines on the curing behavior was monitored by DSC. Fig. 1.14 shows the DSC curves of the PN oligomers and their subsequent prepolymers prepared from various amine-curing agents. As clearly seen from these curves, the PN oligomers only exhibited one endothermic transition and no exotherm was depicted. This confirms the challenging curing character of these PN oligomers. On the other hand, all the prepolymers showed well-defined exothermic peaks related to the polymerization process. The PN prepolymers containing DDS had the largest processing window followed by 4-APN and BDB, respectively.

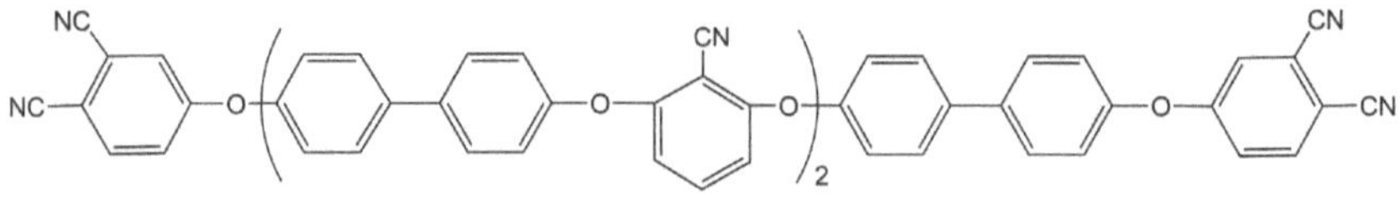

2PEN-BPh

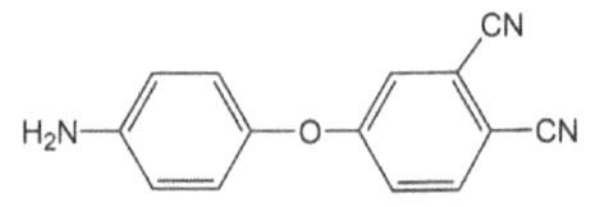

4-aminophenoxyphthalonitrile (4-APN)

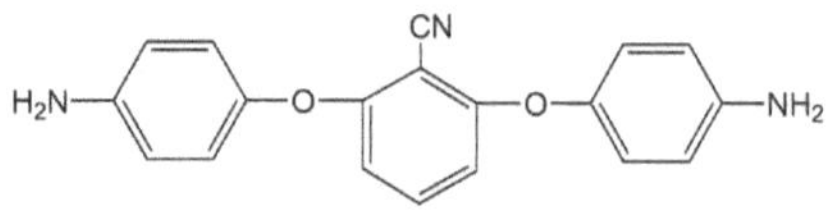

2,6-bis(4-aminobenzoxy)benzonitrile (BDB)

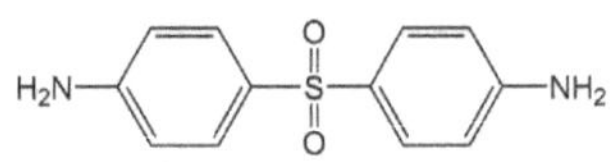

4,4′-diaminediphenyl sulfone (DDS)

Figure 1.13 Chemical structure of the PN oligomer containing biphenyl ethernitrile and the three amines curing agents used by Liu et al. Reprinted from Yang X, Zhang J, Lei Y, Zhong J, Liu X. Effect of different aromatic amines on the crosslinking behavior and thermal properties of phthalonitrile oligomer containing biphenyl ethernitrile. J Appl Polym Sci 2011;121:2331–7.

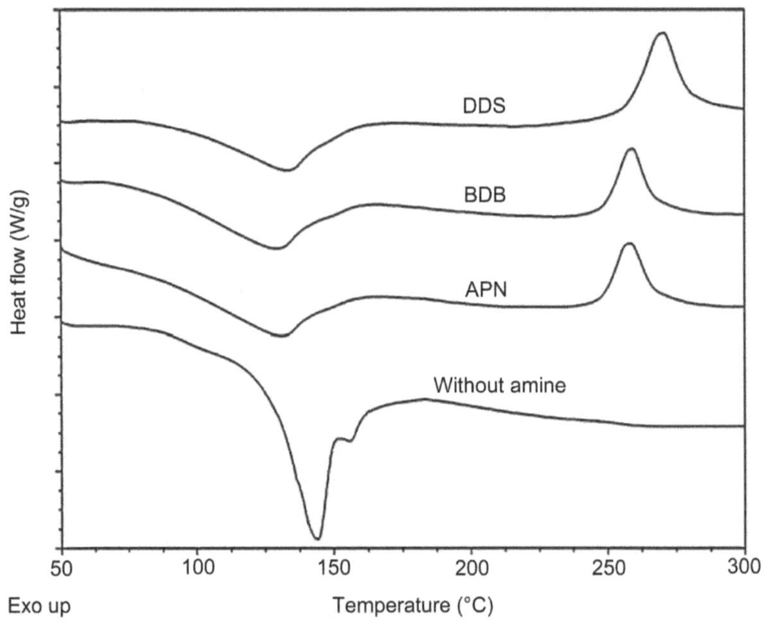

Figure 1.14 Impact of the studied amines on the curing behavior. *APN*, aminophenoxyphthalonitrile; *BDB*, 2,6-bis (4-diaminobenzoxy) benzonitrile; *DDS*, 4,40-diaminediphenyl sulfone. Reprinted from Yang X, Zhang J, Lei Y, Zhong J, Liu X. Effect of different aromatic amines on the crosslinking behavior and thermal properties of phthalonitrile oligomer containing biphenyl ethernitrile. J Appl Polym Sci 2011;121:2331–7.

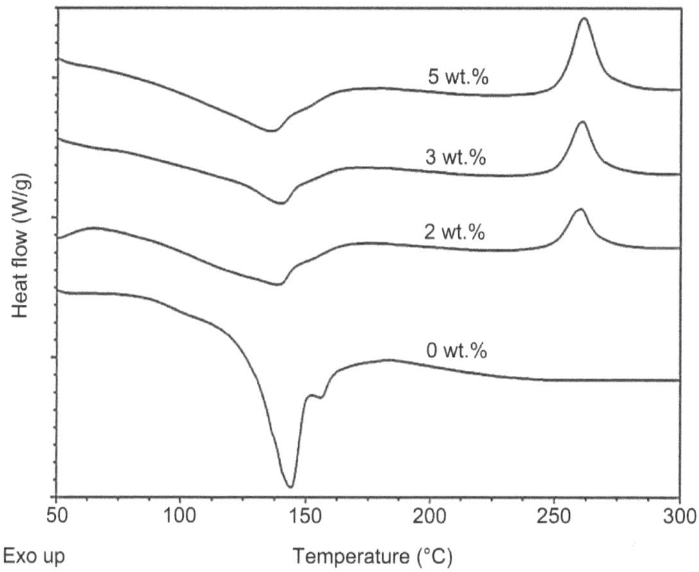

Figure 1.15 DSC curves of the PN prepolymers containing various amount of BDB. *BDB*, 2,6-bis(4-diaminobenzoxy) benzonitrile; *DSC*, differential scanning calorimetry. Reprinted from Yang X, Zhang J, Lei Y, Zhong J, Liu X. Effect of different aromatic amines on the crosslinking behavior and thermal properties of phthalonitrile oligomer containing biphenyl ethernitrile. J Appl Polym Sci 2011;121:2331–7.

The DSC was also performed on the PN prepolymers containing various amount of BDB and the results are plotted in Fig. 1.15. Increasing the amount of BDB resulted in a larger exothermic peak that was accompanied by an increase in the heat polymerization. The exothermic peak maximum remained at about 260°C undependably of the amine amount. Although adding more amine contents increased the curing rate and facilitated the curing process, the increase in the heat of polymerization is not always pursued especially during the preparation of big scale composite materials. Globally, the appropriate amount of the amine-curing agent needs to be carefully chosen to avoid defects in the final products.

The performances of the amine-curing agents proposed by Liu were somehow better than the APB; however, they were largely below those of the *m*-BAPS and *p*-BAPS. Nowadays, the two sulfone-containing diamines are still extensively used as curing agents for numerous PN monomers [29,47–56]. The benefits obtained from these curing additives in terms of processability and curing behavior surpass those obtained with any other amine of phenolic curing agent. Therefore, *m*-BAPS and *p*-BAPS were chosen as the most preferred amine-curing agents for the PN monomers and oligomers.

1.2.2.4 Strong Organic Acid/Amine Salt Curing Agents

From a historical point of view, this class of curing agents has been introduced to overcome the limitations of the early used primary amines. As previously discussed, the first attempts to prepare amine-curing agents resulted in polymers containing voids essentially due to the volatility of the chosen amines. Therefore, Keller and Burchill reported about several possible curing agents based on acids and amine salts [57,58]. The suitable acid-curing agents are those that are proton-donors or electron-acceptors. This comprises organic and inorganic or mineral acids possessing hydrogen atoms which can provide hydrogen ions upon dissociation. A large number of acids having an ionization constant (Ki) preferably ranging between 0.1 and 10^{-4} can also be effectively used. Practically, the acids able to cure the PN monomers range from sulfuric acid to acetic acid. Moreover, as electro acceptors, Lewis acids were also envisaged as PN curing agents. It is also important to point out that regardless of the acid type, the curing agent must be thermally resistant at the curing temperatures or at least one of its decomposed compounds must be able to catalyze the curing reaction.

Suitable inorganic acids encompass sulfuric acid, sulfurous acid, phosphoric acid, phosphorous acid, nitric acid, hydrobromic acid, hydrochloric acid, perchloric acid, boric acid and others [58].

Suitable organic acids include but not limited to *p*-toluenesulfonic acid, aniline-2-sulfonic acid, 8-aniline-1-napthalenesulphonic acid, benzene sulfonic acid, butylsulfonic acid, 10-camphorsulfonic acid, 2,5-diaminobenzenesulfonic acid, 6-dimethylamino-4-hydroxy-2-naphthalene-sulfonic acid, 5-dimethylamino-1-napthalenesulphonic acid, 4-hydroxy-3-nitroso-1-napthalenesulphonic acid tetrahydrate, 8-hydroxyquinoline-5-sulfonic acid, methyl sulfonic acid, phenylboric acid, 1 and 2-napthalenesulphonic acid, 1,5 2,6 and 2,7-napthalenesulphonic acid, picrylsulfonic acid hydrate, 2 and 4-pyridineethanesulfonic acid, 3-pyridinesulfonic acid, 2-pyridinylhydroxylmethanesulfonic acid, 2-sulfobenzoic acid hydrate, 5-sulfosalicylic acid hydrate, 2,4-xylenesulfonic acid, and sulfonic acid containing dyes [58].

Appropriate phosphorous containing acids such as phenylphosphenic acid, diphenylphosphenic acid, propylphosphonic acid, 1-aminoethylphophonic acid, 4-aminophenylphosphonic acid, butylphosphonic acid, 2-carboxyethylphosphonic acid, 2-chloroethylphosphonic acid, dimethylphosphonic acid, ethylphosphonic acid, methylenediphosphonic acid, methylphosphonic acid, phosphonoacetic acid, bis (hydroxymethyl) phosphonic acid, chloromethylphosphonic acid,

1,2-ethylenediphosphonic acid, hystaderylphosphonic acid, hydroxymethylphosphonic acid, *n*-octadecylphosphonic acid, *n*-octylphosphonic acid, phenylphosphonic acid, propylenediphosphonic acid, and *n*-tetradecylphosphonic acid [58].

Lewis acids for possible use as acid-curing agents include aluminum chloride, aluminum bromide, aluminum *n*-butoxide, aluminum *sec*-butoxide, aluminum *sec*-butoxide stearate, aluminum *t*-butoxide, aluminum di(*sec*-butoxide)acetoacetic, aluminum di(isopropoxide)acetoacetic ester chelate, aluminum fluoride, aluminum hydroxide, aluminum iodide, aluminum isopropoxide, aluminum nitrate, aluminum oxide, aluminum 2,4-pentanedionate, aluminum perchlorate, aluminum phenoxide, aluminum phosphate, aluminum sulfide, gallium bromide, gallium chloride, boron oxide, triphenylborane, and methylborane dihydroxide [58].

The necessary amount of the curing agent to effectively catalyze the cure of the PN monomers largely depends upon the acid strength. Trace amounts of strong acids can be enough whereas weak acids need to be added at amounts up to 40 wt.% of the PN monomers. The curing time is also closely related the strength or acidity of the acid and can require fewer minutes to several hours. If a specified acid is chosen as curing agent, the processability and curing behavior of its PN prepolymers needs to be investigated first. The polymerization is initiated by an electrophilic attack of the proton of the acid group with the cyano units of the PN monomer. The PN polymers cured with acid-curing agent display brown color even after postcuring at high temperature. In case of Lewis acid, the acid is believed to initiate the curing reaction by interacting with the nonbonding electron of the cyano groups.

Keller reported on the effect of *p*-toluenesulfonic acid (TsOH) on the properties of oligomeric PN containing multiple aromatic ether linkages [59]. The preparation method involved reacting an excess amount of a bisphenol with 4,4′-difluorobenzophenone in basic environment and in NMP as solvent, the reaction was followed by end-capping with 4-nitrophthalonitrile, the synthetic route and the chemical structure of the resulting PN oligomers are displayed in Scheme 1.7. The TsOH was added in minute amount (2 wt.%) to the melt of the oligomeric PN at 200°C. After stirring for 1 min, the samples were quenched to room temperature and analyzed by DSC. The DSC curve revealed the presence of an endothermic transition at about 100°C, which was ascribed to a T_g and an exothermic transition at around 260°C attributed to the polymerization of the PN oligomers in the presence of TsOH. The DSC was also performed on cured samples at 250°C for 2 h, 280°C for 1 h, and 315°C for 16 h. The resulting polymer had a T_g of 260°C which implied that the reaction effectively occurred in the acidic environment. Moreover, the DSC curve of a sample postcured at 350°C for 2 h and 375°C for 4 h showed no apparent T_g. The polymerization process was also monitored by FTIR spectroscopy. When the PN oligomers were cured in the presence of TsOH only a weak nitrile absorption remained. Further analysis of the spectra indicated the presence of a strong absorption band at 1360 cm^{-1} related to the polytriazine structure. Therefore, it seemed that the TsOH resulted in a PN polymer with polytriazine as the main structure.

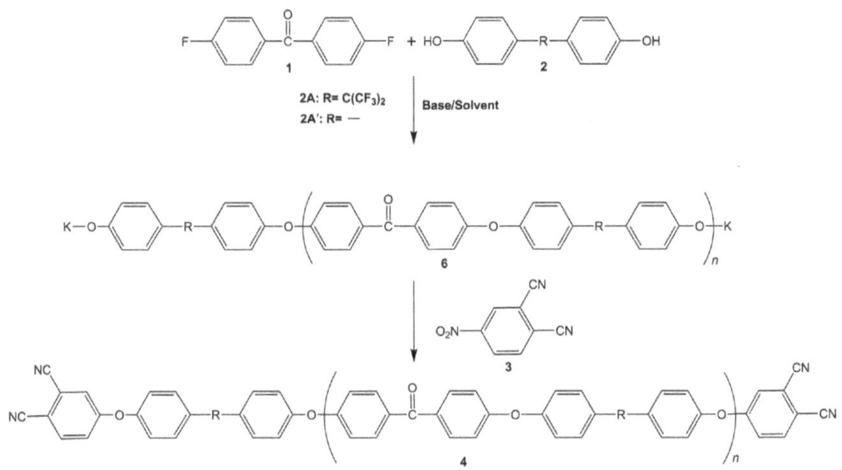

Scheme 1.7 Synthetic route and chemical structure of PN oligomers cured with *p*-toluenesulfonic acid. *PN*, phthalonitrile. Reprinted from Keller TM. Synthesis and polymerization of multiple aromatic ether phthalonitriles. Chem Mater 1994;6:302–5.

The use of amine salts to cure the PN monomers is limited to the work of Burchill in 1994 [43]. The author reported on the formation of a highly cross-linked polytriazine network upon heating the BPh monomers with minute amounts of aromatic amidine or amine salts. The resulting PN thermosets showed similar curing behavior with that of other PN resins obtained by amine-curing agents with one exception: the amine salts promoted the cure reaction more easily and the resulting thermomechanical properties were somehow a stage ahead of the amine-[2]based PN resins. It should be understood that the comparison was made with the first generation of amines additives, since more advanced amines such as APB and BAPS were developed and largely surpass the amines salts in terms of curing behavior and properties.

1.2.3 Self-catalyzed Phthalonitrile Monomers and Oligomers

The previously discussed PN materials were binary compositions composed of PN monomers and curing additives. In those systems, the processability and curing behavior were closely dependent upon the state of dispersion and adhesion between the two phases. Moreover, when small amounts of curing additives were added, extreme conditions needed to be achieved to assure a uniform distribution. Therefore, the necessity of adding a curing agent to cure the PN monomers was an obstacle that needed to be overcome. The main idea behind the development of the self-catalyzed PN monomers is the introduction of amino or hydroxyl functionalities inside the PN monomers structures. In this way, the polymerization reaction can be initiated without adding a second compound.

The first-known, self-catalyzed PN monomers was the 3,5-bis(3,4-dicyanophenoxy)aniline (CPA) (see Scheme 1.8), originally synthetized by Zeng et al. in 2007 [60]. This monomer contains an amino group able to initiate the cure reaction. The DSC and TGA investigations were used to assess the curing behavior and thermal stability of this newly-developed PN monomer, respectively. The DSC results, presented in Fig. 1.16, showed that the PN monomers melting point was around 214°C. Interestingly, these PN monomers also revealed the presence of well-defined exothermic between 257°C and 274°C which confirmed that a self-promoted, curing behavior was effectively obtained. Moreover, the CPA polymer exhibited an excellent thermal resistance with a char yield of 70.6% at 800°C in nitrogen atmosphere. The CPA monomers confirmed that the use of curing additives can be easily circumvented by a simple modification in the molecule design and opened up further research in the field.

1 2 CPA

Scheme 1.8 Synthetic route and chemical structure of the CPA. *CPA*, 3,5-bis (3,4-dicyanophenoxy)aniline. Reprinted from Zeng K, Zhou K, Tang WR, Tang Y, Zhou HF, Liu T, et al. Synthesis and curing of a novel amino-containing phthalonitrile derivative. Chi Chem Lett 2007;18:523–6.

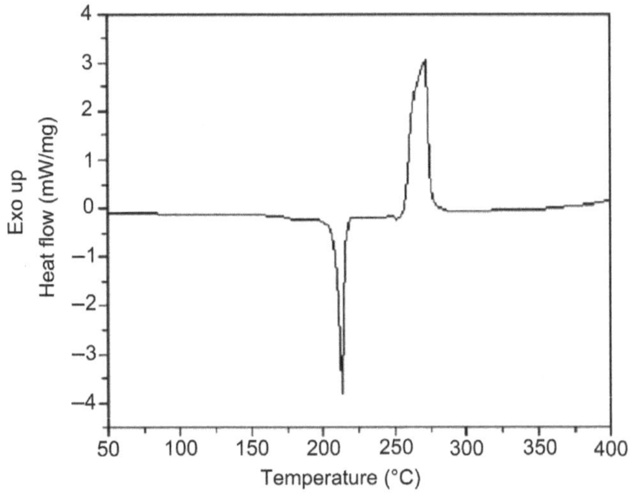

Figure 1.16 DSC curve of the CPA monomers at the heating rate of 10°C/min in N_2. *CPA*, 3,5-bis(3,4-dicyanophenoxy)aniline; *DSC*, differential scanning calorimetry. Reprinted from Zeng K, Zhou K, Tang WR, Tang Y, Zhou HF, Liu T, et al. Synthesis and curing of a novel amino-containing phthalonitrile derivative. Chi Chem Lett 2007;18:523–6.

Zeng and coworkers also prepared series of PN monomers containing hydroxyl and studied their curing behavior by TGA, DSC, and FTIR [61]. The prepared PN monomers are listed in Scheme 1.9 and noted as HPNM-a/d or 3-a/d according to their chemical structure. The self-promoting, curing behavior of the HPNM monomers was investigated under TGA after heating the samples to 225°C at various curing times. The TGA curves of the HPNM-a monomer and its subsequent cured products at 225°C for 15, 30, and 60 min are presented in Fig. 1.17. By analyzing the TGA data, it was evident that curing the HPNM-a monomer for 15 min greatly enhanced its thermal stability. It also appeared that further increase in the curing time, resulted in slight enhancement of the thermal stability. Since the thermal stability of any polymer is mainly enhanced by a formation of a rigid crosslinking network, the previous observations indicated that the HPNM-a heated at 225°C for more than 15 min resulted in the formation of a rigid network. Moreover, the slight increase in the thermal stability after 15 min means that the reaction progressed at various rates.

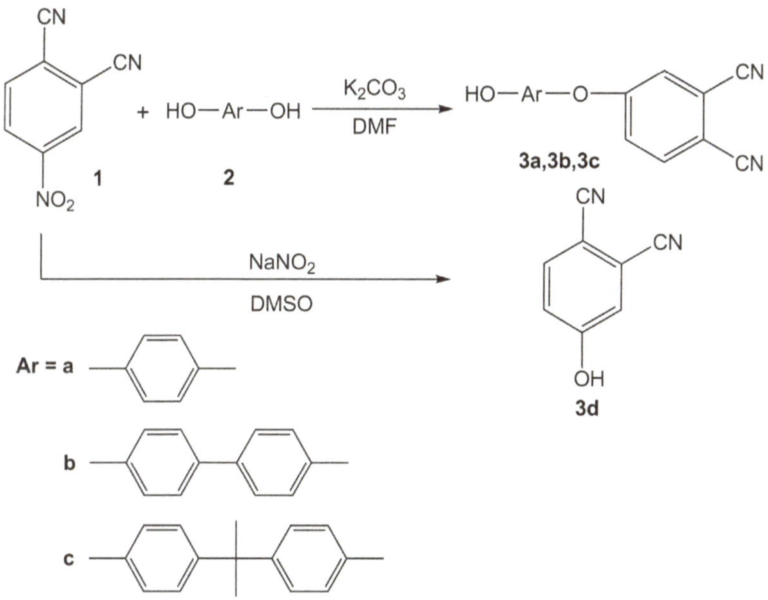

Scheme 1.9 The preparation route and chemical structure of a series of PN monomers containing hydroxyl. *PN*, phthalonitrile. Reprinted from Zeng K, Zhou K, Zhou S, Hong H, Zhou H, Wang Y, et al. Studies on self-promoted cure behaviors of hydroxy-containing phthalonitrile model compounds. Eur Polym J 2009;45:1328–35.

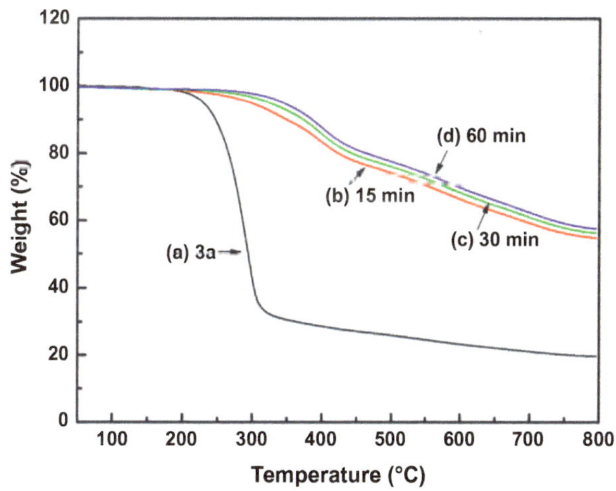

Figure 1.17 TGA curves of the HPNM-a monomer and its subsequent cured products at 225°C for 15, 30, and 60 min. *TGA*, thermogravimetric analysis. Reprinted from Zeng K, Zhou K, Zhou S, Hong H, Zhou H, Wang Y, et al. Studies on self-promoted cure behaviors of hydroxy-containing phthalonitrile model compounds. Eur Polym J 2009;45:1328–35.

The temperature of weight loss at 10 and 5 wt.% for all the HPNM monomers and their cured polymers at different times are depicted in Figs. 1.18 and 1.19, respectively. In contrast with the HPNM-a, the HPNM-b monomers and polymers exhibited a more enhanced thermal stability, which may be due to the high molecular weight and lower volatility of these compounds. On the other hand, the HPNM-c displayed lower thermal stability enhancement probably due to the slower reaction speed of this compound. However, it should be noted that the curing time chosen in the work of Zeng et al. was too short to advance any conclusion concerning the global rate of polymerization of these monomers. More importantly, the TGA investigations confirmed the ability of these new PN monomers to form a more stable structure after heating without the addition of any curing agents.

The authors also attempted to study the cure reaction of the HPNMs by integrating their exothermic peaks; however, the DSC results showed no apparent exothermic transition. This was attributed to either a slow reaction process not detectable by the DSC or to a very fast curing speed, such that the monomers were already cured at 225°C. In either case, no relevant conclusion concerning the effect of the HPNM structures on the curing behavior was possible to be proposed. Meanwhile, the FTIR was used to get some insights into the curing process. The

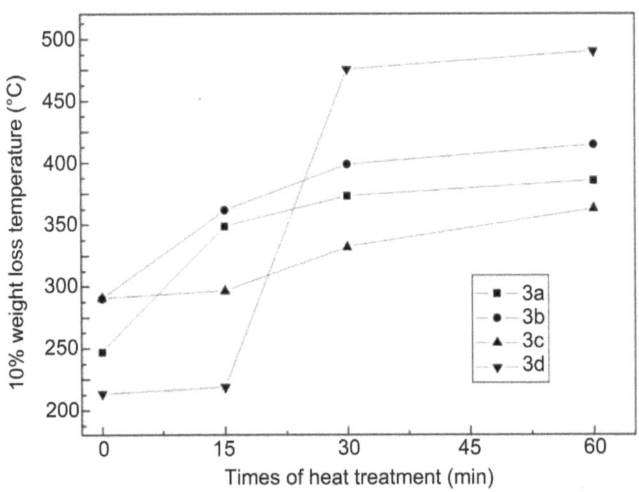

Figure 1.18 Ten percent weight loss temperature of the HPNM before heat treatment (0 min) and cured products versus times of heat treatment at 225°C. Reprinted from Zeng K, Zhou K, Zhou S, Hong H, Zhou H, Wang Y, et al. Studies on self-promoted cure behaviors of hydroxy-containing phthalonitrile model compounds. Eur Polym J 2009;45:1328–35.

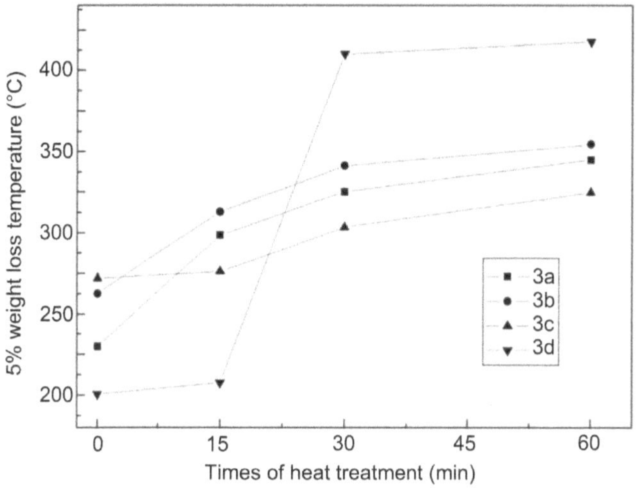

Figure 1.19 Five percent weight loss temperature of HPNM before heat treatment (0 min) and cured products versus times of heat treatment at 225°C. Reprinted from Zeng K, Zhou K, Zhou S, Hong H, Zhou H, Wang Y, et al. Studies on self-promoted cure behaviors of hydroxy-containing phthalonitrile model compounds. Eur Polym J 2009;45:1328–35.

FTIR spectra are presented in Fig. 1.20. Clearly, all the samples showed a reduction in the nitrile absorption at 2230 cm^{-1} and evidences of the formation of triazine structure (1525 and 1521 cm^{-1}) for the HPNM-a and b were found at the early stages of the curing process. The triazine characteristic peaks also increased in intensity as the reaction moved forward in time. Furthermore, the nitrile absorption shifted to lower wave number, similar to that observed when phenolic curing agents were used to cure the PN monomers. For the HPNM-c and d compounds, the absorptions related to phthalocyanine, triazine, and isoindoline were overlapped by other stronger bands. However, their final polymeric networks were expected to contain all the previously cited structures. Generally, the self-catalyzed PN containing hydroxyl compounds show a chemical behavior close to that of PN polymers cured with phenol additives. The main differences were found in the position of the hydroxyl group responsible for the polymerization reaction initiation.

Zhou et al. reported about another self-catalyzed amino containing PN monomers 4-(4-(3, 5-diaminobenzoyl) phenoxy) PN (DAPN) (Scheme 1.10) [62]. The TGA investigations revealed large enhancements in the thermal stability after heating the DAPN monomers for 15 min. Further heating the PN monomers for 30 min showed only slight ameliorations in the thermal stability. Those results are similar to

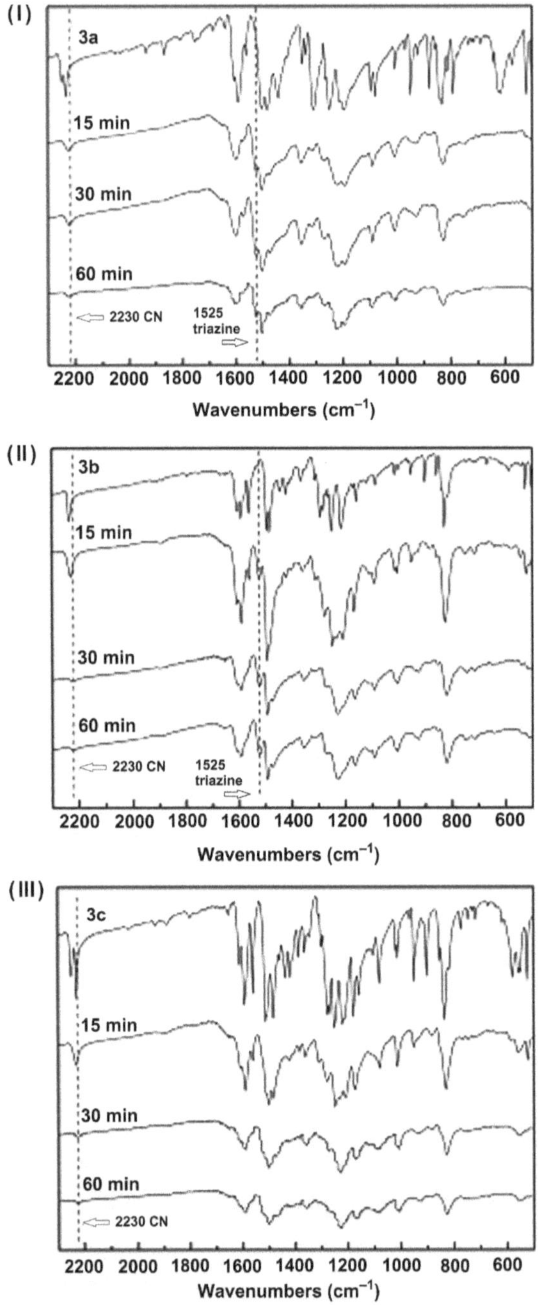

Figure 1.20 FTIR spectra of HPNM 3a (I), 3b (II), 3c (III) and 3d (IV) after heat treatment at 225°C for 15, 30 and 60 min (final cure). *FTIR*, Fourier-transform infrared spectroscopy. Reprinted from Zeng K, Zhou K, Zhou S, Hong H, Zhou H, Wang Y, et al. Studies on self-promoted cure behaviors of hydroxy-containing phthalonitrile model compounds. Eur Polym J 2009;45:1328–35.

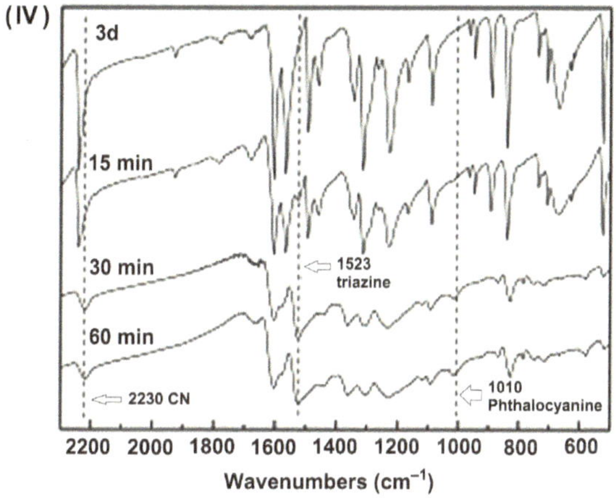

Figure 1.20 (Continued).

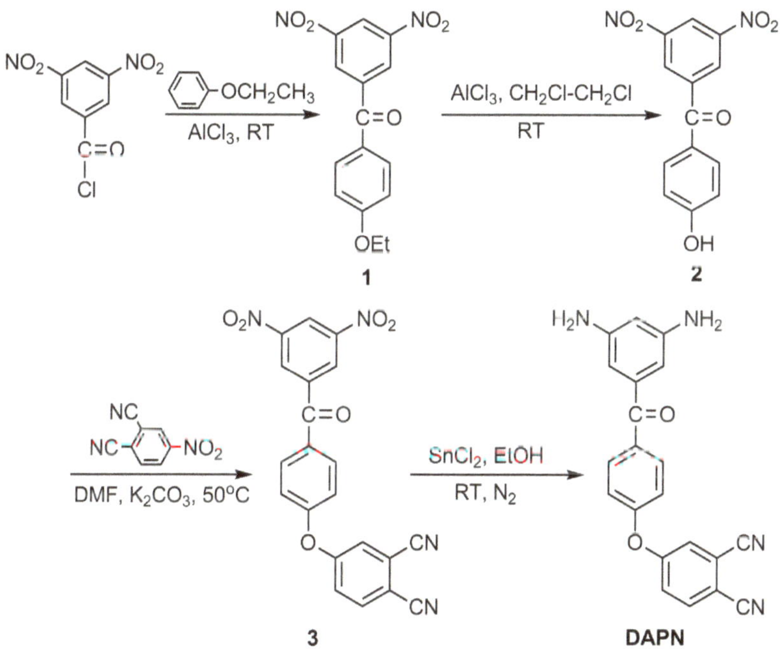

Scheme 1.10 Synthetic route and chemical structure of the self-catalyzed amino containing PN monomers DAPN. *DAPN*, 4-(4-(3, 5-diaminobenzoyl) phenoxy) phthalonitrile; *PN*, phthalonitrile. Reprinted from Zhou S, Hong H, Zeng K, Miao P, Zhou H, Wang Y, et al. Synthesis, characterization and self-promoted cure behaviors of a new phthalonitrile derivative 4-(4-(3, 5-diaminobenzoyl) phenoxy) phthalonitrile. Polym Bull 2009;62:581–91.

those found regarding the previously discussed HPNM compounds. In contrast with the HPNM, the DAPN studied under DSC clearly showed the exothermic peaks (Fig. 1.21) which confirmed the ability of the DAPN to be cured without any external catalyzer. The DAPN polymers final structures analyzed by FTIR also seemed to be very similar to those obtained with an amine-curing agent. For instance, the phthalocyanine, triazine, and isoindoline absorptions were all clearly depicted in the FTIR spectra.

Badshah et al. synthetized some monofuctional PN monomers bearing imide moieties and containing an amino function at different positions (ortho, meta, and para) [63]. The authors investigated the self-promoting behavior of these compounds and studied the influence of the amino position on the processability and curing behavior. The self-catalyzed PN monomers synthesis route and chemical structures studied are drawn in Scheme 1.11. The prepared PN monomers were analyzed by DSC up to 400°C. All the PN monomers exhibited an endothermic transition at 100°C, 204°C, and 242°C for the ortho, meta, and para amino position,

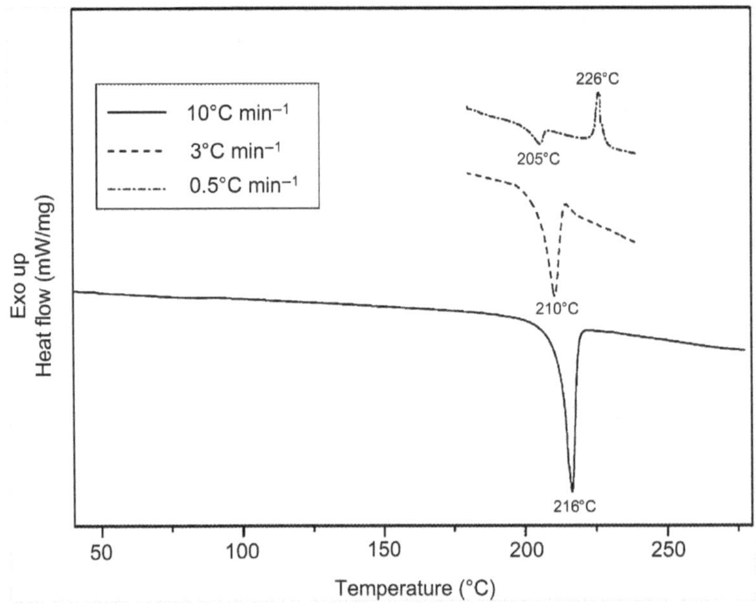

Figure 1.21 DSC curves of the DAPN monomers at various heating rates. *DAPN*, 4-(4-(3, 5-diaminobenzoyl) phenoxy); *DSC*, differential scanning calorimetry. Reprinted from Zhou S, Hong H, Zeng K, Miao P, Zhou H, Wang Y, et al. Synthesis, characterization and self-promoted cure behaviors of a new phthalonitrile derivative 4-(4-(3, 5-diaminobenzoyl) phenoxy) phthalonitrile. Polym Bull 2009;62:581−91.

respectively. As expected, all the PN monomers also presented a clear exothermic transition ranging from 240°C to 300°C. The ortho amino-based PN monomers had the largest processing window, followed by that of the para amino PN monomers. The DSC results confirmed the self-promoting character of these new PN monomers.

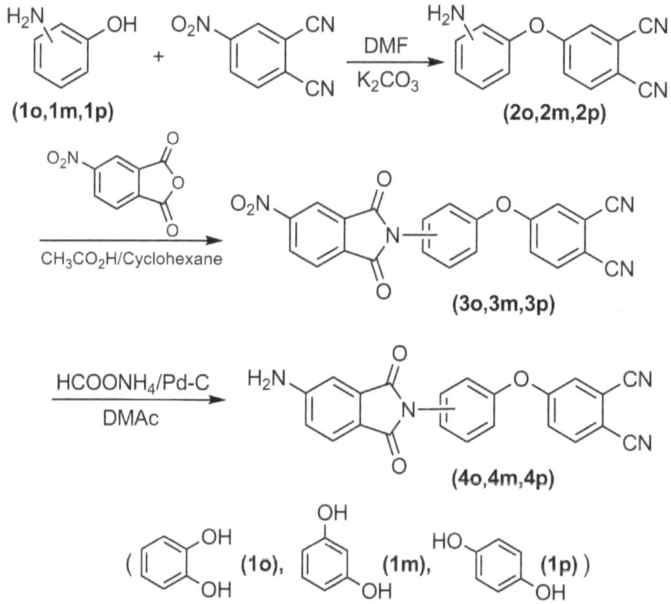

Scheme 1.11 Synthetic route and chemical structures of some monofuctional PN monomers bearing imide moieties. *PN*, phthalonitrile. Reprinted from Amir B, Zhou H, Liu F, Aurangzeb H. Synthesis and characterization of self-catalyzed imide-containing pthalonitrile resins. J Polym Sci: Part A: Polym Chem 2010;48:5916–20

In order to judge the processability of these PN monomers, a rheometric study was conducted to monitor the changes in the viscosity during the polymerization reaction. The viscosity of the ortho compound showed a considerable drop at 130°C and remained at the same value until 235°C where a vertical increase in the viscosity was noted. For the meta compound, the same trend was observed, but with a first viscosity drop at 205°C and a vertical increase at 245°C. On the other hand, the para compound showed a sudden increase in the viscosity after melting. These results are in agreement with those obtained by DSC analyses; the ortho amine-based PN monomers possess the best processability with low viscosity values at moderate temperatures suitable for multiple industrial applications.

The effect of the amino position on the APN monomers curing behavior and processability was investigated by Zhou el al. [64]. The Synthetic route as well as the chemical structures of the self-catalyzed monomers given different amine positions are listed in Scheme 1.12. The monomers were prepared by reacting 4-nitrophthalonitrile with aminophenol (*ortho*, *meta*, or *para*), dimethyl formamide (DMF), and potassium carbonate (K_2CO_3). After heating the reaction to a specified temperature and mechanically stirring for a specified time, the mixtures were cooled to room temperature and purified accordingly.

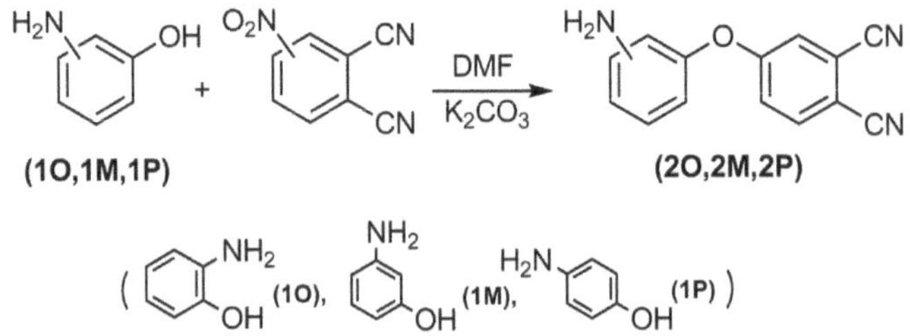

Scheme 1.12 The preparation route and chemical structure of the APN self-catalyzed PN monomers with different amine positions. *APN*, aminophenoxyphthalonitrile; *PN*, phthalonitrile. Reprinted from Zhou H, Badashah A, Luo Z, Liu F, Zhao T. Preparation and property comparison of ortho, meta, and para autocatalytic phthalonitrile compounds with amino group. Polym Adv Technol 2011;22:1459–65.

The curing behavior was monitored by DSC at various heating rates. All the PN monomers exhibited one endothermic transition related to the melting of the monomers and one exothermic peak confirming the autocatalytic ability of these compounds. The melting points and exothermic peaks positions were affected by the amino group position. For instance, the melting points of the *ortho*-, *meta*-, and *para*-based PN monomers were 122°C, 172°C, and 131°C, respectively. Generally, the para position leads to the highest melting point. However, in this case, the meta position, taking advantage of the stronger hydrogen bond, surpassed the para position. Furthermore, the meta position's lower reactivity, due to the steric hindrance, resulted in an exothermic maximum peak located at 258°C. The maximum peak was 20°C higher than those of the meta and para compounds. Therefore, the ortho position was confirmed to exhibit the largest processing window. In contrast with PN monomers/amine-curing agents' systems, the heat of polymerization was found to be higher for all the studied autocatalytic PN compounds regardless of the position of the amino group. This has been explained by the fact that the amounts of the curing agent added to the PN monomers rarely exceed 10 wt.%. However, for the self-catalyzed, PN monomers every unit is bearing an amino group; therefore the excessive number of amino groups is behind the increase in the heat of polymerization.

Badshah et al. recommended an efficient approach to prepare ether and amide-based autocatalytic PN with enhanced processability [65]. The synthesis route of the 4-amino-*N*-[(3,4-dicyanophenoxy)phenyl] benzamides PN monomers is presented in Scheme 1.13. The DSC thermograms of the PN monomers with different positions of the amino group are displayed in Fig. 1.22 and the viscosity changes are plotted in Fig. 1.23. The global trend was similar to that of the previous works presented before. However, the authors successfully enhanced the processing window up to 200°C for the meta amine position. This was the first-ever reported, largest value of the processing window. With such a large processability, numerous additional applications of the PN resins were expected.

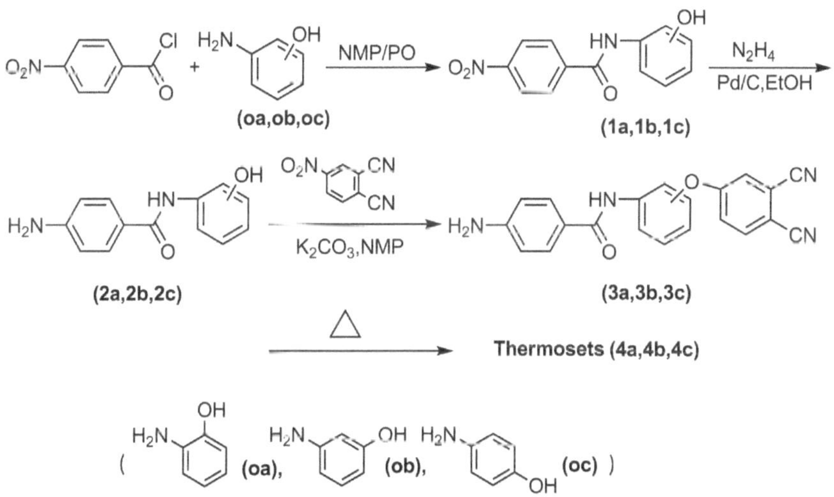

Scheme 1.13 The route of synthesis of the 4-amino-*N*-[(3,4-dicyanophenoxy) phenyl]benzamides PN monomers. *PN*, phthalonitrile. Reprinted from Badshah A, Kessler MR, Heng Z, Zaidi JH, Hameed S, Hasan A. An efficient approach to prepare ether and amide-based self-catalyzed phthalonitrile resins. Polym Chem 2013;4:3617–22.

Another concept of self-promoted, PN monomers was introduced by Zhang et al. The catalytic effect necessary to cure the PN monomers was no longer provided by the hydroxyl or amino groups, but was provided by the silicon units incorporated within the PN monomers

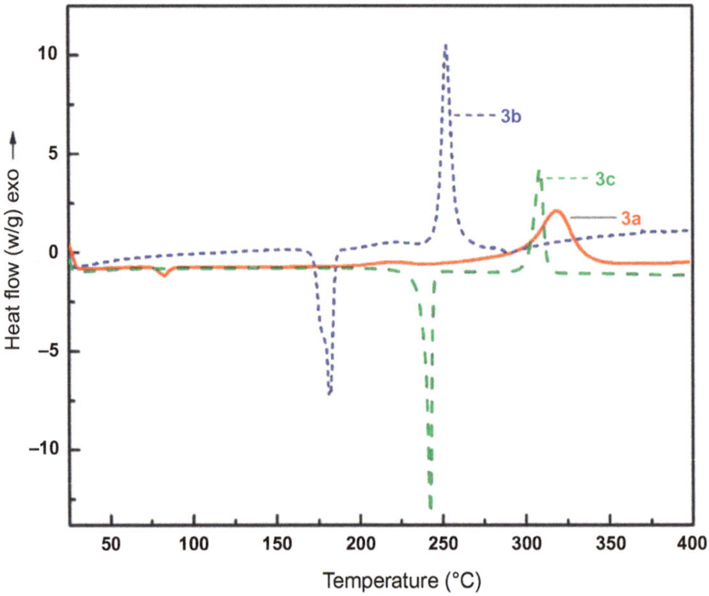

Figure 1.22 The DSC thermograms of the studied PN monomers with different positions of the amino group. *DSC*, differential scanning calorimetry; *PN*, phthalonitrile. Reprinted from Badshah A, Kessler MR, Heng Z, Zaidi JH, Hameed S, Hasan A. An efficient approach to prepare ether and amide-based self-catalyzed phthalonitrile resins. Polym Chem 2013;4:3617–22.

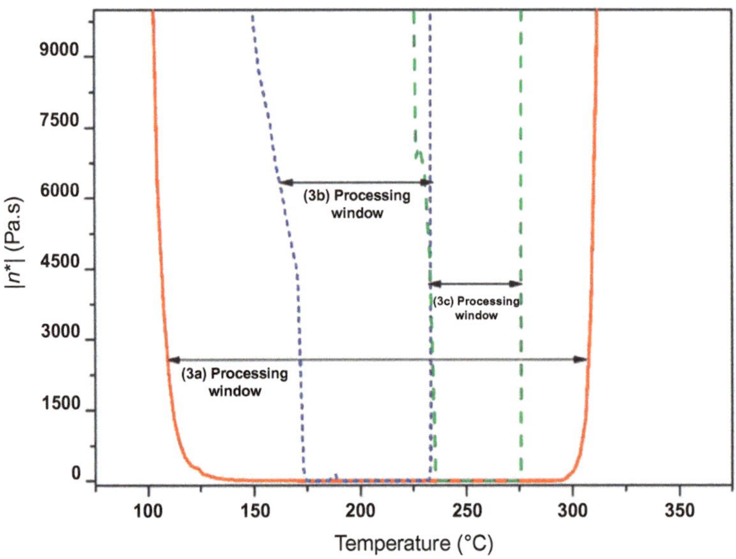

Figure 1.23 Viscosity changes of the studied PN monomers as function of temperature and position of the amino group. *PN*, phthalonitrile. Reprinted from Badshah A, Kessler MR, Heng Z, Zaidi JH, Hameed S, Hasan A. An efficient approach to prepare ether and amide-based self-catalyzed phthalonitrile resins. Polym Chem 2013;4:3617–22.

backbones [66]. These new monomers were synthetized by aminolysis reaction of chlorosilanes with 4-APN. The route of preparation and chemical structures of the studied monomers are presented in Scheme 1.14. The polymerization reaction of the SiPNs and 4-APN monomers were studied and compared by DSC analyses. The DSC curves of all the PN monomers are shown in Fig. 1.24. Obviously, all the monomers exhibited an endothermic transition attributed to their respective melting points. Interestingly, the SiPNs monomers exhibited exceptionally reduced, melting transitions, with melting points temperatures below 60°C. Those results were described as the lowest melting points ever reported in the field of PN resins. Meanwhile an exothermic transition was observed in all the DSC curves of the SiPNs and 4-APN. The SiPNs curing peaks were higher than those of the 4-APN because of the difference in reactivity between these monomers. Indeed the basicity of the 4-APN with an amino group at the para position is higher than that of the SiPNs bearing (-NH-), which means that the 4-APN monomer is more reactive and therefore its curing temperature is lower.

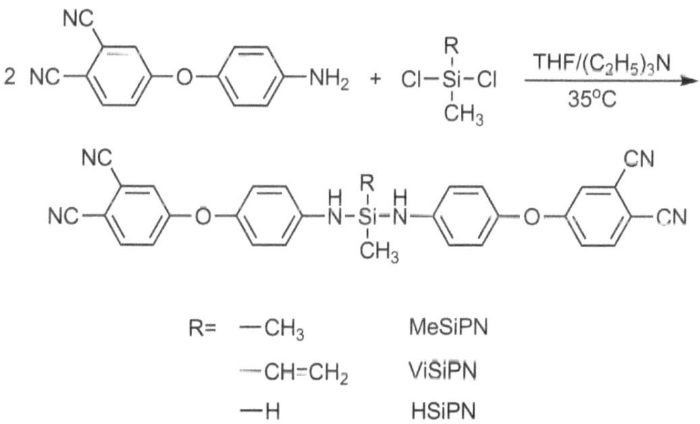

Scheme 1.14 The preparation route and chemical structure of a series of silicon-containing self-catalyzed phthalonitrile derivatives (SiPNs). Reprinted from Zhang Z, Li Z, Zhou H, Lin X, Zhao T, Zhang M, et al. Self-catalyzed silicon-containing phthalonitrile resins with low melting point, excellent solubility and thermal stability. J Appl Polym Sci 2014;131:40919.

In terms of processability, the ViSiPN had the largest processing window of 231°C, more than 30°C higher than that reported by Badshah et al. concerning the 4-amino-*N*-[(3,4-dicyanophenoxy)phenyl] benzamides self-catalyzed PN monomers. The curing exotherm position was also closely related to the basicity of the SiPNs. For example, the HSiPN showed the lowest exothermic peak at around 255°C. This

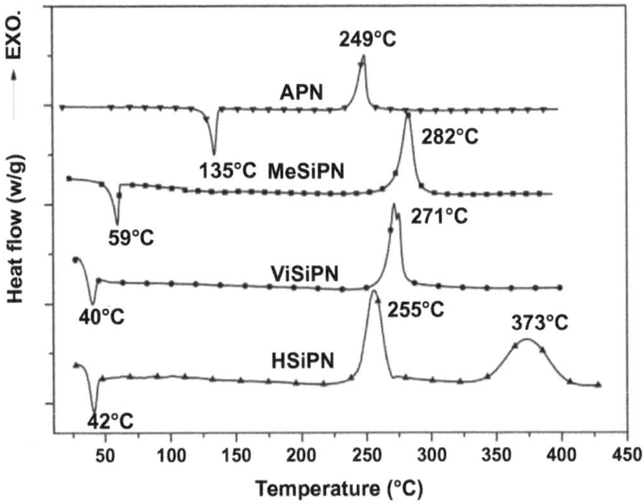

Figure 1.24 DSC thermograms of 4-APN, MeSiPN, ViSiPN, and HSiPN monomers. *APN*, 4-aminophenoxyphthalonitrile; *DSC*, differential scanning calorimetry. Reprinted from Zhang Z, Li Z, Zhou H, Lin X, Zhao T, Zhang M, et al. Self-catalyzed silicon-containing phthalonitrile resins with low melting point, excellent solubility and thermal stability. J Appl Polym Sci 2014;131:40919.

compound basicity takes advantage of the smallest steric hindrance of the hydrogen linked to the Si atom. In contrast with the MeSiPN, the ViSiPN possesses stronger basicity due to the electron-donating (d-p)Π, conjugated effect derived from vinyl structure and Si atom. Therefore, the exothermic peak of the ViSiPN was 11°C lower than that of the MeSiPN. Concerning the heat of polymerization, the ViSiPN clearly surpassed that of the HSiPN and MeSiPN. This was probably due to the overlapping of two additional reactions; one related to the vinyl groups and another to the curing reaction of the cyano groups. Nevertheless, the heat generated by the curing reaction of the ViSiPN monomers was still inferior to that of the 4-APN. This implies that the SiPNs have lower reactivity and better processability than the 4-APN.

The FTIR was recorded for the monomers and polymers of the SiPNs. The obtained spectra are presented in Fig. 1.25. The nitrile absorption depicted at 2232 cm^{-1} for the uncured SiPNs decreased in intensity as the curing reaction progressed. Meanwhile, two new absorption bands materialized at 1354 and 1048 cm^{-1} and were attributed to the triazine and phthalocyanine structures, respectively. The FTIR was also used to confirm the participation of the vinyl group of the ViSiPN in the thermal crosslinking reaction. Although the absorptions bands of the vinyls ($Si\text{-}CH{=}CH_2$) were overlapped by those of the phenyls, a comparative study between the spectra of the SiPNs revealed that the vibration of the (=C-H) located at 1010 cm^{-1} was stronger for the ViSiPN before curing. However, after the curing, all the SiPNs presented the same intensity for the 1010 cm^{-1} band which clearly indicated that the vinyl groups effectively participated in the curing process.

Zhang et al. took advantage of the excellent mechanical and thermal properties of the fluorenyl structure and prepared self-catalyzed PN monomers containing fluorine structure bearing two amino groups, namely 9,9-bis(4-aminophenyl)-2,7-(3,4-dicyanophenoxy)-fluorene (AFPN) [67]. The Scheme 1.15 illustrates the route of preparation of the AFPN monomers. For the sake of comparison, the authors introduced a nonautocatalytic fluorene-based PN monomer, 9,9-bis[(3,4-dicyanophenoxy) phenyl]fluorene (BPF), and studied its curing behavior and processability along with the AFPN. Investigations of the cure behavior were performed using DSC and FTIR analyses. The DSC curves for both AFPN and BPF

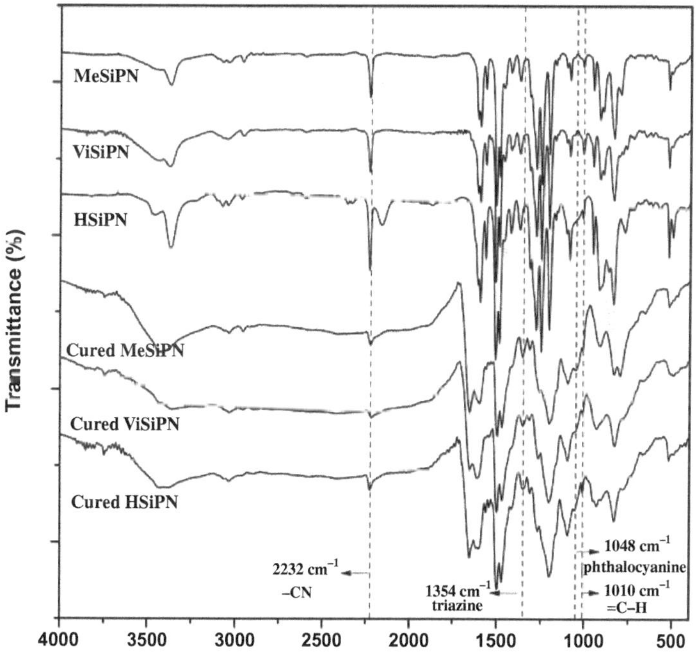

Figure 1.25 FTIR spectra of the uncured and cured SiPNs. *FTIR*, Fourier-transform infrared spectroscopy. Reprinted from Zhang Z, Li Z, Zhou H, Lin X, Zhao T, Zhang M, et al. Self-catalyzed silicon-containing phthalonitrile resins with low melting point, excellent solubility and thermal stability. J Appl Polym Sci 2014;131:40919.

are shown in Fig. 1.26. The BPF showed only one endothermic transition at around 262°C corresponding to its melting point. As expected, the BPF as a nonautocatalytic PN monomer failed to exhibit a clear exothermic transition, which means that the curing of this monomer without any curing agent is a sluggish process. Conversely, the introduction of two amino groups led to a different curing behavior. Beside the evident melting transition at about 157°C, the AFPN confirmed its autocatalytic polymerization process with an acceptable processing window. Taking into account the high molecular weight of the AFPN and the evident steric hindrance caused by the presence of multiple benzene rings, its melting point was surprisingly lower than those of the conventional PN monomers (such as BPh and BAPh with melting points about 232°C and 195°C, respectively).

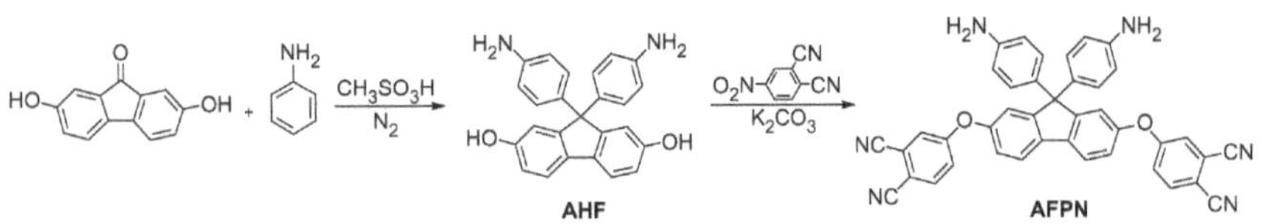

Scheme 1.15 Synthetic route and chemical structure of the AFPN. *AFPN*, amino-containing fluorene-based bisphthalonitrile. Reprinted from Zhang T, Wang J, Derradji M, Ramdani N, Wang H, Lin ZW, et al. Synthesis, curing kinetics and thermal properties of a novel self-promoted fluorene-based bisphthalonitrile monomer. Thermochim Acta 2015;602:22–9.

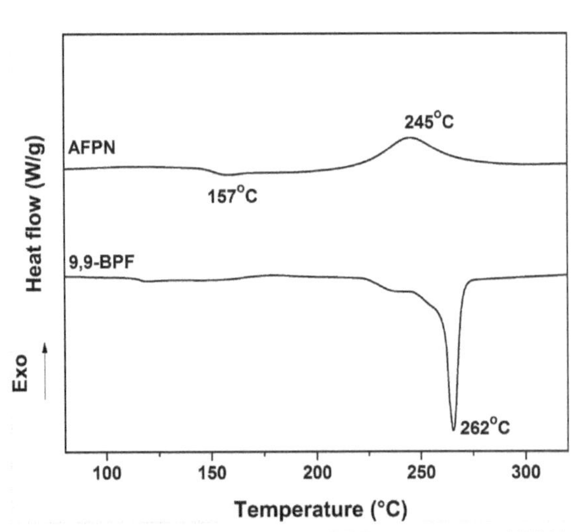

Figure 1.26 DSC curves for both AFPN and BPF monomers. *AFPN*, 9,9-bis(4-aminophenyl)-2,7-(3,4-dicyanophenoxy)-fluorene; *BPF*, 9,9-bis[(3,4-dicyanophenoxy) phenyl]fluorene; *DSC*, differential scanning calorimetry. Reprinted from Zhang T, Wang J, Derradji M, Ramdani N, Wang H, Lin ZW, et al. Synthesis, curing kinetics and thermal properties of a novel self-promoted fluorene-based bisphthalonitrile monomer. Thermochim Acta 2015;602:22–9.

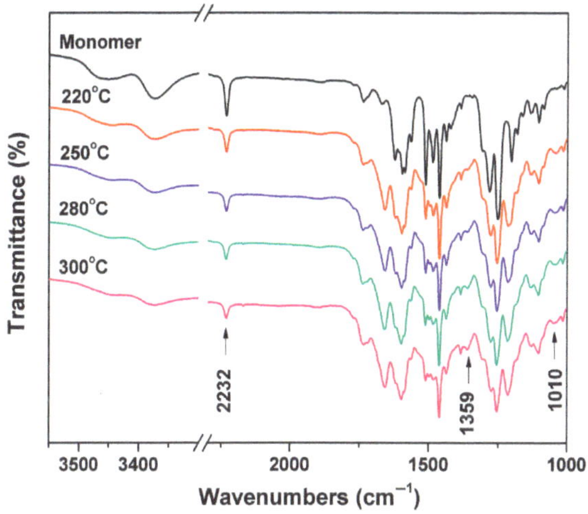

Figure 1.27 FTIR spectra of the AFPN at different curing stage. *AFPN*, 9,9-bis(4-aminophenyl)-2,7-(3,4-dicyanophenoxy)-fluorene; *FTIR*, Fourier-transform infrared spectroscopy. Reprinted from Zhang T, Wang J, Derradji M, Ramdani N, Wang H, Lin ZW, et al. Synthesis, curing kinetics and thermal properties of a novel self-promoted fluorene-based bisphthalonitrile monomer. Thermochim Acta 2015;602:22–9.

The FTIR spectra of the AFPN at different curing stages are illustrated in Fig. 1.27. By analyzing the spectra, it clearly appears that the intensity of the nitrile absorption at 2232 cm^{-1} was reduced at each curing stage. This decrease was accompanied by the appearance of new absorptions related to the formation of triazine such as the band at 1359 cm^{-1} corresponding to the vibration of (-C = N-). However, not all the nitriles participated in the polymerization reaction due to the high steric hindrance of the AFPN monomers. The FTIR spectra further confirmed the autocatalytic behavior of the AFPN. The variations of complex viscosity of the AFPN and BPF between 160°C and 300°C are displayed in Fig. 1.28. The AFPN showed a sharp decrease in viscosity at 180°C and maintained very low viscosity values until 220°C. After 220°C a vertical increase in the viscosity was noticed implying the starting of the curing reaction. Thus, the rheometric study confirmed the DSC results concerning the good processability of the AFPN. On the other hand, the drop and increase in the BPF viscosity were very close, indicating an obvious poor processability.

Another interesting feature of the self-catalyzed PN monomers is their ability to catalyze the cure of other nonautocatalytic PN monomers. Indeed, the self-catalyzed PN monomers contain in their backbone enough initiating functions to catalyze their cure as well those of

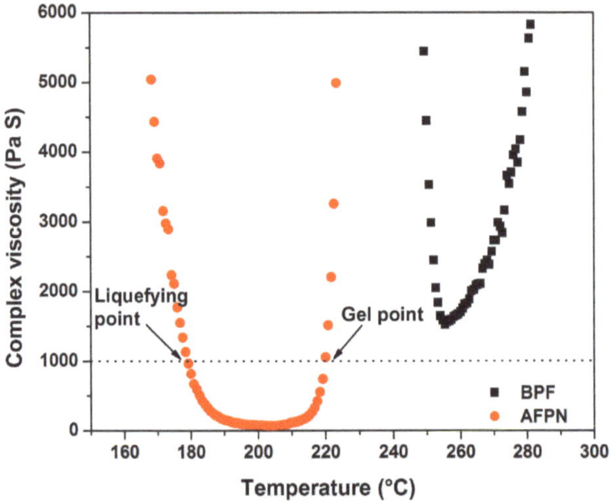

Figure 1.28 Variations of the complex viscosity of the AFPN and BPF between 160°C and 300°C. *AFPN*, 9,9-bis(4-aminophenyl)-2,7-(3,4-dicyanophenoxy)-fluorene; *BPF*, 9,9-bis[(3,4-dicyanophenoxy) phenyl]fluorene. Reprinted from Zhang T, Wang J, Derradji M, Ramdani N, Wang H, Lin ZW, et al. Synthesis, curing kinetics and thermal properties of a novel self-promoted fluorene-based bisphthalonitrile monomer. Thermochim Acta 2015;602:22–9.

other PN monomers. As seen in the sections related to the curing additives, the preferred amount of the curing agent to catalyze the cyano addition reaction rarely surpasses 10 wt.%. On the other hand, each self-catalyzed PN monomer contains its own initiating functions which generally results in enhanced reactivity. Thus, using self-catalyzed PN monomers to cure the PN monomers is a promising area to be addressed. For instance, Sheng et al. used 4-APN to cure the RPh and BAPh monomers [68]. The 4-APN self-catalyzed behavior is shown in Fig. 1.29. The DSC curves of RPh and BAPh with various amounts of 4-APN are presented in Fig. 1.30. Clearly, all the mixtures exhibited an exothermic peak confirming the autocatalytic behavior of these blends. Moreover, adding more 4-APN contents enlarged the processing window up to 96°C. These results imply that both PN monomers can be effectively cured with 4-APN.

To acknowledge about the curing process and the final structure of these PN resins, the FTIR spectra of the prepolymers and polymers of RPh and BAPh containing different amounts of 4-APN are presented in Fig. 1.31. By analyzing these spectra, it appears that all the prepolymers and polymers presented the absorption band at 1384 cm^{-1} related of triazine structure. The nitrile absorption intensity was lowered by adding more 4-APN, which once more confirmed the ability of the 4-APN to

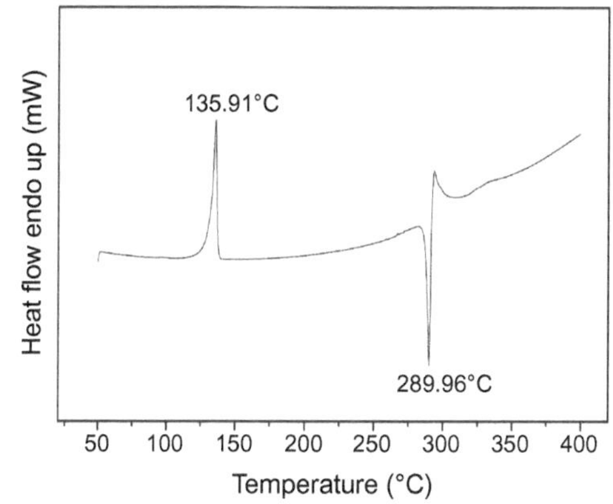

Figure 1.29 DSC curve of 4-APN. *APN*, aminophenoxyphthalonitrile; *DSC*, differential scanning calorimetry. Reprinted from Sheng H, Peng X, Guo H, Yu X, Naito K, Qu X, et al. Synthesis of high performance bisphthalonitrile resins cured with self-catalyzed 4-aminophenoxy phthalonitrile. Thermochim Acta 2014;577:17–24.

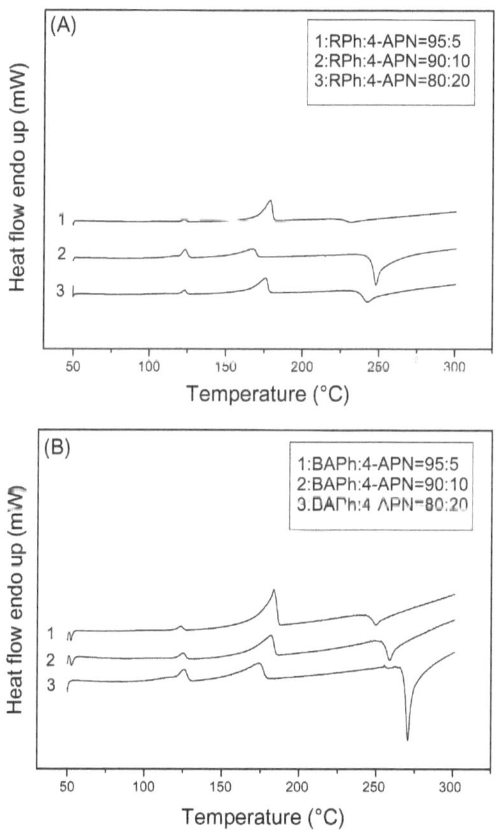

Figure. 1.30 DSC curves of (A) RPh and (B) BAPh with various amount of 4-APN. *BAPh*, BPA-based PN resin; *DSC*, differential scanning calorimetry; *RPh*, resorcinol-based PN. Reprinted from Sheng H, Peng X, Guo H, Yu X, Naito K, Qu X, et al. Synthesis of high performance bisphthalonitrile resins cured with self-catalyzed 4-aminophenoxy phthalonitrile. Thermochim Acta 2014;577:17–24.

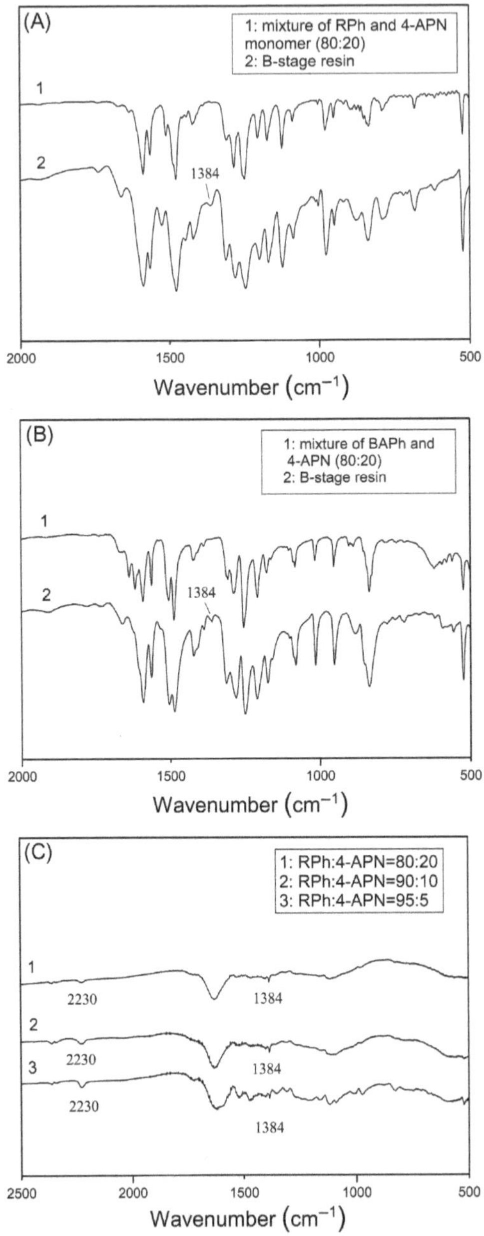

Figure 1.31 FTIR spectra of the prepolymers and polymers of (A) RPh and (B) BAPh containing different amounts of 4-APN. *BAPh*, BPA-based PN resin; *FTIR*, Fourier-transform infrared spectroscopy; *RPh*, resorcinol-based PN. Reprinted from Sheng H, Peng X, Guo H, Yu X, Naito K, Qu X, et al. Synthesis of high performance bisphthalonitrile resins cured with self-catalyzed 4-aminophenoxy phthalonitrile. Thermochim Acta 2014;577:17–24.

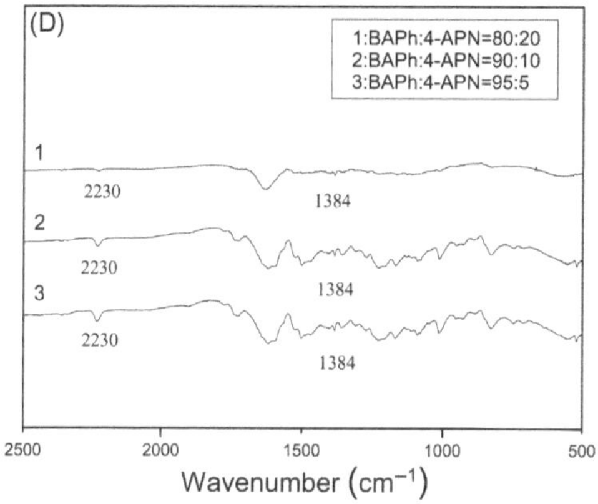

Figure 1.31 (Continued).

act as an advanced curing agent for the RPh and BAPh monomers. Interestingly, no bands related to phthalocyanine or isoindoline were seen in these spectra. This may be ascribed to the fact that 4-APN, as PN monomers, not only catalyze the curing reaction, but also actively participate in the cyano addition. Therefore the polytriazine was the major structure constituting the polymeric structures.

Derradji et al. during the preparation of several PN composites also reported about the use of APN to cure the BAPh resin with different kinds of ceramic micro and nanoparticles [69]. For instance, the curing process was investigated by DSC for the Baph containing 10 wt.% 4-APN and various amounts of silicon carbide particles. The neat BAPh/4-APN matrix and all the composites showed two distinct endothermic transitions and a clear exothermic peak. The composites were easily cured in the presence of 4-APN with a reduced heat of polymerization (not more than 25 J/g). The FTIR results obtained by Derradji et al. were also in agreement with those obtained by Sheng and coworkers, concerning the final structure of the polymeric network.

It should be pointed out that the self-curing behavior can also achieved by the preparation of X-functional PN monomers/oligomers where X can include oxazine, allyl, alkyne, propargyl, and so on. The X-functional PN materials are discussed in Chapter 3, X-Functional Phthalonitrile Monomers and Polymers. Although the concept of self-catalyzed, PN monomers is new, many researchers are now actively working on developing new, autocatalytic, PN monomers to be used alone or as high-performance, curing agent. The numerous improvements in the curing behavior and processability obtained with the self-catalyzed PN-based monomer and prepolymers are also accompanied by multiple enhancements in the thermal and mechanical properties. A detailed discussion about these improvements can be found in Chapter 2, Phthalonitrile Resins' Properties and Chapter 3, X-functional phthalonitrile monomers and polymers.

1.3 Conclusions

The present chapter gives a clear idea about the tremendously interesting properties and applications of the PN resins. The curing behavior and processability, as the two main factors influencing the development and properties of the PN-based materials, are also fully discussed. Finally, this chapter provides the basics information about the PN resins and prepares the way for further discussions.

References

[1] Linstead R. 212. Phthalocyanines. Part I. A new type of synthetic colouring matters. J Chem Soc (Resumed) 1934;1016–17.

[2] Byrne G, Linstead R, Lowe A. 213. Phthalocyanines. Part II. The preparation of phthalocyanine and some metallic derivatives from o-cyanobenzamide and phthalimide. J Chem Soc (Resumed) 1934;1017–22.

[3] Linstead R, Lowe A. 214. Phthalocyanines. Part III. Preliminary experiments on the preparation of phthalocyanines from phthalonitrile. J Chem Soc (Resumed) 1934;1022–7.

[4] Dent C, Linstead R. 215. Phthalocyanines. Part IV. Copper phthalocyanines. J Chem Soc (Resumed) 1934;1027–31.

[5] Linstead R, Lowe A. 216. Phthalocyanines. Part V. The molecular weight of magnesium phthalocyanine. J Chem Soc (Resumed) 1934;1031–3.

[6] Ross SD, Fineman M. The trimer of o-phthalonitrile1. J Am Chem Soc 1950;72:3302–4.

[7] Marvel C, Martin MM. Polymeric phthalocyanines. II1. J Am Chem Soc 1958;80:6600–4.

[8] Griffith JR, O'Rear JG, Walton TR. Phthalonitrile resins. Adv chem; Am Chem Soc 1975;142:458–64.

[9] Griffith JR, Jacques G. Bisorthodinitriles. United States Patent US 4057569. 1977 Nov 8.

[10] Griffith JR, Jacques G.N., N′-bis (3, 4-dicyanophenyl) alkanediamides. United States Patent US 4056560. 1977 Nov 1.

[11] Griffith JR, Jacques G. Electrical conducting phthalonitrile polymers. United States Patent US 4102873. 1978 Jul 25.

[12] Keller TM, Griffith JR. Fluorinated phthalonitriles. United States Patent US 4209458. 1980 Jun 24.
[13] Keller TM, Griffith JR. The synthesis of highly fluorinated phthalonitrile resins and cure studies. J Fluorine Chem 1979;13:315–24.
[14] Keller TM, Price TR, Griffith JR. Synthesis of phthalonitriles by nitro displacement. Synth 1980 1980; 613-.
[15] Keller T, Griffith J. The synthesis of a new class of polyphthalocyanine resins. Resins for aerospace ACS Symp Ser 1980;25–34.
[16] Keller TM, Griffith JR. Polyphthalocyanine resins. United States Patent US 4234712. 1980 Nov 18.
[17] Keller TM, Griffith JR. Perfluorinated aliphatic phenoxy bisorthodinitriles and polyphthalocyanines therefrom. United States Patent US 4238601. 1980 Dec 8.
[18] Keller TM, Griffith JR. Aliphatic phenoxy polyphthalocyanine. United States Patent US 4223123. 1980 Sep 16.
[19] Keller TM, Griffith JR. Polyphenylether-bridged polyphthalocyanine. United States Patent US 4259471. 1981 Mar 31.
[20] Keller TM, Griffith JR. Polyphthalocyanine resins. United States Patent US 4304896. 1981 Dec 8.
[21] Ting RY, Keller TM, Price TR, Poranski Jr CF. Characterization of the cure of diether-linked phthalonitrile resins. ACS Sym Ser 1982;195:337–50.
[22] Keller TM, Price TR. Amine-cured bisphenol-linked phthalonitrile resins. J Macromol Sci Chem 1982;18:931–7.
[23] Sastri SB, Armistead JP, Keller TM. Phthalonitrile-carbon fiber composites. Polym Compos 1996;17:816–22.
[24] Sastri SB, Armistead JP, Keller TM. Phthalonitrile-glass fabric composites. Polym Compos 1997;18:48–54.
[25] Boghozian T, Stackpoole MM, Gasch M. Cyanate ester and phthalonitrile impregnated carbon ablative TPS. NASA Technical Reports. 2016.
[26] Snow AW, Griffith JR, Marullo N. Syntheses and characterization of heteroatom-bridged metal-free phthalocyanine network polymers and model compounds. Macromolecules 1984;17:1614–24.
[27] Keller TM. Phenolic-cured phthalonitrile resins. United States Patent US 4410676. 1983 Oct 18.
[28] Chen Z, Guo H, Tang H, Yang X, Xu M, Liu X. Preparation and properties of bisphenol A-based bis-phthalonitrile composite laminates. J Appl Polym Sci 2013;129:2621–8.
[29] Laskoski M, Dominguez DD, Keller TM. Synthesis and properties of a bisphenol A based phthalonitrile resin. J Polym Sci: Part A: Polym Chem 2005;43:4136–43.
[30] Xu M, Jia K, Liu X. Effect of bisphenol-A on the structures and properties of phthalonitrile-based resin containing benzoxazine. Express Polym Lett 2015;9.

[31] Luo Y, Xu M, Pan H, Jia K, Liu X. Effect of ortho-diallyl bisphenol A on the processability of phthalonitrile-based resin and their fiber-reinforced laminates. Polym Eng Sci 2016;56:150–7.
[32] Keller TM. Phthalonitrile resin from diphthalonitrile monomer and amine. United States Patent US 4408035. 1983 Oct 4.
[33] Keller TM. Synthesis and polymerization of phthalonitrile monomers containing multiple phenoxy and sulfone linkages. United States Patent US 4409382. 1983 Oct 11.
[34] Keller TM. Processable and stable conductive polymers from diether-linked bisorthodinitrile monomers. United States Patent US 4587325. 1986 May 6.
[35] Keller TM. Phthalonitrile-based conductive polymer. J Polym Sci: Part A: Polym Chem 1987;25:2569–76.
[36] Keller TM. Phthalonitrile-Based High Temperature Resin. J Polym Sci: Part A: Polym Chem 1988;26:3199–212.
[37] Keller TM. Amino phenyl containing curing agent for high performance phthalonitrile resin. United States Patent US 5003039. 1991 Mar 26.
[38] Keller TM, Price TR. Polymer of dithioether-linked phthalonitrile. United States Patent US 5004801. 1991 Apr 2.
[39] Keller TM. Imide-containing phthalonitrile resin. Polymer 1993;34:952–5.
[40] Keller TM. Synthesis of phthalonitrile resins containing ether and imide linkages. United States Patent US 5159054. 1992 Oct 27.
[41] Warzel M, Keller T. Tensile and fracture properties of a phthalonitrile polymer. Polymer 1993;34:663–6.
[42] Keller TM, Price TR. Synthesis and polymerization of dithioether-linked phthalonitrile monomers. United States Patent US 5350828. 1994 Sep 27.
[43] Burchill PJ. On the formation and properties of high-temperature resin from a bisphthalonitrile. J Polym Sci: Part A: Polym Chem 1994;32:1–8.
[44] Sastri SB, Keller TM. Phthalonitrile cure reaction with aromatic diamines. J Polym Sci: Part A: Polym Chem 1998;36:1885–90.
[45] Keller TM, Dominguez DD. High temperature resorcinol-based phthalonitrile polymer. Polymer 2005;46:4614–18.
[46] Yang X, Zhang J, Lei Y, Zhong J, Liu X. Effect of different aromatic amines on the crosslinking behavior and thermal properties of phthalonitrile oligomer containing biphenyl ethernitrile. J Appl Polym Sci 2011;121:2331–7.
[47] Laskoski M, Dominguez DD, Keller TM. Processable Phthalonitrile Resins with High-Thermal and Oxidative Stability. ACS Symp Ser 2005;922:378–88.
[48] Laskoski M, Dominguez DD, Keller TM. Synthesis and properties of aromatic ether phosphine oxide containing oligomeric phthalonitrile resins with improved oxidative stability. Polymer 2007;48:6234–40.
[49] Badshah A, Kessler MR, Heng Z, Hasan A. Synthesis and characterization of phthalonitrile resins fromortho-linked aromatic and heterocyclic monomers. Polym Inter 2014;63:465–9.

[50] Laskoski M, Dominguez DD, Keller TM. Alkyne-containing phthalonitrile resins: Controlling mechanical properties by selective curing. J Polym Sci: Part A: Polym Chem 2013;51:4774–8.
[51] Laskoski M, Neal A, Keller TM, Dominguez D, Klug CA, Saab AP. Improved synthesis of oligomeric phthalonitriles and studies designed for low temperature cure. J Polym Sci: Part A: Polym Chem 2014;52:1662–8.
[52] Laskoski M, Schear MB, Neal A, Dominguez DD, Ricks-Laskoski HL, Hervey J, et al. Improved synthesis and properties of aryl ether-based oligomeric phthalonitrile resins and polymers. Polymer 2015;67:185–91.
[53] Laskoski M, Neal A, Schear MB, Keller TM, Ricks-Laskoski HL, Saab AP. Oligomeric aliphatic-aromatic ether containing phthalonitrile resins. J Polym Sci: Part A: Polym Chem 2015;53:2186–91.
[54] Keller TM, Dominguez DD, Laskoski M. Oligomeric bisphenol A-based PEEK-like phthalonitrile-cure and polymer properties. J Polym Sci: Part A: Polym Chem 2016;54:3769–77.
[55] Laskoski M, Clarke JS, Neal A, Harvey BG, Ricks-Laskoski HL, Hervey WJ, et al. Sustainable high-temperature phthalonitrile resins derived from resveratrol and dihydroresveratrol. ChemistrySelect 2016;1:3423–7.
[56] Laskoski M, Clarke JS, Neal A, Ricks-Laskoski HL, Hervey WJ, Keller TM. Synthesis of bisphenol A-free oligomeric phthalonitrile resins with sulfone and sulfone-ketone containing backbones. J Polym Sci: Part A: Polym Chem 2016.
[57] Burchill PJ, Keller TM. Curing phthalonitrile resins with acid and amine. United States Patent US 5237045. 1993 Aug 17.
[58] Keller TM. Curing phthalonitriles with acid. United States Patent US 5247060. 1993 Sep 21.
[59] Keller TM. Synthesis and polymerization of multiple aromatic ether phthalonitriles. Chem Mater 1994;6:302–5.
[60] Zeng K, Zhou K, Tang WR, Tang Y, Zhou HF, Liu T, et al. Synthesis and curing of a novel amino-containing phthalonitrile derivative. Chi Chem Lett 2007;18:523–6.
[61] Zeng K, Zhou K, Zhou S, Hong H, Zhou H, Wang Y, et al. Studies on self-promoted cure behaviors of hydroxy-containing phthalonitrile model compounds. Eur Polym J 2009;45:1328–35.
[62] Zhou S, Hong H, Zeng K, Miao P, Zhou H, Wang Y, et al. Synthesis, characterization and self-promoted cure behaviors of a new phthalonitrile derivative 4-(4-(3, 5-diaminobenzoyl) phenoxy) phthalonitrile. Polym Bull 2009;62:581–91.
[63] Amir B, Zhou H, Liu F, Aurangzeb H. Synthesis and characterization of self-catalyzed imide-containing pthalonitrile resins. J Polym Sci: Part A: Polym Chem 2010;48:5916–20.
[64] Zhou H, Badashah A, Luo Z, Liu F, Zhao T. Preparation and property comparison of ortho, meta, and para autocatalytic phthalonitrile compounds with amino group. Polym Adv Technol 2011;22:1459–65.

[65] Badshah A, Kessler MR, Heng Z, Zaidi JH, Hameed S, Hasan A. An efficient approach to prepare ether and amide-based self-catalyzed phthalonitrile resins. Polym Chem 2013;4:3617–22.

[66] Zhang Z, Li Z, Zhou H, Lin X, Zhao T, Zhang M, et al. Self-catalyzed silicon-containing phthalonitrile resins with low melting point, excellent solubility and thermal stability. J Appl Polym Sci 2014;131 40919-7.

[67] Zhang T, Wang J, Derradji M, Ramdani N, Wang H, Lin ZW, et al. Synthesis, curing kinetics and thermal properties of a novel self-promoted fluorene-based bisphthalonitrile monomer. Thermochim Acta 2015;602:22–9.

[68] Sheng H, Peng X, Guo H, Yu X, Naito K, Qu X, et al. Synthesis of high performance bisphthalonitrile resins cured with self-catalyzed 4-aminophenoxy phthalonitrile. Thermochim Acta 2014;577:17–24.

[69] Derradji M, Ramdani N, Zhang T, Wang J, Feng TT, Wang H, et al. Mechanical and thermal properties of phthalonitrile resin reinforced with silicon carbide particles. Mater Des 2015;71:48–55.

2 Phthalonitrile Resins' Properties

2.1 Introduction

During the last few decades, phthalonitrile (PN)-based materials have been extensively studied and have emerged as a new class of high-performance polymers. In addition to their exceptional and unique thermal resistance performances, the PN resins can be merely processed into void-free shaped structures and provide a nonesuch combination of physical properties, particularly appealing for the preparation of advanced composites. The high-thermal and thermo-oxidative stabilities, superior flame resistance, low water uptake, and absence of a viscoelastic transition before the thermal decomposition, are the main features distinguishing the PN polymers from other high-performance thermosets. Nowadays, the PN resins have reached a high degree of maturity thanks to the devoted works of many talented researchers around the world. The major studies available in the open literature describe the PN resins as good candidates for replacing the currently used polymers, such as epoxies, polyimides, and cyanate esters in many applications. Some of the potential applications of these materials include adhesive, electronic, and structural components for use in extremely exigent applications, such as aerospace, aircraft, and marine fields.

The common types of PN monomers display a sluggish curing behavior, necessitating temperatures in excess of 300°C and extended time. Thus the development of the PN resins properties has been in close contact with the advances in the curing behavior and processability of the monomeric precursors. As previously discussed in the first chapter of this book, the curing can be achieved in a timely manner by adding a proper amount of a catalyst or by designing PN monomers able to self-catalyze. The researched curing agents included metals and metallic salts, phenols, organic amines, and strong organic acid/amine salt. Meanwhile the self-catalyzed, PN monomers can be afforded by the introduction of amino or hydroxyl functionalities inside the PN monomers backbones. As a result of these multiple curing paths, one PN polymer may show slight or great variations in its final properties according to the choice of the type and amount of the curing agent, as well as the choice of the curing schedule.

Phthalonitrile Resins and Composites.
DOI: https://doi.org/10.1016/B978-0-12-812966-1.00002-0

Hence, it becomes important to acknowledge the influence of each parameter on all properties to fully take advantage of the exceptional features of the PN resins.

The PN resins can be grouped into either the first-generation resins or the oligomeric second-generation PN resins. The first-generation PN are obtained by reacting 4-nitrophthalonitrile with the diphenolate salts of bisphenols in a one-step reaction process. On the other hand, the second-generation PN resins consist of oligomeric aromatic ether linkages between the two terminal PN units. These resins are obtained by a two-step reaction process. This process begins with a nucleophilic displacement reaction involving a dihaloaromatic or dihaloaliphatic compound and an appropriate bisphenol, and finishes by end-capping with 4-nitrophthalonitrile. This classification does not suggest in any way that one type surpasses the other. In fact, the need to tailor the properties of the PN resins to meet the requirements of many applications resulted in these changes in the synthetic routes.

In the course of this chapter, the thermal, mechanical, and water uptake properties of the PN resins are thoroughly discussed. The objective is to reveal the outstanding performances of these materials and to highlight the areas that need further investigation. This chapter first details the properties of the early PN resins, which can be considered fundamental in the PN field. Subsequently, the performances of the first and second-generation PN polymers are presented by focusing upon the main findings and by putting each result in close relationship to the initial conditions. Indeed, the initial conditions, such as the choice of the type and amount of the curing agent as well as the adopted curing schedule, are important parameters that define the final polymer properties.

2.2 Early Phthalonitrile Resins

The first successful attempt to prepare high-molecular-weight phthalocyanine was achieved by Griffith et al. in 1975 [1]. The researchers estimated that the weak polymerization process reported by Marvel and Martin was essentially due to the sterical hindrance of the tetranitrile structure. Therefore, they proposed the inclusion of flexible moieties between the PN ends to increase the freedom of movements of the chains, which allowed a proper orientation for the inclusion of an end into the phthalocyanine structure. Based on this principle, the authors synthetized series of PN monomers containing aliphatic diamides with various chain lengths, varying from 4 to 18. Overall, two synthetic routes were envisaged, including those through the tetraamids and through the reaction of 4-aminophthalonitrile with a chosen aliphatic diacid chloride. The first method was quickly rejected considering the purification difficulties arising from the formation of many intermediate compounds, in addition to the necessity that all the structures should be stable under the reaction conditions. Although the second method was promising, at that time difficulties regarding the preparation of the 4-nitrophthalonitrile through the direct placement of a nitro group on PN, obliged the authors to alter the procedure. Hence, it was proposed to synthetize the 4-nitrophthalonirile by diamide formation and dehydration of 4-nitrophthalimide. Subsequently, the 4-aminophthalonitrile was afforded by the reduction of the nitro group in the presence of the cyano units. Further reacting the 4-aminophthalonitrile with an appropriate aliphatic diacid chloride allowed the preparation of the desired PN monomers.

The polymerization reaction was achieved by, first, adding to the melt of these PN monomers a certain amount of clean copper flake and, then, the obtained mixtures were heated at temperatures above 170°C. Based upon the Fourier-transform infrared spectroscopy (FTIR) and the physical strength measurements, the formation of a fully cured network was confirmed, after heating the mixtures at 220°C for 48 h. The thermal stability evaluations for the cured polymers revealed that at 300°C a gradual weight loss of 2 wt.% per day was noticed. Meanwhile, the structural integrity of these PN polymers was found to be maintained for several days at this temperature. Although the thermal decomposition process was not studied, it seemed that the amide linkages as well as the unreacted compounds were behind the early weight losses. If compared with the epoxy cured polymers, these new resins showed huge improvements in the thermal resistance. Moreover, in terms of mechanical performances, these PN polymers showed strength levels satisfying the needs of many practical applications. On the other hand, the adhesive properties of these PN resins were found to be somehow lower than those of epoxies. To sum up, the work performed by Griffith et al. successfully presented the PN resins as new, high-performance polymers with many interesting features, some of them better than those of the epoxies.

In a similar way, Keller and Griffith, aiming to prepare more performant PN polymers, attempted the preparation of highly-fluorinated, PN resins by reacting fluorinated diols with tetrafluoro-PN compounds [2]. The choice of the fluoro-aromatic compounds was driven by their ability to engage in the nucleophilic aromatic substitution reaction. In fact, these compounds are extremely susceptible to attacks by even weak nucleophile. Additionally the cyano groups are also well-known for their powerful activating effect on nucleophilic substitution reactions. Although the authors successfully prepared the targeted PN monomers, their conversion to polymeric structures was not achieved even at high temperatures and in the presence of strong curing agent. Further investigations revealed that the low reactivity of these PN monomers was once more due to the inflexibility of their backbones. Therefore, more flexible fluorinated aliphatic spacers were developed to circumvent these limitations. The oxidative resistance of the resins was improved compared to the hydrocarbon spacer-linked PN resins, even when the used curing agents were metallic salts for both types of the PN resins. No weight loss was detected at temperatures below 285°C and the weight loss was around 1 wt.% per day in the temperature range of 285–300°C. In contrast with the previous PN polymers, the thermal resistance was markedly improved by the incorporation of fluoro-aromatic structures.

Considering the appealing properties of the early PN resins, Nash and Ting investigated their potential use as matrices for carbon fiber-reinforced composites [3]. The researchers chose the PN monomers containing aliphatic diamides with varying chain lengths, from 4 to 36, as potential matrices. Prior to the composites fabrications, some mechanical tests were performed on the cured polymers to select the PN matrix exhibiting the best match in terms of overall performances. Typically, four polymers issued from various PN monomers having the chain lengths of 6, 10, 22, and 36 were evaluated. The curing schedule adopted was 220°C for 24 h except for the chain length of 36, which required a 3-day cure. Interestingly, these monomers were cured without the introduction of any type of curing agent.

Torsion pendulum analyses were performed on the samples to study the structure/properties relationship of these PN polymers. Table 2.1 summarizes the glass transition temperature (T_g) and the room temperature shear modulus variations as a function of the interaromatic aliphatic chain length (number of carbon units) for the four studied PN resins. The obtained data first confirmed the remarkable performances of these PN materials. Indeed, the T_g was as high as 350°C for the chain length of 4 with a shear modulus of about 1.6 GN/m^2. Meanwhile, increasing the chain length allowed more flexibility but induced reductions in the mechanical performances. For instance, the T_g was reduced by almost 65°C when moving from the chain length of 6–10 and the same trend was observed for the shear modulus. Considering that the T_g

Table 2.1 Shear Modulus and T_g Variations as a Function of the PN Interaromatic Aliphatic Chain Length

Chain Length (n)	Shear Modulus (GN/m^2)	T_g (°C)
6	1.6	350
10	1.2	286
22	0.7	110
36	0.3	90

Data were extracted from Fig. 2 in Ting R, Nash H. The development of C-10 phthalocyanine for composite matrix applications. Polym Eng Sci 1981;21:441–8.

determines the use temperature of any polymeric material and the PN resins are designed to meet the requirement of exigent applications, one is limited to resins with interaromatic chain length of no more than 14 carbon atoms.

By examining the properties of the early PN resins, it clearly appears that this new class of thermosets possessed many advantages over the traditional high-performance polymers. In addition, the ease of preparation and the possibilities of tailoring the molecular structure and the resulting properties suggested room for larger improvements was still possible. In fact, only few years after the work of Nash and Ting was published a new class of PN materials with superior properties, known as the first-generation of PN resins, emerged as an advanced class of high-performance thermosets and the number of researchers on the subject exponentially increasing.

2.3 First-Generation Phthalonitrile Resins

As previously discussed, the early PN resins were confirmed to be excellent candidates for high-performance, fiber-reinforced composites. On the other hand, the cost of the preparation was quite expensive considering the high cost of the precursors and the low reaction yield. Therefore, considerable efforts at the Naval Research Laboratory (NRL) were focused upon the preparation of PN resins combining high performances and low cost. Thus, continuous works led to the development of a new class of PN with diether linking structure, known as the first--generation of PN resins. These PN are obtained by reacting 4-nitrophthalonitrile with the diphenolate salts of bisphenols in a one-step, reaction process. The used bisphenols include but are not limited to bisphenol A (BPA), bisphenol S, dihydroxybiphenyl, resorcinol, hydroquinone, or bisphenol-A6F (BPF). The main advantages of the bisphenols are their highly aromatized structure and more importantly their low costs. In the open literature, the first-generation, PN resins are also known as the bisphenol-based PN resins or the typical PN resins.

In contrast with the early PN monomers, the bisphenol-based PN ones can be easily obtained via a simple nucleophilic displacement reaction of a nitro-substituent from 4-nitrophthalonitrile by the alkali salt of a bisphenol in a dry dipolar aprotic, preferably under inert atmosphere. The chosen solvent or mixture of solvents needs to allow a proper dissolution of the reactants without any interference with the base. On the other hand, both weak and strong bases can be used, if they do not compete with the diphenoxide during the nitro displacement. For instance, if strong bases are used, such as sodium hydroxide or potassium hydroxide, the formation of the diphenoxide may be accompanied with water formation as a by-product. Knowing that the cyano groups can react with alkali metal hydroxide in an aqueous medium, it becomes necessary to remove the water before the addition of 4-nitrophthalonitrile. The synthetized PN monomers can be achieved with high yield and lower cost. Examples of the common chemical molecular structures of the first-generation. PN monomers and their respective abbreviations are shown in Fig. 2.1.

The polymerization reaction of these PN monomers is an addition reaction occurring through the cyano groups present at the end of the structures. However, the real problem is the activation of the first cyano groups that will further expend the reaction. While the first attempts to cure the first-generation PN monomers failed to achieve the initial goal of cost-effective materials, large efforts were dedicated to both monomer design and curing agent choice. It is important to remind that the curing agents/cure behavior relationship was previously discussed in Chapter 1, Introduction to the Phthalonitrile Resins and Their Curing Behavior, along with the impact of the monomer structure on the curing conditions. Herein, we focus on the properties and performances on various first-generation PN resins cured in the presence or not of a curing agent.

2.3.1 Thermal Resistance Properties

The first-generation PN resins are known for their outstanding thermal and thermo-oxidative stabilities. It is noteworthy to point out that a proper comparison between the PN polymers needs to be done

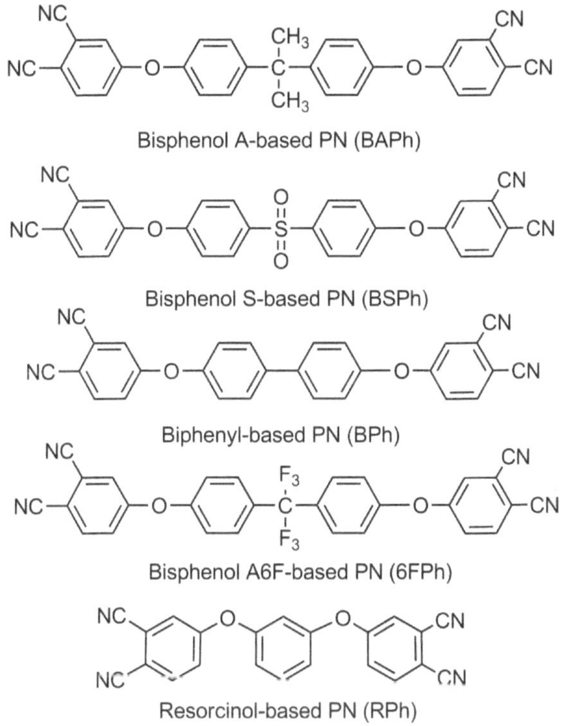

Figure 2.1 Chemical molecular structures of the typical first-generation PN monomers and their respective abbreviations.

with respect to the curing schedule, the used type and amount of curing agent, and postcure conditions. When the first-generation PN monomers, seen in Fig. 2.1, were melted and subsequently cured without any catalyst, fully polymerized structures were only obtained after a 6-day continuous cure at 280°C [4]. Thermo-oxidative stability tests were performed by isothermally curing the samples at a specific temperature and measuring the weight losses at various times. In all instances, no weight loss was depicted after heating at 250°C for 1000 h. Meanwhile, at 280°C the biphenyl-based PN (BPh) polymer only lost 1.6 wt.% of its initial weight after 3200 h. The second-most, thermally-resistant, PN polymer was the RPh followed by bisphenol S-based PN (BSPh), BAPh, and bisphenol A6F-based PN (6FPh) polymers. Moreover the thermogravimetric analysis (TGA) results under nitrogen atmosphere, presented in Fig. 2.2, revealed that no weight loss was observed up to 300°C. Further increasing the temperature initiated the weight losses in rates dependent upon the type of the polymer. Nevertheless, the previous classification was maintained and the BPh once more confirmed its superior thermal properties.

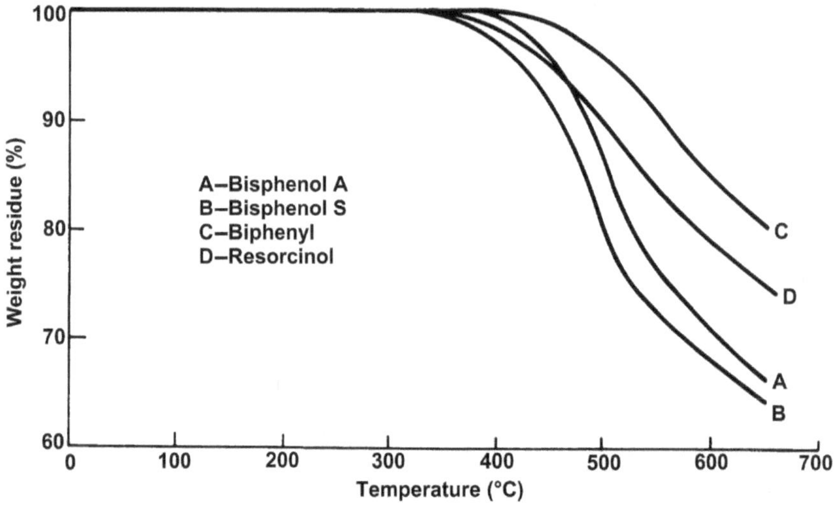

Figure 2.2 TGA curves of the BAPh, BSPh, BPh, and RPh under nitrogen atmosphere. *TGA*, thermogravimetric analysis. Reprinted from Ting RY, Keller TM, Price TR, Poranski JCF. Characterization of the cure of diether-linked phthalonitrile resins. ACS Symp Ser 1982;195:337–50.

Although the first-generation PN resins showed excellent thermal stability, curing at high temperatures, and extended time, do not qualify these materials for practical applications. Therefore there has been an unceasing need to improve the stubborn curing characteristic of the PN monomers by various means, all of them discussed in the first chapter of this book. The impact of these curing agents on the performance of the first-generation PN resins was studied since the earliest eighties. For instance, Keller and Price studied the effect of various amine-curing agents on the thermal properties of the BAPh resin [5]. The BAPh monomers and the various curing agent were melted, mixed, and heated at temperatures above the melting point till solidification. Afterwards, BAPh polymers were obtained after curing the samples between 280 and 310°C in a relatively short time.

The thermal stability of the BAPh polymers cured with and without the amine-curing agents was studied by TGA, in both nitrogen and air atmospheres, and the obtained results are summarized in Table 2.2. The BAPh polymer cured without any catalyst showed a temperature of 10 wt.% weight loss ($T_{10\%}$) of 467 and 440°C in nitrogen and air, respectively. Although these were the lowest results among all the tested samples, when compared with other, high-performance polymers, the superiority of the PN resin was still maintained. On the other hand,

Table 2.2 Thermal Properties of the BAPh Polymer as a Function of the Type and Amount of Several Amine-Curing Agents

Type of the Curing Agent	Amount of the Curing Agent (wt.%)	$T_{10\%}$ in Air (°C)	$T_{10\%}$ in Nitrogen (°C)
4-Aminophenyl sulfone	5	473	482
4-Aminophenyl sulfone	10	440	460
4,4′-(*p*-Phenylenedioxy) dianiline	3	455	485
4-Aminophenyl ether	3	490	478
4,4′-Methylene dianiline	3	480	488
–	0	440	467

Reprinted from Keller TM, Price TR. Amine-cured bisphenol-linked phthalonitrile resins. J Macromol Sci Chem 1982;18:931–7.

the amine-cured BAPh polymers exhibited improved thermal resistance, which also varied according to the type, amount, and reactivity of each amine-curing agent. Indeed, when highly reactive amines are used, only smaller amounts can greatly improve the thermal performances. Meanwhile, with less reactive amines, higher amounts are needed to achieve the targeted improvements. For instance, when 4-aminophenyl sulfone was used, which is a less reactive amine, the addition of 5 wt.% was not enough to achieve consequent ameliorations in thermal and thermo-oxidative stabilities. In fact, the reactivity of the curing agent plays an important role in the degree of conversion of the nitrile groups. With respect to the chosen curing schedule, the highly reactive amines allow greater cross-linking degrees, which give the resulting polymers improved thermal resistances. In this case, the addition of 3 wt.% of 4-aminophenyl ether was found to impart the best $T_{10\%}$ of 478 and 490°C in nitrogen and air atmospheres, respectively.

In a different report, Keller studied the effect of various, postcure conditions on the thermal resistance properties of the BPh resin cured in the presence of 1 wt.% of 4,4′-methylene dianiline (MDA) [6]. The curing agent was added to the melt of the BPh monomers followed by a vigorous stirring. The adopted curing program was as follows: 240–280°C for 5–18 h and 280–300°C for 15–20 h. Subsequently, various postcure conditions were applied from 316 to 375°C under inert atmosphere. The cured samples of BPh presented an outstanding thermal stability under both nitrogen and air atmospheres. For instance, under nitrogen atmosphere, the temperature of 5 wt.% weight loss ($T_{5\%}$)

reached about 490°C with no postcure, while postcuring up to 375°C for 24 h increased the $T_{5\%}$ to around 550°C. Under air atmosphere, the polymer started to lose weight at 400°C when cured up to 316°C for 16 h and the char yield at 800°C was measured to be 11%. Postcuring the polymer at 350°C for 4 h and at 375°C for 4 h shifted the initial weight loss temperature to 490°C. Further postcuring at 375°C for an additional 20 h also increased the initial weight loss temperature to 500°C. In addition, it was noticed that the catastrophic decomposition occurring at higher temperatures was aggravated with enhanced, postcure conditions. For the sake of comparison, another high-performance, polyimide resin (PMR-15) was cured in the same way as that of BPh (260°C for 16 h, 280°C for 1 h, and postcured at 316°C for 16 h) and the resulting thermal stability performances were compared. It was found that the temperature of initial decomposition was higher for the BPh and the rate of decomposition was also slower. Therefore this BPh polymer has been proven to be a good candidate for aerospace applications, since it surpassed the currently in-use polymers, including the PMR-15. Thermal aging studies were also conducted by heating the BPh samples at various temperatures for 100 h under an air flow of 50 cm^3/min. The tested BPh polymers were cured at 260–280°C for 15 h, at 316°C for 16 h, and at 350°C for 2 h. The results indicated that no weight loss was noticed for temperatures under 316°C, while at 357 and 372°C the BPh polymers only lost 33% and 34% of their initial weight.

During the previously discussed researches, the used amine-curing agents were found to cause void problems, resulting from the evaporation of these compounds at elevated temperatures, which negatively impacted the overall performances of the resulting PN polymers. Therefore most of the efforts were focused upon developing new kinds of organic amines. Thus, in 1991, Keller reported the preparation of an advanced amine that can cure some PN monomers at temperatures lower than their melting points. The new organic amine is known as 1,3-bis(3-aminophenoxy)benzene (*m*-APB) [7]. Later in 1998, Sastri et al. studied the impact of this curing agent on the performances of three first-generation, PN resins, namely the BPh, BAPh, and 6FPh [8]. The polymers were afforded by first adding to the melt of the PN monomers 2.56 mol.% of *m*-APB prior to curing the samples at 260, 280, and 315°C for 8 h each.

The TGA was performed on samples postcured at 375°C under nitrogen atmosphere and the resulting curves are displayed in Fig. 2.3. In all instances, the first-generation PN polymers cured with *m*-APB showed

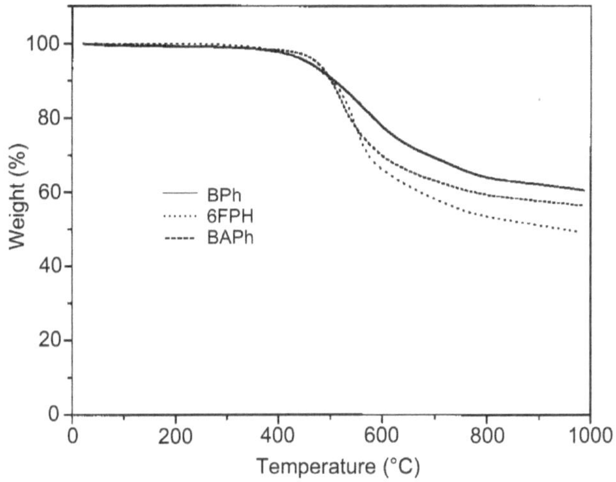

Figure 2.3 TGA curves of the BPh, 6FPh, and BAPh polymers cured in the presence of 2.56 mol.% of *m*-APB and postcured up to 375°C. *TGA*, thermogravimetric analysis; *m-APB*, 1,3-bis(3-aminophenoxy)benzene. Reprinted from Sastri SB, Keller TM. Phthalonitrile polymers: cure behavior and properties. J Polym Sci A: Polym Chem 1998;37:2105–11.

excellent thermal stability up to 450°C and upon pyrolysis the char yield at 100°C was around 60%. The superiority of the BPh was also maintained by changing the amine-curing agent. The 6FPh exhibited the lowest performances. Under air atmosphere, the polymers started to catastrophically lose weight between 500 and 600°C and, in that case, the BAPh had a slightly lower, oxidative stability than the other PN polymers.

The thermal aging was studied at both short and long terms. The short-term oxidative aging was carried out by TGA by measuring the weight losses occurring after 5 h dwell at various temperatures ranging from 250 to 375°C under an air flow rate of 100 cm^3/min. Globally, all the studied polymers showed outstanding thermal resistances with minor weight losses. The long-term oxidative aging was investigated for larger samples heated at 315 and 343°C for 100 h in an oven at the same air flow rate of 100 cm^3/min. The obtained data along with those of the PMR-15 are listed in Table 2.3. Obviously, all the first-generation PN polymers surpassed the PMR-15 with weight losses not exceeding 0.003 and 0.016 g/cm^2 at 315 and 343°C, respectively. Once again, the PN resins showed promises to replace the currently in-use, high-performance polymers in extremely exigent applications.

Table 2.3 Thermal Aging Resistance and Flammability Properties of the BPh, 6FPh, BAPh, and PMR-15 Polymers

	Weight Loss in g/cm^2 at				
Polymer	315°C	343°C	PHR (kJ/g)	THR (kJ/g)	Char Yield at 900°C (%)
BAPh	0.003	0.016	29.7	4.8	79.8
BPh	0.002	0.019	13.1	2.1	87.3
6FPh	0.002	0.007	7.2	1.5	67.5
PMR-15	0.014	0.027	–	–	–

Data from Table 1 and 2 in Sastri SB, Keller TM. Phthalonitrile polymers: cure behavior and properties. J Polym Sci A: Polym Chem 1998;37:2105–11.

The flammability properties of the BPh, BAPh, and 6FPh polymers were studied by microscale calorimeter at an incident heat flux of 50 kW/m^2. The data extracted from the experimental investigations included the peak heat release (PHR), total heat release (THR), and char yield at 900°C, all these parameters are also listed in Table 2.3. The data related to the char yield corroborated the previous TGA investigations by confirming the superiority of the BPh. As far as the heat release was concerned, the general trend was altered. In fact, the 6FPh showed the lowest PHR and THR of 7.2 kJ/g and 1.5 kJ/g, respectively. In contrast with the BPh and BAPh, the formation of noncombustible fluorocarbons for the 6FPh at elevated temperatures was behind its slightly enhanced flame resistance. Interestingly, all the PN polymers were far more superior to other, high-performance, polymers such as polyimide (kapton), polybenzimidazole, phenolic triazine (PT-30).

Although the APB curing agent allows good results, some issues regarding the loss of the amine from volatility impact on the polymerization process. This problem is more accentuated when thick sections are cured in open molds. Therefore, there was a need to develop amines with higher boiling points and decomposition temperatures. Moreover, improving the processability of the PN prepolymers by reducing the initial melt viscosities and enlarging the processing windows are important challenges to be addressed. These need to be addressed because not all high-performance, polymeric materials can be amenable to processing by resin transfer molding (RTM), vacuum assisted RTM, or filament winding. In an attempt to overcome these limitations, Sastri et al. suggested the use of two sulfone containing diamines (*m*-Bis [(aminophenoxy) phenyl]sulfone (BAPS) and *p*-BAPS) as an ameliorated amine-curing agents [9]. The effect of these curing agents on the curing behavior and processability of the PN monomers were discussed in Chapter 1, Introduction to the Phthalonitrile Resins and Their Curing Behavior and herein we report about their effect on the thermal properties.

Examples of the effect of *m*-APB and *p*-BAPS on the thermal resistance performances of a PN resin were studied by Keller and Dominguez [10]. The investigated PN resin was the first-generation RPh resin which was synthetized according to the conventional nucleophilic displacement of a nitro-substituent from 4-nitrophthalonitrile in the presence of a dipolar aprotic solvent. The polymers were prepared by first mixing the RPh monomers with various amounts of *m*-APB and *p*-BAPS and then heating the resulting prepolymers under nitrogen atmosphere following the curing schedule of 250°C for 2 h, 325°C for 2 h, 350°C for 4 h, and 375°C for 8 h.

The TGA curves of the RPh polymers under nitrogen and air atmospheres and with two different amounts of curing agents are shown in Fig. 2.4. Prior TGA investigations of the neat RPh resin indicated that the initial weight losses started at around 300°C with a thermal failure between 350 and 440°C. Thus it confirmed the importance of achieving a high degree of conversion at the initial curing temperatures. Thereby, the RPh prepolymers were first cured at 250°C to avoid their degradation during the cure reaction. As seen in Fig. 2.4A, cured samples with 2.4 mol.% of *m*-APB and *p*-BAPS, under nitrogen atmosphere, showed char yields at 1000°C of 70% and 66%, respectively. The enhanced reactivity of the *m*-APB in contrast with that of the *p*-BAPS imparted a slightly improved char yield due to the difference in the crosslink degree. In addition, increasing the amounts of the curing agents to 5.8 mol.%, resulted in RPh polymers with similar char yield at 1000°C (74 and 73 for the *m*-APB and *p*-BAPS-based RPh polymers, respectively). In fact, with more contents of catalysts, the extent of nitrile groups conversion was increased to a higher degree in which the initial differences in the reactivity did slightly affect the final cross-linking degree. Moreover, with higher amounts of catalysts, large numbers of nitrile groups were involved in the reaction resulting in less volatilization of unreacted units. In Fig. 2.4B are shown the TGA curves of the RPh polymers cured with 2.4 mol.% of *m*-APB and *p*-BAPS under air atmosphere. While both polymers exhibited similar performances up to 450°C, further increasing the temperature outlined some better thermo-oxidative stability for the *p*-BAPS-based RPh polymer. To sum up, it

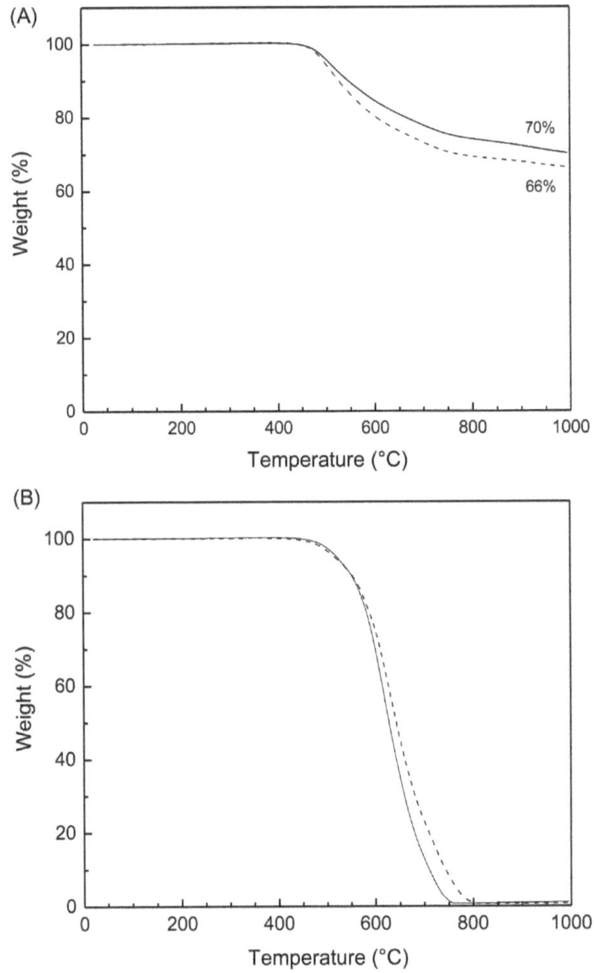

Figure 2.4 Thermal and thermo-oxidative stabilities of RPh polymers cured with 2.4 mol% of *m*-APB (solid line) and *p*-BAPS (dash line) and heated to 1000°C in nitrogen (A) and air (B) atmospheres. *m-APB*, 3-bis(3-aminophenoxy)benzene. Reprinted from Keller TM, Dominguez DD. High temperature resorcinol-based phthalonitrile polymer. Polymer 2005;46:4614–18.

was deduced that the *m*-APB and *p*-BAPS curing agents have a similar impact on the overall thermal properties of the RPh and also to the other bisphenols-based PN polymers by extrapolation.

The thermal properties of the first-generation PN resins were also investigated for monomers cured in the presence of phenolic curing agents. In a similar way as the amine-curing agent, the active hydroxyl groups attack some of the nitrile groups forming an isoindoline structure, which further reacts with other nitriles to form a cross-linked polymer structure. Ting et al. reported about the properties of the BAPh resin cured in the presence of small amounts of BPA. The molar ratio of 90/10 BAPh/BPA was found to be the most appropriate in terms of curing behavior and processability. The resulting polymer was afforded by curing the prepolymer at 210°C for 48 h, followed by a postcure at 250°C for 46 h. The TGA results indicated that the postcured samples had a better thermal stability than that of the neat PN resins cured at 280°C for 6 days.

Chen et al. also studied the effect of various amounts of BPA on the thermal and thermo-oxidative stabilities of the BAPh polymer [11]. Based on the cure behavior study, it was highlighted that the BAPh/BPA prepolymers can be cured at relatively lower temperatures. Indeed, cured polymers were achieved by only heating the samples at 200°C for 10 h, in addition to a postcure at 240°C for another 10 h. The prepared polymers were investigated by TGA and the obtained data corresponding to the $T_{5\%}$ and char yield at 800°C were compared. From the various studied molar ratios, the 2:0.5 BAPh/BPA showed the highest $T_{5\%}$ of 443 and 420°C under nitrogen and air, respectively. Further increasing the BPA contents resulted in a negative impact on both $T_{5\%}$ and char yield. The decrease in these parameters was related to the evaporation of the unreacted BPA, causing voids and weakening the polymeric structure. Nevertheless, considering the relatively lower curing temperature, the BAPh polymers cured BPA were found to exhibit excellent thermal resistances.

In all the previous reviewed reports, the major shortcoming of the used curing agents was their volatilizations at higher temperatures. Thus, dedicated efforts were targeted toward the development of more advanced curing agents which could be fully incorporated in the polymeric structures. The solution came from the PN monomers themselves. Indeed, with the development of the PN monomers bearing amine of hydroxyl groups in their backbones, many possibilities emerged. In this context, it was proposed to cure the first-generation PN monomers in the presence of other self-catalyzed, PN monomers to achieve void-free, self-catalyzed, and highly-performant, PN materials. In those cases, the introduced, self-catalytic, PN monomers were viewed as curing agents.

One of the most preferred, self-catalyzed, PN monomers is the aminophenoxyphthalonitrile (APN) which is a monofunctional PN monomer bearing an amino group at one end and a cyano group at the other end. The many interesting features of this monomer allow it to be used as stand-alone, PN polymer and more importantly as a curing agent, not only for the other PN monomers, but also for various other monomers such as epoxies and bismaleimides [12,13]. In this context, an interesting work was published by Sheng et al., in which the authors studied the effect of the 4-APN of the overall performances of two first-generation PN resins (RPh and BAPh) [14]. The RPh and BAPh were mixed with 4-APN in the weight ratios of 80:20, 90:10, and 95:5 of PN to 4-APN. The obtained PN prepolymers at their B-stage were cured in a muffle furnace following two different curing procedures. The RPh prepolymers were cured at 220°C for 2 h, 250°C for 4 h, 280°C for 6 h, and 320°C for 6 h. Meanwhile, the BAPh prepolymers were first cured at 240°C for 2 h, 280°C for 6 h, and 300°C for 6 h and further postcured 325°C for 6 h, 350°C for 6 h, and 375°C for 6 h. The differences in the cure programs are imparted to discrepancies in the curing behavior and processability.

The effect of the 4-APN on the thermal resistance properties of the RPh and BAPh polymers was studied by TGA. The $T_{5\%}$ and $T_{10\%}$ temperatures as well as the char yield at 1000°C were extrapolated from the obtained TGA curves. All these thermal stability parameters are listed in Tables 2.4 and 2.5. The first obvious observation was that increasing the 4-APN contents gradually improved the thermal performances of both first-generation PN resins. For instance, the $T_{5\%}$ and $T_{10\%}$ were measured to be 503 and 546°C, respectively. The char yield at 1000°C reached 69.8% for the 95:5 RPh/4-APN copolymer under nitrogen atmosphere. Increasing the proportion of the 4-APN to 20% improved the $T_{5\%}$, $T_{10\%}$, and char yield by 15°C, 12°C, and 1.4%, respectively.

Table 2.4 Thermal Properties of the RPh Polymers Cured in the Presence of Various Amounts of 4-APN

	RPh/4-APN	$T_{5\%}$ (°C)	$T_{10\%}$ (°C)	Char Yield at 1000°C (%)
In N_2	95:5	503	546	69.8
	90:10	512	551	70.4
	80:20	518	558	71.2
In air	95:5	498	551	–
	90:10	512	557	–
	80:20	522	564	–

Reprinted from Sheng H, Peng X, Guo H, Yu X, Naito K, Qu X, et al. Synthesis of high performance bisphthalonitrile resins cured with self-catalyzed 4-aminophenoxy phthalonitrile. Thermochim Acta 2014;577:17–24.

Table 2.5 Thermal Properties of the BAPh Polymers Cured in the Presence of Various Amounts of 4-APN

	BAPh/4-APN	$T_{5\%}$ (°C)	$T_{10\%}$ (°C)	Char Yield at 1000°C (%)
In N_2	95:5	455	484	64.5
	90:10	461	490	65.0
	80:20	472	502	70.2
In air	95:5	471	514	–
	90:10	477	521	–
	80:20	479	519	–

Reprinted from Sheng H, Peng X, Guo H, Yu X, Naito K, Qu X, et al. Synthesis of high performance bisphthalonitrile resins cured with self-catalyzed 4-aminophenoxy phthalonitrile. Thermochim Acta 2014;577:17–24.

The same trend was observed under air atmosphere and for the BAPh-based polymers. The enhancement seen upon increasing the 4-APN contents was attributed to its successful activation of the cyano addition reaction, as well as its participation in the formation of the polymeric structure, which was mainly composed of polytriazine structure in this case. Moreover, the participation of the 4-APN in the polymerization reaction ensued void-free materials. If the RPh and BAPh are compared, it seems that the latter showed slightly lower thermal resistance, which was potentially due to the early breakage of the alkyl molecular structures. In order to improve the initial BAPh polymers thermal stability, the authors postcured the samples at high temperatures as described above. The results indicated that the whole thermal resistance properties were markedly improved. For example, the $T_{5\%}$ and $T_{10\%}$ of the 80:20 BAPh/4-APN copolymer moved from 472 and 502°C to 499 and 547°C, respectively. Globally, the 4-APN was confirmed to be an advanced curing agent for the first-generation PN resins by allowing polymers with high performances and ameliorated structural integrity.

2.3.2 Mechanical and Thermomechanical Properties

The mechanical and thermomechanical properties of the first-generation PN polymers depend on many factors such as the monomer structure, curing schedule, presence or not of a curing additive, and so one. As for the thermal properties, in each case, we first define the initial conditions to allow the readers to better understand the effect of each parameter on the overall mechanical properties of the first-generation PN resins.

Ting et al. first reported about the mechanical properties of many first-generation PN resins afforded by curing their monomeric precursors without any curing agent at 280°C for 6 days [4]. The studied mechanical properties included the room temperature measurements of the fracture energy, bending modulus, shear modulus, and T_g. The obtained data for the BAPh, BSPh, BPh, and 6FPh polymers were compared with those of the tetrafunctional epoxy Narmco 5208, as seen in Table 2.6. By excluding the 6FPh, all the other, first-generation PN were marginally tougher than epoxy. Meanwhile in terms of stiffness, both PNs and epoxy were quite similar. On the other hand, the advantage of the PN polymers was clearly shown through their higher use temperatures. These results revealed that even cured without any catalyst, the first-generation PN materials still showed better properties than epoxy.

The thermomechanical properties of the first-generation BPh resin cured in the presence of 1 wt.% of MDA were investigated by dynamic mechanical analysis (DMA) as reported by Keller [6]. The thermomechanical properties including the storage modulus and damping response were studied for the BPh samples cured by various heat programs. The aim of the study was to investigate the influence of the postcure conditions on the thermomechanical properties. Sample A refers to the polymer cured at 260°C for 16 h, 300°C for 15 h, and postcured at 316°C for 16 h. Sample B were obtained by further heating the sample A at 350°C for 4 h. Similarly, sample C and D were afforded by heating the sample B for an additional 4 and 24 h at 375°C, respectively. The storage modulus variations indicated that the sample A

Table 2.6 Room Temperature Mechanical Properties of Various First-Generation PN Polymers and 5208-Type Epoxy

Sample	Fracture Energy (J/m^2)	Bending Modulus (GN/m^2)	Shear Modulus (GN/m^2)	T_g (°C)
BAPh	99	3.92	1.20	320
BSPh	202	2.25	1.32	300
BPh	140	4.35	1.22	300
6FPh	53	4.02	1.25	225
Epoxy 5208	76	3.89	1.25	260

Reprinted from Ting RY, Keller TM, Price TR, Poranski JCF. Characterization of the cure of diether-linked phthalonitrile resins. ACS Symp Ser 1982;195:337–50.

exhibited a storage modulus at 40°C of about 3.3 GPa, however, this value was reduced by about 90% at 375°C. On the other hand, toughening the postcure conditions reduced the initial values of the storage modulus to 2–2.5 GPa, but most importantly, the samples maintained a large proportion of their initial storage modulus values at high temperatures. Therefore, it was advanced that with high temperatures and extended postcure times, the rigidity of the polymers was enhanced, while the chain mobility was reduced. Those two phenomena were the direct results of an increase in the cross-linking degree. Moreover, by analyzing the damping response, it appeared that the sample A and B showed T_g values of 290 and 332°C, respectively, while the sample C and D did not show any transition to the rubbery state indicating that their respective T_g values were superior to 375°C. To sum up, this work revealed the properties of the first-generation PN resin that can be easily tailored by changes in the curing procedures.

Another interesting work was published by Warzel et al. in which the authors investigated the tensile and fracture properties of the first-generation BPh resin cured in the presence of 1.5 wt.% of *m*-APB [15]. The measurements also included the study of the effect of various cure/postcure conditions on these important mechanical properties. Preliminarily, all the sample were cured in air at 240°C for 6 h, 280°C for 16 h, and 315°C for 24 h. Subsequently, free-standing samples were postcured in an argon atmosphere at 350 and 375°C for 12 h each. Some of the resulting samples also underwent a long-term oxidative aging at 315°C for 100 h in an air-circulating oven.

The effects of the cure/postcure conditions on the room temperature tensile strength of the BPh polymer are presented in Table 2.7. The obtained data revealed that the samples cured up to 315°C without any additional postcure had the highest tensile strength of about 94 MPa. For the sake of comparison, the tensile strength of the PMR resins ranges between 48 and 83 MPa. Although some other polymers may have comparable or even better tensile strength, these polymers generally fail to maintain their initial performances at high temperatures. On the contrary, the PN resins are well-known for their outstanding thermal stability, which allows them to maintain their initial mechanical properties even at high temperatures. Indeed, the BPh polymers further postcured at 350°C for 12 h did not show any change in the tensile strength. Pushing the postcure to even higher temperatures (375°C for 12 h) only slightly affected the tensile properties. In addition, the long-term aging study of the 375°C postcure samples only reduced the initial tensile strength by 23.4%. The tensile modulus determination showed that the

Table 2.7 Effects of the Cure/Postcure Conditions on the Tensile Strength of the BPh Polymers Cured in the Presence of 1.5 wt.% of *m*-APB

	Cure/Postcure Conditions		
Temperature (°C)	Time (h)	Atmosphere	Tensile Strength (MPa)
315	24	Air	94 ± 17
350	12	Argon	94 ± 21
375	12	Argon	80 ± 7
315	100	Air	72 ± 5

Reprinted from Warzel M, Keller T. Tensile and fracture properties of a phthalonitrile polymer. Polymer 1993;34:663–6.

BPh specimens cured up to 315 and 375°C had modulus values ranging from 3.6 to 2.4 GPa, respectively. As far as the T_g was concerned, it was noticed that samples postcured up to 375°C and those long-term aged at 315°C for 100 h did not show any evidence of glass/rubber transition up to 375°C.

The fracture measurements were carried out according to the double torsion technique for the BPh specimens cured up to 315°C for 24 h. The fracture toughness (K_{ic}) of the as such cured samples ranged between 0.61 and 0.63 MN $m^{-3/2}$ with a fracture energy of 120–130 J m^{-2} and a crack speed of 70–80 mm/min. In similar conditions, the poly(methyl methacrylate) polymer exhibited a K_{ic} of about 1.3 MN $m^{-3/2}$. The first-generation PN resins are indeed high-performance thermosets displaying highly cross-linked and brittle structures. The brittleness of the PN resins is somehow akin to that of epoxies. For instance, the K_{ic} values of unmodified epoxies range between 0.5 and 0.8 MN $m^{-3/2}$.

The thermomechanical properties of the BPh resin were also compared in terms of impact of the used amine-curing agent. For instance, Sastri and Keller studied the effect of *m*-APB and *p*-BAPS on the storage modulus and T_g of the BPh polymer [9]. The study also included the investigations of the postcure conditions' effects on the same thermomechanical properties. It should be understood that all the samples underwent an initial heating at 260°C for 8 h and 280°C for 8 h. As far as the storage modulus was concerned, both *m*-APB and *p*-BAPS cured BPh showed the same behavior when it came to the postcure effect. In fact, it was once more confirmed that postcuring the PN resins at higher temperatures and extended time ensued minor changes in the storage modulus initial values at elevated temperatures. Meanwhile, the tan delta variations also confirmed the formation of highly rigid polymeric network characterized by the absence of any T_g for specimens postcured over 350°C. In terms of impact of the curing agent on the overall performances, it was noticed that the samples cured up to 315°C for 8 h in the presence of *m*-APB afforded a T_g superior by 50°C to that of the *p*-BAPS-based polymer. Nevertheless, for higher curing cycles no differences in the T_g values were noticed. These initial slight differences were attributed the higher reactivity of the *m*-APB, which allowed a rapid conversion of the nitrile groups at relatively low curing temperatures.

The flexural strengths of three first-generation PN resins were also reported by Sastri et al. [8]. The studied PN resins, namely, the BPh, BAPh, and 6FPh, were cured in the presence of 2.56 mol.% of *m*-APB following the curing schedule of 260, 280, and 315°C for 8 h each. The flexural strengths of these polymers evaluated at two different postcure conditions are presented in Fig. 2.5. After heating the specimens at 325°C for 8 h, the BAPh, BPh, and 6FPh exhibited flexural strengths of 130, 84, and 94 MPa, respectively. For an additional heating at 350 and

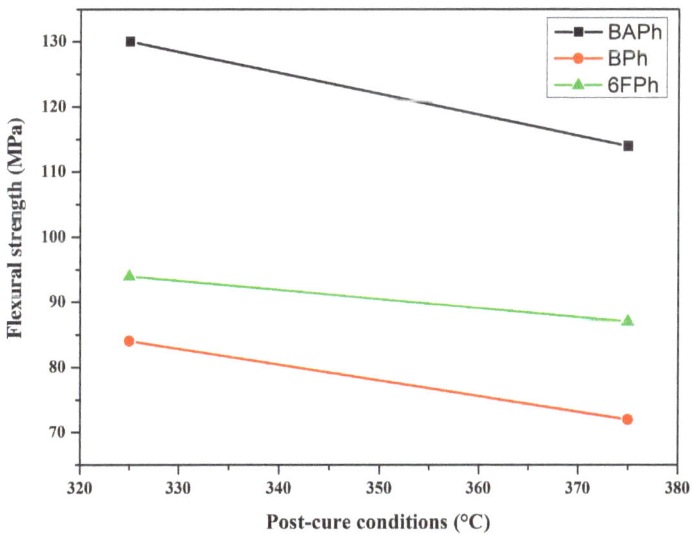

Figure 2.5 Flexural strength variations of the BPh, BAPh, and 6FPh polymers cured in the presence of 2.56 mol.% of *m*-APB as a function of the postcure conditions. *m-APB*, 3-bis(3-aminophenoxy)benzene. Data extracted from Table 3 in Sastri SB, Keller TM. Phthalonitrile polymers: cure behavior and properties. J Polym Sci A: Polym Chem 1998;37:2105–11.

375°C for 16 h each, the BAPh, BPh, and 6FPh showed a lowering in the flexural strength by 16, 7, and 12 MPa, respectively. Therefore, it was confirmed that the studied first-generation resins maintained their initial performances at high temperatures. Such kind of materials is highly appreciated in the aerospace and aircraft industries, especially for the preparation of composites designated to replace some of the heavy metallic components.

In 2005, Keller and Dominguez studied the thermomechanical properties of the first-generation RPh polymers distinctively cured in the presence of *m*-APB and *p*-BAPS [10]. The adopted cure schedule was 250°C for 2 h, 325°C for 2 h, 350°C for 4 h, and 375°C for 8 h. The storage modulus and loss factor (tan delta) variations over temperatures up to 500°C are shown in Fig. 2.6. Globally, there were no differences between the thermomechanical properties of the *m*-APB and *p*-BAPS cured RPh polymers. For instance, at 40°C both polymers exhibited storage modulus values around 1 GPa. As far as the T_g was concerned, both polymers also showed similar behavior with T_g values obviously above the range of the studied temperatures. Moreover, the RPh polymers also successfully maintained a large proportion of their initial storage modulus values at elevated temperatures. These outstanding results were justified by the formation of a highly cross-linked network with an enhanced rigidity.

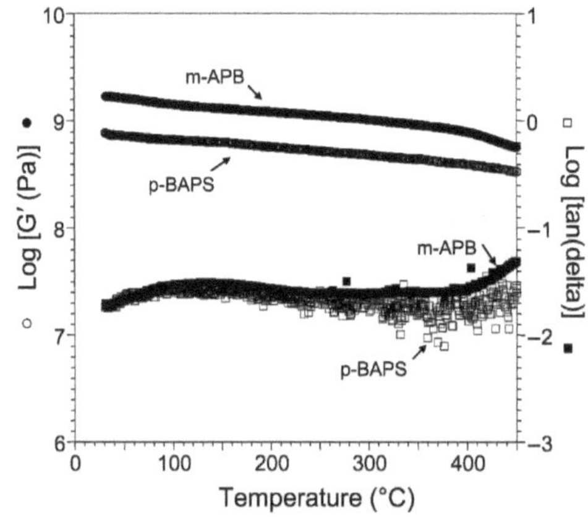

Figure 2.6 Storage modulus and loss factor (tan delta) variations for the RPh polymers distinctively cured in the presence of *m*-APB and *p*-BAPS. *m-APB*, 3-bis(3-aminophenoxy)benzene. Reprinted from Keller TM, Dominguez DD. High temperature resorcinol-based phthalonitrile polymer. Polymer 2005;46:4614–8.

The effect of another phenolic curing agent (BPA) on the thermomechanical properties of the first-generation, BAPh polymer was researched by Chen et al. [11]. In the course of their work, the authors prepared various molar ratios of BAPh/BPA and heated the resulting prepolymers up to 240°C for 10 h for DMA measurements. In all instances, the storage modulus values at low temperature were higher than those obtained with the amine-cured first-generation PN polymers. For instance, the lowest storage modulus values of about 2.7 GPa was depicted for the 2:1 BAPh/BPA polymer. Meanwhile, the best match was achieved with the 2:0.5 BAPh/BPA polymer, which not only showed a high initial modulus (about 3.9 GPa at 50°C), but also maintained the highest percentage of this value at relatively elevated temperatures. In terms of T_g values, 2:0.5 BAPh/BPA polymer exhibited the highest value of 299°C, which later decreased by increasing the contents of BPA. These observations were consistent with those of thermal stability measurements, especially when it came to the volatilization of the unreacted BPA at high temperatures and its impact on the final properties.

Similarly to what has been previously presented in the thermal resistance properties section; we herein further discuss the effect of the advanced 4-APN curing agent on the thermomechanical properties of the RPh and BAPh polymers [14]. For further information related to the preparation method, readers are referred to the previous mentioned section. The curves related to the storage modulus as well the tan delta variations for the RPh and BAPh polymers cured in the presence of various weight ratios of 4-APN are displayed in Figs. 2.7 and 2.8. The global trend indicated that increasing the contents of the 4-APN allowed an improvement in the overall performances. For instance, the 80:20 weight ratio of RPh/4-APN exhibited one of the highest low temperature storage modulus in the field of PN materials with a value of 4.3 GPa at 30°C. Such a high value was only achieved with particle-reinforced PN composites as can be seen in the works of Derradji et al. presented in Chapter 6, Ceramic-Based Phthalonitrile Micro- and Nanocomposites. The T_g value for the same weight ratio was also confirmed to be superior to 375°C. If the RPh and BAPh polymers were to be compared, the RPh polymers markedly surpassed the BAPh ones in terms of thermomechanical properties. Nonetheless, the BAPh polymers still exhibited excellent performances with a storage modulus at 30°C of about 2.5 GPa and a T_g of 325°C for the polymer issued from the 80:20 weight ratio. The ability of the 4-APN to activate and participate in the polymerization reaction was the main reason behind the enhancements seen in the thermomechanical properties.

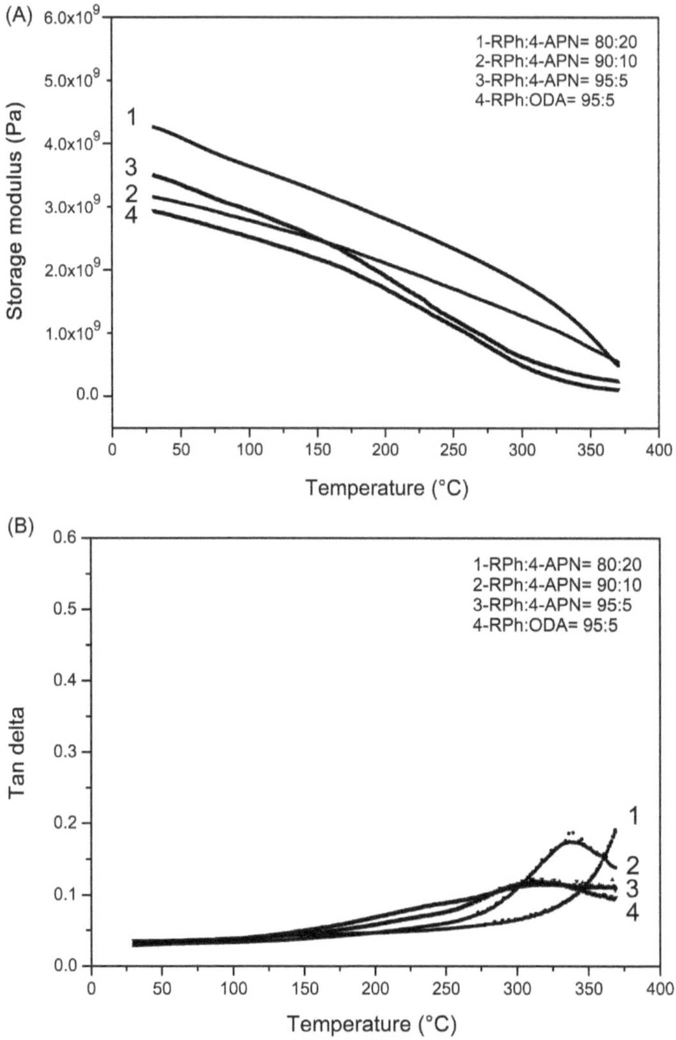

Figure 2.7 Storage modulus (A) and loss factor (tan delta) (B) variations for the RPh polymers cured in the presence of various weight ratios of 4-APN. Reprinted from Sheng H, Peng X, Guo H, Yu X, Naito K, Qu X, et al. Synthesis of high performance bisphthalonitrile resins cured with self-catalyzed 4-aminophenoxy phthalonitrile. Thermochim Acta 2014;577:17–24.

2.3.3 Water Absorption Properties

Determining the amount of absorbed water in polymeric materials is of great importance especially when it comes to marine applications where polymers are generally used as coatings to prevent against the

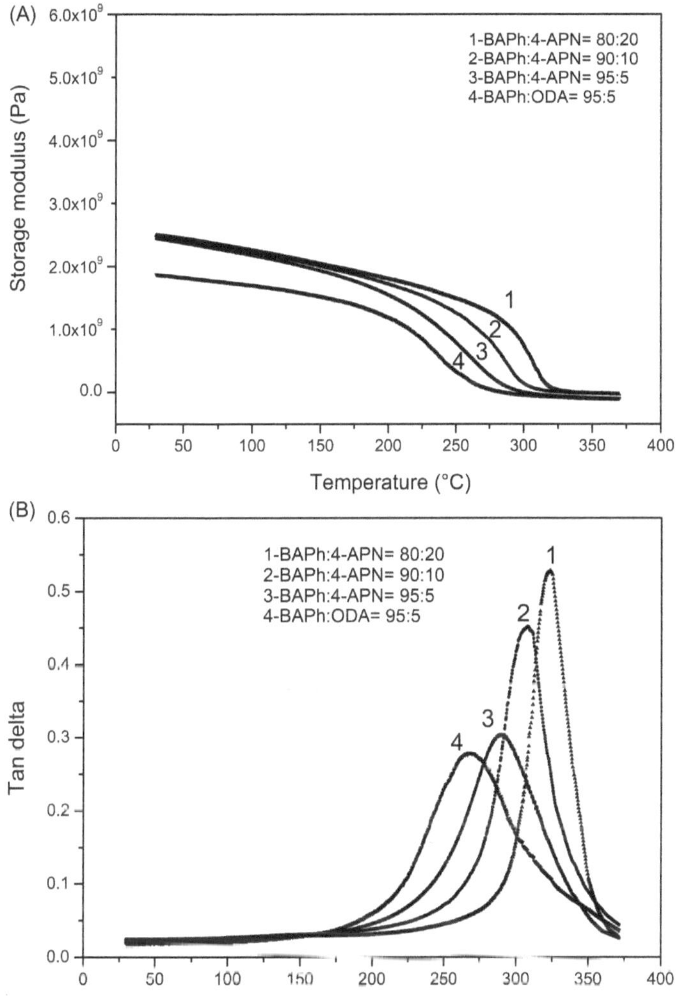

Figure 2.8 Storage modulus (A) and loss factor (tan delta) (B) variations for the BAPh polymers cured in the presence of various weight ratios of 4-APN. Reprinted from Sheng H, Peng X, Guo H, Yu X, Naito K, Qu X, et al. Synthesis of high performance bisphthalonitrile resins cured with self-catalyzed 4-aminophenoxy phthalonitrile. Thermochim Acta 2014;577:17–24.

corrosion of the metallic structures. The PN resins and precisely the first-generation of them are well-known for their water repellent behavior. The amount of absorbed water generally depends on the same factor that affects the thermal and mechanical properties.

When the first-generation PN resins cured without any catalyst at 280°C for 6 days were immersed up to 2800 h in water under ambient conditions, the results indicated that the BAPh and 6FPh only absorbed less than 1.5% [4]. On the other hand, the BSPh absorbed about 3.5% which may be attributed to the water affinity of the polar sulfone group. The introduction of curing catalyst affected the water absorption in various ways. For instance, the BAPh polymer cured with 3 wt.% of MDA showed a maximum water absorption of 0.7% after 2 months of immersion [5]. Under the same conditions, the BAPh cured with 3 wt.% of 4-aminophenyl ether absorbed more water (about 1.5%). Other investigations revealed that the BPh, BAPh, and 6FPh cured and postcured up to 375°C in the presence of 2.56 mol.% of *m*-APB showed an outstanding water repellent behavior with less than 3% of water absorption after 7 months of immersion [8]. Through these examples, it clearly appears that the PN resins are suitable for extremely exigent, marine applications. The ability of the PN resins to be used as coating materials to protect against the marine corrosion was investigated by Derradji et al. [16]. The findings showed that the PN resins resulted in better corrosion protective properties than the epoxy- and polybenzoxazine-based coatings.

2.4 Second-Generation Phthalonitrile Resins

In spite of the outstanding features of the first-generation PN polymers, their monomeric precursors have high melting points. For instance, the BPh and BAPh monomers melt at around 235 and 200°C, respectively. With such higher melting points, the polymerization reactions start immediately after the monomers are melted, resulting in short processing windows. As an example, the BPh monomers cured in the presence of *m*-APB exhibit the small processing window of about 20°C. This behavior directly impacts the preparation of large-scale, PN materials and also reduces their competitiveness with the other, high-performance polymers. Therefore there has been a need to develop cost-effective low-melting, PN resins with larger processing windows, which may be easily processed into composites using the conventional fabrication techniques. The challenge to be addressed is to improve the processability, while maintaining the other desirable properties of the first-generation PN materials.

The approach toward enhancing the processability is based on the oligomer preparation and the insertion of flexible segments into the monomer backbone. The aim of these modifications is to achieve PN oligomers that may be either liquid at room temperature or solid with relatively lower melting points. The PN materials able to meet these criteria are known as the second-generation PN resins. In contrast with the first-generation PN monomers, the second-generation PN resins are composed of oligomeric aromatic/aliphatic ether linkages between the terminal PN units. These PN resins are prepared through a two-step, one-pot, nucleophilic displacement reaction involving activated or unactivated dihaloaromatic/dihaloaliphatic compounds and an appropriate bisphenol, followed by end-capping with 4-nitrophthalonitrile. The average length of these PN oligomers can be easily tailored by modifying the molar ratio between the bisphenol (in excess) and the dihaloaromatic/dihaloaliphatic compounds. Furthermore the availability of various bisphenol and dihaloaromatic/dihaloaliphatic compounds ensures the versatility of the synthetic route. The general chemical structure of the studied second-generation PN oligomers can be found in Fig. 2.9. Meanwhile, according to the composition of each oligomer, a proper labellization has been elaborated as also depicted in Fig. 2.9. In addition, the number *n* of the repeating units inside the PN backbone will be specified accordingly.

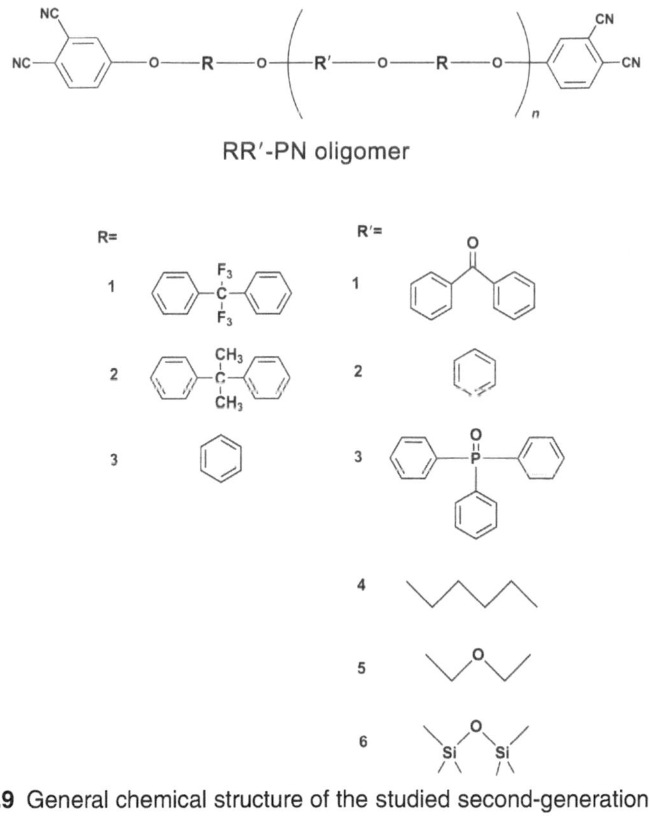

Figure 2.9 General chemical structure of the studied second-generation PN oligomers. *PN*, phthalonitrile.

The curing behavior and processability of the second-generation PN resins are contingent on the oligomer structure as well as the type and amount of the used curing agent. Indeed, as for the first-generation PN monomers, the second-generation PN oligomers need to be cured in the presence of a curing additive to achieve fully cured polymers in a timely manner. In addition, the polymers performances further depend upon the adopted curing schedule. Therefore to allow a proper comparison between these polymers, it should be understood that each result is closely related to the set of the previously listed parameters. Hence, any changes into those parameters will certainly affect the final properties. In the following parts, all the researched, second-generation PN resins are reviewed and compared with their first-generation counterparts.

2.4.1 Thermal Resistance Properties

One of the firstly investigated second-generation PN resin was the one obtained through the typical two-step, one-pot nucleophilic displacement reaction involving BPF and difluorobenzophenone compounds in the presence of either weak or strong base in *n*-methyl-2-pyrrolidone solvent [17]. The reaction was terminated by end-capping with 4-nitrophthalonitrile to afford the targeted PN oligomers. According to the proposed nomenclature, the resulting oligomers are referred to as the 11-PN. These synthetized PN oligomers were cured in the presence of 2 wt.% of either *m*-APB or TsOH. In both cases, the adopted curing schedule first included curing in air atmosphere at 250°C for 2 h, 280°C for 1 h, and 315°C for 16 h, and then the resulting samples were additionally postcured in argon atmosphere at 350°C for 2 h and 375°C for 4 h. The thermal resistance properties of the 11-PN polymers with a number $n = 1$ of repeating units were investigated by TGA analyzes under air atmosphere. To acknowledge the effect of the curing agent on these properties, the temperatures corresponding to the initial weight losses as well as those where the catastrophic thermal decompositions occurred were compared for both polymers. When cured up to 315°C without any further postcure conditions, both *m*-APB and TsOH-based 11-PN polymers showed similar thermo-oxidative stabilities. By applying the postcure, the initial decomposition temperatures were slightly different. In fact, it was observed that the *m*-APB-based 11-PN polymer started to lose weight at higher temperatures than that of the TsOH-based 11-PN polymer. In terms of catastrophic decomposition temperatures, both polymers exhibited a similar behavior.

Laskoski et al. reported about the exceptional high-thermal resistance properties of two aromatic ether-linked second-generation PN polymers. Following the typical two-step, one-pot, nucleophilic displacement reaction, the authors synthetized the two second-generation, PN oligomers (21-PN and 31-PN) by reacting BPA or resorcinol with difluorobenzophenone [18]. It should be pointed out that the number *n* of the repeating units was fixed to one for both PN oligomers. The consolidation of these oligomers to fully cured polymers was achieved by heating the oligomers in the presence of 3 wt.% *p*-BAPS under an inert atmosphere at 270°C for 12 h, 300°C for 3 h, 350°C for 6 h, 375°C for 8 h, and 425°C for 8 h. As far as the processability was concerned, the 21-PN and 31-PN oligomers were found to be completely free flowing at around 150 and 130°C, respectively, and their polymerization reaction in presence of *p*-BAPS started at about 250°C. Therefore, the processing windows of the 21-PN and 31-PN oligomers were about 100 and 120°C, which was a considerable improvement in contrast with the narrow processing windows of the first-generation PN resins.

The thermal and thermo-oxidative properties of the studied polymers were analyzed by TGA and the obtained curves are shown in Fig. 2.10.

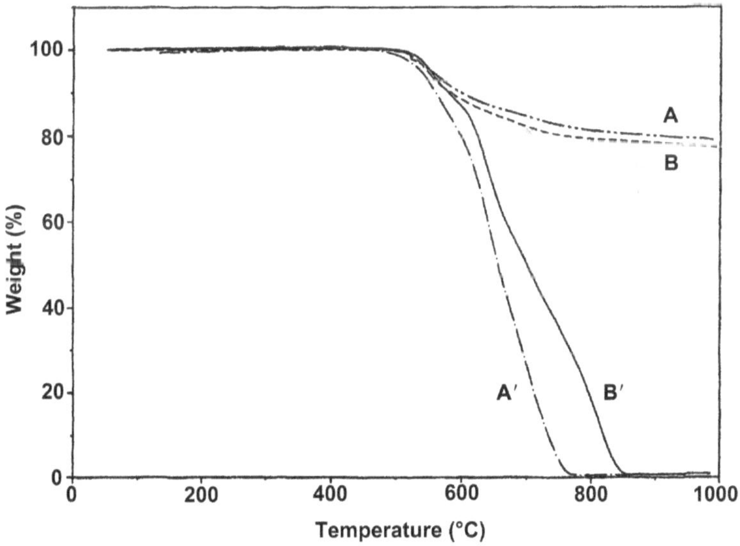

Figure 2.10 TGA curves of the 21-PN and 31-PN polymers under inert and oxidizing environments: 21-PN under nitrogen (A) and air (A′) and 31-PN under nitrogen (B) and air (B′). *TGA*, thermogravimetric analysis; *PN*, phthalonitrile. Reprinted from Laskoski M, Dominguez DD, Keller TM. Processable phthalonitrile resins with high-thermal and oxidative stability. ACS Symp Ser 2005;922:378–88.

Table 2.8 Oxidative Aging Data for the 21-PN and 31-PN Polymers

Segment	Temperature (°C)	Total Weight Loss for the 21-PN Polymer (%)	Total Weight Loss for the 31-PN Polymer (%)
A	250	0.2	0.1
B	300	0.4	0.3
C	325	1.0	0.7
D	350	3.0	1.7
E	375	8.0	5.0

Reprinted from Laskoski M, Dominguez DD., Keller TM. Processable phthalonitrile resins with high-thermal and oxidative stability. ACS Symp Ser 2005;922:378–88.

Under nitrogen atmosphere, a $T_{5\%}$ value of 545°C was measured for both 21-PN and 31-PN polymers, while the char yield at 1000°C was 79% and 77%, respectively. Under air atmosphere, the $T_{5\%}$ values were slightly different with 530 and 550°C for the 21-PN and 31-PN polymers, respectively, with thermal failures occurring for both polymers between 600 and 900°C. The long-term aging test was also performed by heating the cured polymers at various temperatures up to 375°C under air atmosphere in 8 h temperature intervals. After each isothermal step, the percentage of total weight loss was measured and the obtained results are listed in Table 2.8. In both cases, the investigations revealed that no major weight loss occurred up to 350°C, while further heating up to 375°C resulted in 8 and 5 wt.% losses for the 21-PN and 31-PN polymers. The slightly lower performances of the 21-PN polymer in contrast with those of the 31-PN one were attributed to the decomposition of the aliphatic moieties in the 21-PN polymer. Overall, both studied second-generation PN resins exhibited outstanding thermal resistance properties.

Laskoski et al. also compared the thermal properties of the second-generation 21-PN polymer with those of the first-generation BAPh one [19]. Both polymers were achieved by curing their precursors in the presence of 3 wt.% *p*-BAPS at 270°C for 12 h, 300°C for 3 h, 350°C for 6 h, and 375°C for 8 h under nitrogen atmosphere. The obtained TGA curves related to the thermal and thermo-oxidative behavior are displayed in Fig. 2.11. Globally, the BAPh polymer showed better thermal resistance performances than those of the 21-PN. For instance, under nitrogen atmosphere, the $T_{10\%}$ values for the 21-PN and BAPh

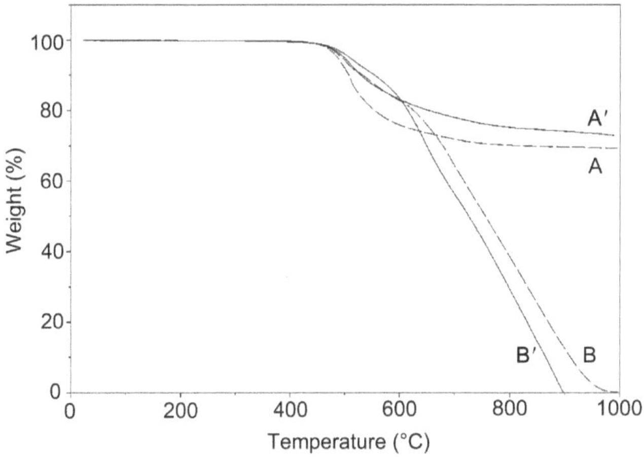

Figure 2.11 TGA curves of the 21-PN and BAPh polymers under inert and oxidizing environments: 21-PN under nitrogen (A) and air (B) and BAPh under nitrogen (A′) and air (B′). *TGA*, thermogravimetric analysis; *PN*, phthalonitrile. Reprinted from Laskoski M, Dominguez DD, Keller TM. Synthesis and properties of a bisphenol A based phthalonitrile resin. J Polym Sci A: Polym Chem 2005;43:4136–43.

polymers were measured to be about 505 and 520°C, respectively. In addition, the char yields at 1000°C were 70% and 72% for the 21-PN and BAPh polymers, respectively. It was also confirmed that improving the thermal stability of the 21-PN can be easily achieved by performing a postcure. Indeed, when postcured for an additional 4 h at 375°C and 8 h at 425°C, the 21-PN polymer exhibited a $T_{10\%}$ of 565°C and a char yield at 1000°C of 79%. When the specimens of the 21-PN and BAPh were heated in air without postcuring, they showed the $T_{10\%}$ values of about 530 and 550°C, respectively, with catastrophic decomposition occurring between 600 and 1000°C. To sum up, it can be concluded that the gain in the processability obtained with the second-generation, PN resins was achieved with minor sacrifices in the thermal resistance performances.

Dominguez et al. reported about the effect of various parameters on the thermal resistance performances of the second-generation 32-PN polymer [20]. The studied parameters included the type and amount of the used curing agent, the adopted curing schedule and the chain length between the terminal PN units. First of all, the TGA investigations were performed for the 32-PN polymers issued from oligomers containing a number $n = 2$ of repeating units and cured in the presence of various amounts of *m*-APN or *p*-BAPS. The adopted curing program was

Table 2.9 Thermal Stability Parameters of the 32-PN Polymers Issued From Oligomers Containing a Number $n = 2$ of Repeating Units and Cured in the Presence of Various Amounts of *m*-APN or *p*-BAPS

	p-BAPS		*m*-APB	
Curing Agent Amount (mole %)	$T_{5\%}$ (°C)	Char Yield at 1000°C (%)	$T_{5\%}$ (°C)	Char Yield at 1000°C (%)
2.5	494	61	489	58
5	509	68	493	62
10	513	71	509	68

250°C for 1 h, 325°C for 3 h, 350°C for 2 h, 375°C for 8 h. The data related to the $T_{5\%}$ and char yield at 1000°C are summarized in Table 2.9. By analyzing the data, it appeared that increasing the amount of the curing agent resulted in an improvement in the overall thermal performances. For instance, when increasing the amount of *p*-BAPS from 2.5 to 10 mol.%, the $T_{5\%}$ moved from 494 to 513°C and the char yield went from 61% to 71%. In fact, with a larger amount of curing agent, the extent of nitrile groups conversions was ameliorated, resulting in a higher cross-linking degree and better thermal resistance. From another perspective, it was observed that the *p*-BAPS-based polymers showed slightly better performances than those of the *m*-APB-based ones. These differences were attributed to the high-thermal stability of the *p*-BAPS in contrast with the *m*-APB.

Further investigations related to the effects of various heating procedures were carried out for the 32-PN polymers with a number $n = 2$ of repeating units and cured in the presence of 5 mol.% of *p*-BAPS. The TGA investigations revealed that increasing both the curing temperature and time positively affected the thermal resistance. For example, specimens cured at 200°C for 8 h showed a $T_{5\%}$ and char yield at 1000°C of 370°C and 41%, respectively, while an additional heating until 425°C for 8 h improved these parameters up to 528°C and 75%. These results suggested that higher heat treatments were needed to achieve fully cured, PN polymers because the extended chain lengths of the oligomeric spacer reduced the density of the functional end groups in these molecules. Therefore higher curing temperatures were needed to increase the mobility and put the reactive ends in contact.

Indeed, the chain length of the oligomers was another important parameter that needed to be investigated. Therefore the authors compared the thermal resistance properties of the 32-PN polymers issued from oligomers with a number *n* of repeating units of 1 and 2. The obtained data related to the $T_{5\%}$ and char yield at 1000°C are listed in Table 2.10. In the same table are listed the data from the first-generation RPh and BPh polymers for comparison purposes. All the studied PN polymers were thermally stable up to 500°C in both nitrogen and air atmospheres. For specimens cured up to 375°C for 8 h, the measured $T_{5\%}$ temperatures were in the ranges of 487–530°C and 479–533°C under nitrogen and air atmospheres, respectively. The first-generation PN resins were more thermally stable than the second-generation ones due to the higher concentrations of the cyano groups in those systems. To further improve the thermal stability of the 32-PN polymers an additional heating at 425°C for 8 h was performed to allow a higher degree of cyano groups conversion. The procedure was effective since consequent ameliorations were noticed upon further heating the 32-PN polymers. For example, the measured $T_{5\%}$ temperatures were in the ranges of 557–567°C and 547–551°C under nitrogen and air atmospheres, respectively. In addition, the char yields at 1000°C were also markedly improved upon heating to 425°C for 8 h. Interestingly, the second-generation 32-PN polymers were more

Table 2.10 Thermal and Thermo-Oxidative Stability of *p*-BAPS- and *m*-APB-Cured 32-PN Polymers Heated to Different Final Temperatures

	375°C for 8 h		425°C for 8 h	
	$T_{5\%}$ (°C)	Char Yield at 1000°C (%)	$T_{5\%}$ (°C)	Char Yield at 1000°C (%)
p-BAPS cured polymers				
32-PN ($n = 2$)	509	68	557	70
32-PN ($n = 1$)	501	67	567	73
BPh	511	69	–	–
RPh	530	72	–	–
m-APB-cured polymers				
32-PN ($n = 2$)	493	62	560	69
32-PN ($n = 1$)	487	55	565	72
BPh	519	72	–	–
RPh	527	73	–	–

thermally stable than the first-generation RPh and BPh when cured up to 425°C for 8 h. Hence, the previous results showed that the thermal resistance performances of the second-generation 32-PN polymers can be easily tailored to even surpass those of the first-generation PN polymers. Such PN polymers combining good processability and excellent thermal stability are promising materials with many possible applications. In summary, the work achieved by Dominguez et al. proved that scarifying the thermal stability of the PN resins for better processability was not longer necessary with the 32-PN resins.

In another interesting work, Laskoski et al. prepared series of second-generation PN oligomers bearing an aromatic ether phosphine oxide group [21]. The PN oligomers, labeled as 23-PN and 33-PN, were synthetized from the reaction of an excess amount of BPA or resorcinol with bis(4-fluorophenyl)phenylphosphine oxide in the presence of potassium carbonate (K_2CO_3) in a dimethylformamide/toluene solvent mixture, followed by reacting with 4-nitrophthalonitrile in a two-step, one-pot reaction. The number n of repeating unit in each PN oligomer was set to 1. The polymerization of these oligomers was achieved in the presence of 3 wt.% of p-BAPS. In terms of processability, the 23-PN and 33-PN oligomers were completely free flowing at around 150°C and had a processing window of about 100°C.

The TGA investigations were carried out in both nitrogen and air atmospheres for samples cured at 270°C for 12 h, 300°C for 4 h, 350°C for 4 h, and 375°C for 8 h. From the obtained TGA curves shown in Figs. 2.12 and 2.13, it appeared that the polymers exhibited outstanding thermal and thermo-oxidative stabilities. Under nitrogen atmosphere, both 23-PN and 33-PN polymers showed a similar $T_{5\%}$ of about 490°C with weight retentions at 1000°C of 56% and 65%, respectively. Under flowing air, the $T_{5\%}$ temperatures were depicted at 495 and 490°C, respectively, with catastrophic thermal decompositions happening between 600 and 800°C for both polymers.

The long-term oxidative aging properties were also explored for the 23-PN and 33-PN polymers heated from 250 to 400°C in 8 h temperature segments under air atmosphere. Table 2.11 summarizes the total weight losses after each isothermal step. The data related to the second-generation PN polymers containing benzophenone (21-PN and 31-PN) were also included in Table 2.11 for comparison purpose. The results indicated that the incorporation of phosphorus atoms in the second-generation PN oligomeric backbones significantly improved the long-term, oxidative aging performances. For instance, by curing to 375°C, the total weight losses for the 23-PN and 33-PN polymers were 4.2%

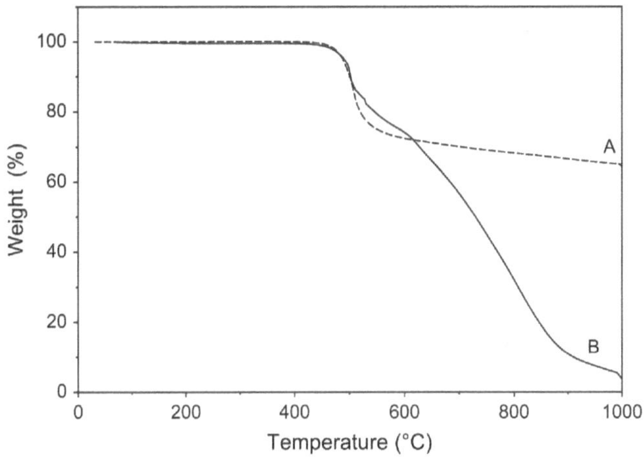

Figure 2.12 TGA curves of the 23-PN polymer under nitrogen (A) and air (B) atmospheres. *TGA*, thermogravimetric analysis; *PN*, phthalonitrile. Reprinted from Laskoski M, Dominguez DD, Keller TM. Synthesis and properties of aromatic ether phosphine oxide containing oligomeric phthalonitrile resins with improved oxidative stability. Polymer 2007;48:6234–40.

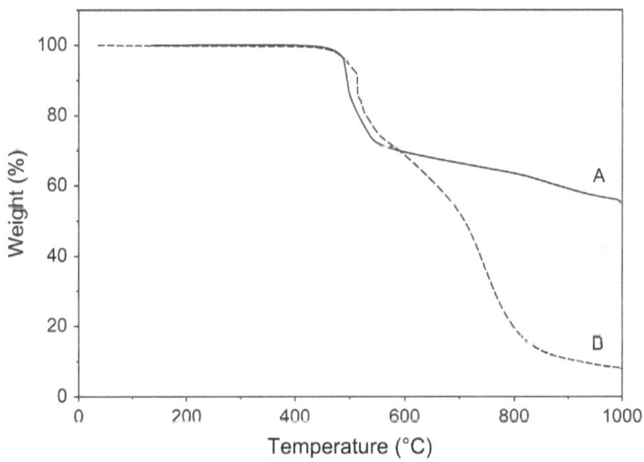

Figure 2.13 TGA curves of the 33-PN polymer under nitrogen (A) and air (B) atmospheres. *TGA*, thermogravimetric analysis; *PN*, phthalonitrile. Reprinted from Laskoski M, Dominguez DD, Keller TM. Synthesis and properties of aromatic ether phosphine oxide containing oligomeric phthalonitrile resins with improved oxidative stability. Polymer 2007;48:6234–40.

and 3.9%, respectively, while the 21-PN and 31-PN showed the elevated total weight losses of 7.2% and 4%, respectively. The same observations were applied for the specimens cured up to 400°C where the 23-PN and 33-PN polymers showed even better weight retention.

Table 2.11 Oxidative Aging Data for the 21-PN, 23-PN, 31-PN and 33-PN Polymers Heated from 250 to 400°C in 8 h Temperature Segments

	Total Weight Loss (%)			
Temperature (°C)	33-PN	31-PN	23-PN	21-PN
250	0.0	0.0	− 0.4	− 0.1
300	0.1	0.1	− 0.4	0.5
325	0.3	0.5	0.3	1.6
350	1.1	1.3	1.7	3.3
375	3.9	4.0	4.2	7.2
400	7.9	17.1	10.8	20.4

Reprinted from Laskoski M, Dominguez DD, Keller TM. Synthesis and properties of aromatic ether phosphine oxide containing oligomeric phthalonitrile resins with improved oxidative stability. Polymer 2007;48:6234–40.

Generally, it was confirmed that the inclusion of phosphorus units in the PN backbone was effectively able to improve the overall thermal resistance properties.

Other second-generation PN oligomers were synthesized by reacting in an appropriate way BPF or BPA with dibromobenzene [22]. These prepared oligomers with a number $n = 1$ of repeating units are referred to as 12-PN and 22-PN. Following the same objective of improving the processability with minor effects of the thermal stability, the authors chose to replace the generally used *p*-BAPS with *m*-BAPS. This choice was dictated by the relatively lower melting point of the latter (about 132°C) in contrast with that of the *m*-BAPS (about 192°C). The TGA investigations were performed on samples cured in the presence of 10 mol.% of *m*-BAPS at 270°C for 12 h, 300°C for 3 h, 350°C for 6 h, and 375°C for 8 h. From the obtained TGA curves seen in Fig. 2.14, it appeared that both polymers showed excellent thermal resistance properties. Under inert conditions, the 12-PN and 22-PN exhibited $T_{5\%}$ temperatures of 510 and 505°C with char yields at 1000°C of 74% and 60%, respectively. Under air atmosphere, the $T_{5\%}$ temperatures were 520 and 505°C, respectively, with catastrophic thermal decomposition starting from 600°C for both polymers. It should be pointed out that better thermal properties could be achieved for the 12-PN and 22-PN polymers, if further cured at higher temperatures. The oxidative aging studies were carried out by heating the specimens stepwise over an extended period at various temperatures in air at 8 h temperature intervals. The measured total weight

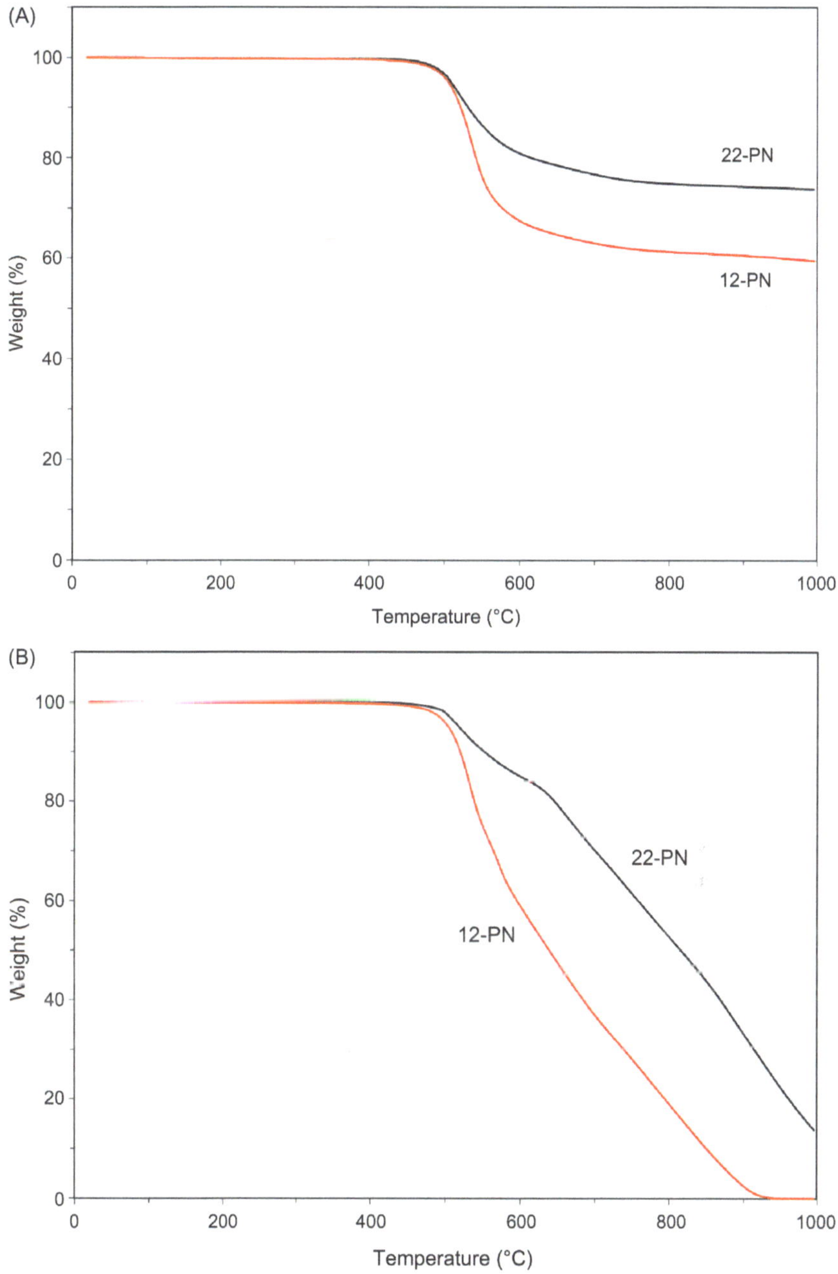

Figure 2.14 TGA curves of the 12-PN and 22-PN polymers under nitrogen (A) and air (B) atmospheres. *TGA*, thermogravimetric analysis; *PN*, phthalonitrile. Reprinted from Laskoski M, Schear MB, Neal A, Dominguez DD, Ricks-Laskoski HL, Hervey J, et al. Improved synthesis and properties of aryl ether-based oligomeric phthalonitrile resins and polymers. Polymer 2015;67:185–91.

Table 2.12 Oxidative Aging Data for the 12-PN and 22-PN Polymers

	Weight Loss (%)	
Temperature	**12-PN**	**22-PN**
250	0	0
300	0.2	0.1
325	0.6	0.8
350	1.4	2.3
375	2.7	6.8
Total	4.9	9.9

Reprinted from Laskoski M, Dominguez DD, Keller TM. Processable phthalonitrile resins with high-thermal and oxidative stability. ACS Symp Ser 2005;922:378–88.

loss after each isothermal step is presented in Table 2.12. The obtained results revealed that only minor weight losses occurred at temperatures below 375°C. Meanwhile, by heating up to 375°C for 8 h, the 12-PN and 22-PN polymers lost 4.9% and 9.9% of their initial weights, respectively. If compared with other second-generation PN polymers, the 12-PN and 22-PN polymers showed slightly lower oxidative aging stability, which may be ascribed to the shortened curing schedule. Hence, it is believed that higher curing temperatures may enhance the overall thermal resistance performances of these polymers.

The previously discussed second-generation PN resins achieved their initial objective to combine good processability with excellent thermal performances. Nonetheless, the brittleness of their fully cured polymers still remained an inherent shortcoming that required further attention. The reasons behind the lack of toughness of the PN resins are due to the high rigidity and high cross-linking degree of their polymeric structures. One way to overcome these shortcomings is through the incorporation of flexible, aliphatic linkages in the PN backbones. Meanwhile the introduction of such flexible moieties could negatively affect the thermal properties, which greatly benefit from the high rigidity and high cross-linking degree. Therefore the choice of the aliphatic moiety to be inserted in the PN backbone as well as the set of other conditions, such as the type and amount of the curing agent and curing procedure, need to be carefully chosen to avoid losing the tight balance between the thermal and mechanical properties.

Laskoski et al. first synthetized series of aliphatic-aromatic ether containing PN resins and studied their properties [23]. The studied second-generation PN oligomers were afforded by following the typical two-step, one-pot nucleophilic displacement reaction starting with the reaction between BPA and various dihaloaliphatic compounds. According to the adopted referencing system, the synthetized PN oligomers are labeled as 24-PN, 25-PN, and 26-PN. Further information regarding the chemical structures of the oligomers are provided in Fig. 2.9. The cure behavior investigations unraveled that these PN oligomers displayed lower melting points (below 60°C), while the exothermic transitions corresponding to the reaction between the nitrile groups and the chosen curing agent (in this case 10 mol.% of *m*-BAPS) were depicted at around 250°C. Therefore, the PN oligomers possessed large processing windows in the range of 190–215°C.

Based on the dynamic scanning calorimetry and rheological measurements, the curing schedule of 200°C for 16 h (overnight) and 300°C for 8 h was adopted to cure the oligomers in an adequate manner. The TGA tests were performed under both nitrogen and air atmospheres as seen in Fig. 2.15. As expected, the introduction of weak thermal linkages in the PN backbone reduced the thermal stability of the resulting polymers. Typically, the 24-PN, 25-PN, and 26-PN polymers exhibited the $T_{5\%}$ values of 430, 410, and 420°C and showed char yields of 52%, 55%, and 58%, respectively, under inert conditions. Under air atmosphere, the three polymers showed weight retention of 95% at about 420°C with thermal failures occurring between 450 and 1000°C. The long-term oxidative aging investigations shown in Table 2.13 revealed that the polymers showed no weight loss at 250°C. Further increasing the temperature to 350°C resulted in greater weight losses. For instance, the 24-PN, 25-PN, and 26-PN polymers lost 12.8%, 18.4%, and 8.0% of their initial weights at the final aging step of 350°C for 8 h. Although these values are higher than those of the multiple aromatic-based, second-generation PN polymers, these polymers still exhibited interesting thermal resistance for long-term uses below 300°C.

2.4.2 Mechanical and Thermomechanical Properties

The storage modulus and tan delta variations were studied for the 21-PN and 31-PN polymers and the results are shown in Fig. 2.16. The cured thermosets were obtained by curing their oligomeric precursors in the presence of 3 wt.% *p*-BAPS under an inert atmosphere at 270°C for 12 h, 300°C for 3 h, 350°C for 6 h, 375°C for 8 h, and 425°C for 8 h

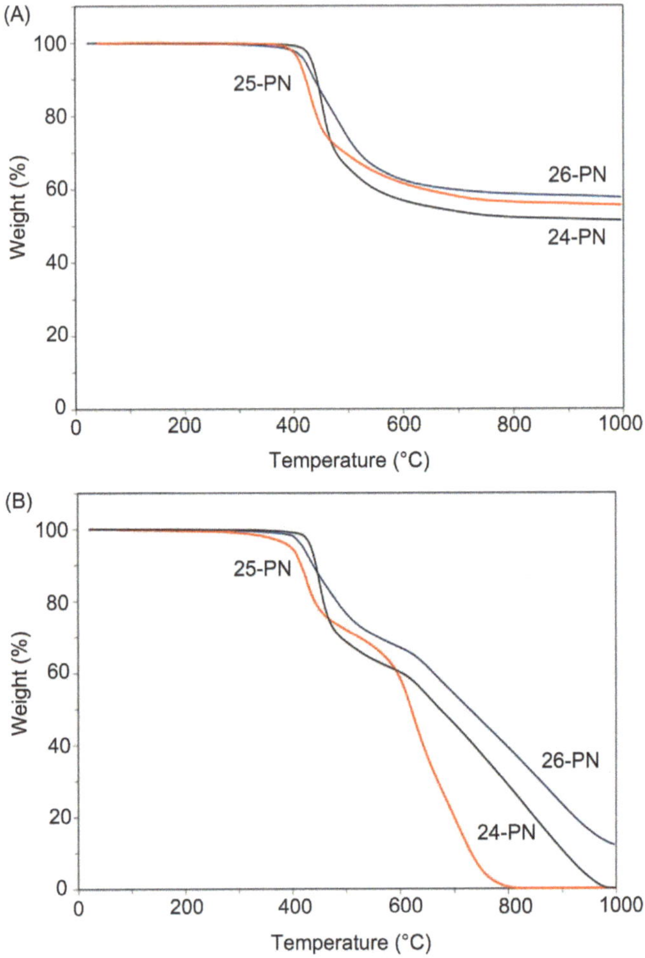

Figure 2.15 TGA curves of the 24-PN, 25-PN, and 26-PN polymers under nitrogen (A) and air (B) atmospheres. *TGA*, thermogravimetric analysis; *PN*, phthalonitrile. Reprinted from Laskoski M, Neal A, Schear MB, Keller TM, Ricks-Laskoski HL, Saab AP. Oligomeric aliphatic-aromatic ether containing phthalonitrile resins. J Polym Sci, A: Polym Chem 2015;53:2186–91.

[18]. The storage modulus of the 21-PN polymer moved from 1290 to 725 MPa when the temperature was increased from 30 to 500°C, respectively. The 31-PN polymer exhibited higher storage modulus values with a 1625 MPa at 30°C and 830 MPa at 500°C. The slightly better performances of the 31-PN were ascribed to its structure in which the PN ends were relatively much closer allowing the curing agent to easily initiate the reaction. As a result, the 31-PN polymer cross-linking

Table 2.13 Long-Term Oxidative Aging Data for the 24-PN, 25-PN, and 26-PN Polymers

	Weight Loss (%)		
Temperature (°C)	**24-PN**	**25-PN**	**26-PN**
250	0	0	0
300	0.9	2.3	1.2
325	3.3	6.7	2.5
350	8.6	9.4	4.3
Total	12.8	18.4	8

Reprinted from Laskoski M, Neal A, Schear MB, Keller TM, Ricks-Laskoski HL, Saab AP. Oligomeric aliphatic-aromatic ether containing phthalonitrile resins. J Polym Sci A: Polym Chem 2015;53:2186–91.

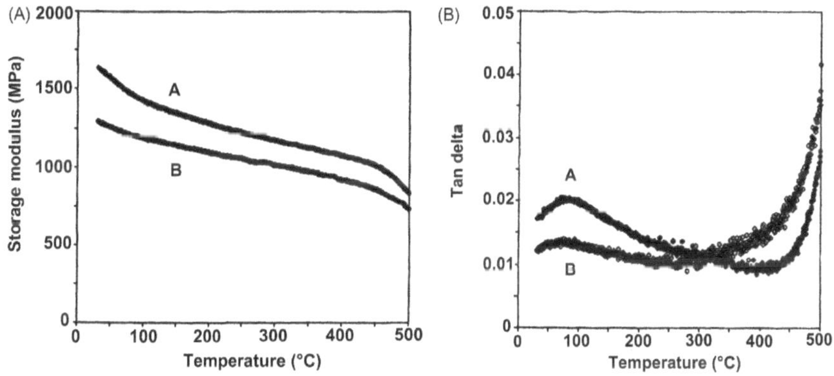

Figure 2.16 Storage modulus (A) and loss factor (tan delta) (B) variations for the 21-PN B and 31-PN A polymers. *PN*, phthalonitrile. Reprinted from Laskoski M, Dominguez DD, Keller TM. Processable phthalonitrile resins with high-thermal and oxidative stability. ACS Symp Ser 2005;922:378–88.

density was also higher than that of the 21-PN polymer. In terms of tan delta variations, both polymers showed peaks starting from 400°C with no well-defined maximum up to 500°C confirming that the T_g values of these polymers were higher than 500°C. What is also interesting is that the 21-PN and 31-PN polymers maintained 56% and 51% of their initial storage modulus at 500°C. Such behavior confirmed that these second-generation PN polymers can maintain their structural integrity at elevated temperatures.

The mechanical performances of the BPA-based first and second-generation PN polymers (BAPh and 21-PN) were also compared by Laskoski et al. [19]. The polymers were afforded by curing their respective precursors in the presence of 3 wt.% of *p*-BAPS to 375°C under akin conditions. The storage modulus and tan delta variations over temperature are displayed in Fig. 2.17. By analyzing the data, it was observed that both polymers had the same initial storage modulus value of 1260 MPa at 30°C. By increasing the temperature to 450°C, the BAPh polymer retained 240 MPa of its initial storage modulus, while the

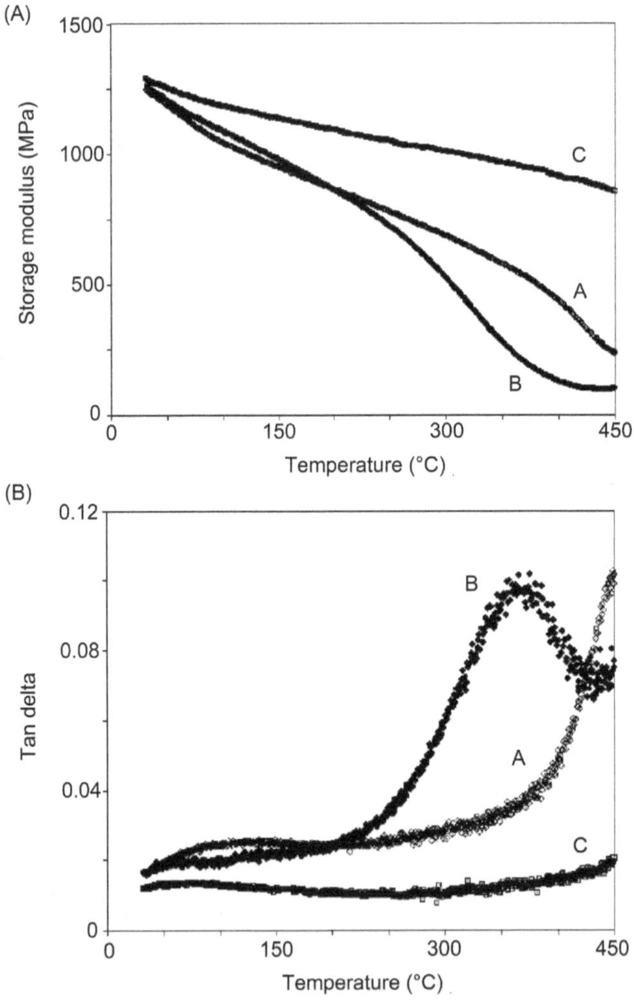

Figure 2.17 Storage modulus (A) and loss factor (tan delta) (B) variations for the 21-PN and BAPh polymers. *PN*, phthalonitrile. Reprinted from Laskoski M, Dominguez DD, Keller TM. Synthesis and properties of a bisphenol A based phthalonitrile resin. J Polym Sci, A: Polym Chem 2005;43:4136–43.

21-PN polymer only maintained 100 MPa. The same explanations given for the 21-PN and 31-PN are also valid in this case. Indeed, with their reduced chain length the BAPh monomers are more easily to be cured to a larger extent, resulting in higher cross-linking density and better stiffness. In order to increase the cross-linking density of the 21-PN polymers, samples were postcured for an additional 4 h at 375°C and 8 h at 425°C. The obtained results, also shown in Fig. 2.17, revealed a significant amelioration in the storage modulus which reached about 850 MPa at 450°C. Hence, postcuring the 21-PN polymers was a good alternative to achieve better properties and improve the structural integrity at high temperatures. As far as the T_g was concerned, the tan delta variations for samples cured to 375°C further confirmed that the high rigidity and high cross-linking density of the BAPh polymer played a crucial role in the final results. For instance, the BAPh and 21-PN polymers had T_g values of 450 and 365°C, respectively. Meanwhile, it was once more confirmed that postcuring the samples at higher temperatures allowed consequent improvement in the T_g which was superior to 450°C for the 21-PN samples heated up to 425°C.

The thermomechanical properties of the second-generation PN resins bearing phosphine oxide moieties were also investigated [21]. The 33-PN oligomers were cured with 3 wt.% of *p*-BAPS at 270°C for 12 h, 300°C for 4 h, 350°C for 4 h, and 375°C for 8 h. The DMA were carried out from 30 to 400°C, and the resulting storage modulus and tan delta variations as a function of temperature are presented in Fig. 2.18. Interestingly, no visible changes characteristic of a viscoelastic transition to a rubbery state were depicted. Similarly, to other PN polymers, the 33-PN storage modulus decreased by increasing the temperature. In fact, the storage modulus went from 2780 to 580 MPa when the temperature was increased from 30 to 400°C. What is more important is that the introduction of the phosphine linkages markedly increased the stiffness of the resulting PN polymer. Indeed, the benzophenone- and benzene-based second-generation PN polymers showed initial storage modulus values lower than 1500 MPa. The tan delta variations revealed the presence of two minor transitions at 98 and 177°C, in addition to a broader transition starting from 300°C. The two first transitions were ascribed to the some mobility in the backbone. Similarly, other phosphine oxide-based second-generation PN polymers such as the 23-PN showed quite similar thermomechanical behavior.

Globally, the mechanical properties of the second-generation PN resins were limited to the dynamic mechanical properties' investigations and no report can be found on any other mechanical test. The reason behind this lack of researchers is only a matter of time and consideration of the

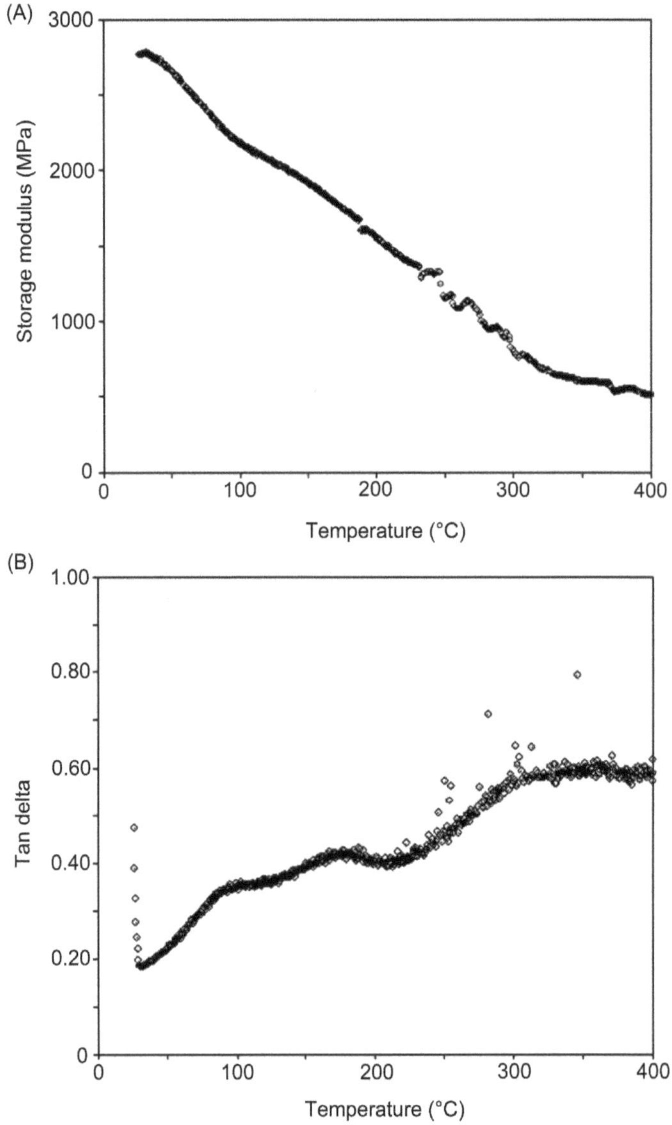

Figure 2.18 Storage modulus (A) and loss factor (tan delta) (B) variations for the 33-PN polymer. *PN*, phthalonitrile. Reprinted from Laskoski M, Dominguez DD, Keller TM. Synthesis and properties of aromatic ether phosphine oxide containing oligomeric phthalonitrile resins with improved oxidative stability. Polymer 2007;48:6234–40.

many interesting features of these resins, we expect an exponential increase in the number of researchers on the subject. Indeed, it is one of the objectives of this book to introduce the PN-based materials hoping to attract as many talented researchers as possible.

2.4.3 Water Absorption Properties

The water absorption properties of the second-generation PN resins are regrouped in Table 2.14. First of all, all the tested polymers showed excellent water repellent properties with no more than 2.3 wt.% of total water absorption. Therefore, the second-generation PN resins can easily

Table 2.14 Water Uptake of the Second-Generation PN Polymers as a Function of the Curing and Test Conditions

PN Designation	Curing Conditions	Test Conditions	Total Water Absorbed (wt.%)	Reference
21-PN 31-PN	270°C for 12 h, 300°C for 3 h, 350°C for 6 h, 375°C for 8 h, and 425°C for 8 h, in the presence of 3 wt.% *p*-BAPS	45 days at ambient temperature in distilled water	1.3 1.4	[18]
21-PN	270°C for 12 h, 300°C for 3 h, 350°C for 6 h, and 375°C for 8 h, in the presence of 3 wt.% *p*-BAPS	50 days at ambient temperature in distilled water	2.3	[19]
21-PN	270°C for 16 h, 300°C for 4 h, 350°C for 6 h, 375°C for 6 h, and 415°C for 2 h, in the presence of 5 wt.% *p*-BAPS or Cu(AcAc)/TsOH (3.75/1.5 wt.%)	1 day at 100°C in distilled water	1.5	[24]
12-PN 22-PN	270°C for 12 h, 300°C for 3 h, 350°C for 6 h, and 375°C for 8 h, in the presence of 10 mol.% *m*-BAPS	1 day at 100°C in distilled water	0.3 0.9	[22]
24-PN 25-PN 26-PN	200°C for 16 h, and 300°C for 8 h, in the presence of 10 mol.% *m*-BAPS	30 days at ambient temperature in distilled water	1.5 1.5 1.5	[23]

meet the requirements of marine applications, especially for use as advanced polymeric coatings. In addition, these polymers can also be used in any other application requiring materials with high water resistance and excellent thermal and mechanical properties. By further analyzing the data, it also appeared that the adopted curing schedule greatly affected the total water absorption. For instance, the 21-PN polymer cured up to 375°C had a total water absorption of 2.3%, while the same polymer cured up to 425°C showed the lowest value of 1.3%. Similar to the explanations given in the thermal and mechanical sections, pushing the curing toward higher temperatures allows more rigidity and an enhanced cross-linking degree, which directly affects the overall polymer performances. Moreover it was confirmed that changing the curing agent had little effects on the water absorption if the polymers were cured until full transformation of the nitrile groups. For example, the 21-PN polymers cured in the presence of 5 wt.% *p*-BAPS or Cu(AcAc)/TsOH (3.75/1.5 wt.%) up to 415°C displayed the same total water absorption of 1.5%. In terms of molecular compositions, the PN polymers containing benzene in the repeating unit ensued better water repellent properties than the benzophenone-based PN polymers. What is more interesting is that the aliphatic-aromatic ether containing PN resins (24-PN, 25-PN, and 26-PN), cured at relatively lower temperatures, exhibited water repellent properties similar to those of the highly aromatized PN polymers.

2.5 Actual and Potential Applications of the Phthalonitrile Resins

Owing to their combination of excellent properties, the PN resins have entered the field of industry and claimed the leading place as advanced, high-temperature polymers for extreme environmental applications. Generally, the PN-based materials are designated to meet the requirements of extremely exigent military and civilian applications (see Fig. 2.19). For instance, in the aerospace industry, the PN resins can be incorporated in the internal components of commercial aircraft, space vehicles, supersonic military aircraft, missiles, rockets, turbine engine applications, second-generation shuttle, satellite components, and engine components on aircraft [25]. In addition, the PN resins can be used as advanced coatings or as a part of structural components in the surface ships and submarines. In fact, the PN resins are the only

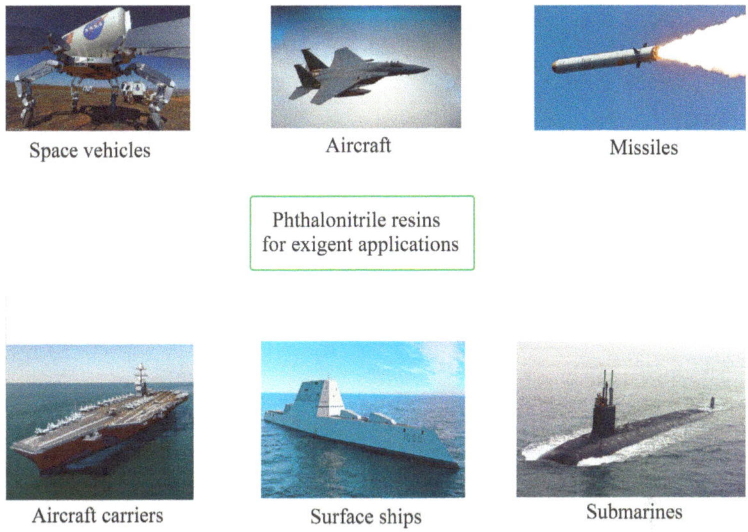

Figure 2.19 Some of the PN resins applications in aerospace and military. *PN*, phthalonitrile.

polymeric material that meets the MIL-STD-2031 flame test requirements for usage inside submarines [26]. Other applications include but are not limited to high-temperature adhesives, oil rig components, high-temperature sizing or coating material for fibers, tooling, car battery casing, geothermal tooling, microelectronic device applications, and printed circuit boards. Meanwhile, considering the high technological impact of the PN resins, their actual use is often classified. Therefore, the potential and actual applications of these polymers may be overlapped. One of the recent reports from the NRL indicated that the second-generation poly(ether ether ketone) (PEEK)-like PN or herein referred to as the 21-PN can be used to manufacture composites for various applications (see Fig. 2.20), such as lightweight automobiles, ships, oil rigs, aircraft, wind blades, high-temperature bearings, valves, battery and electronic casings, fire-resistant textiles, robotic and autonomous firefighting on ships, ammunition casings and storage containers, and fire-resistant building materials [27]. What is more important is the ability to tailor the properties of the PN resins to various needs and applications conditions.

One of the most interesting applications of the PN resins arises from the ease of their conversion into carbon nanotubes. As a matter of facts, the PN resins have been successfully used to synthesize multiwalled, carbon nanotubes in bulk solid compositions [28]. The synthetic route

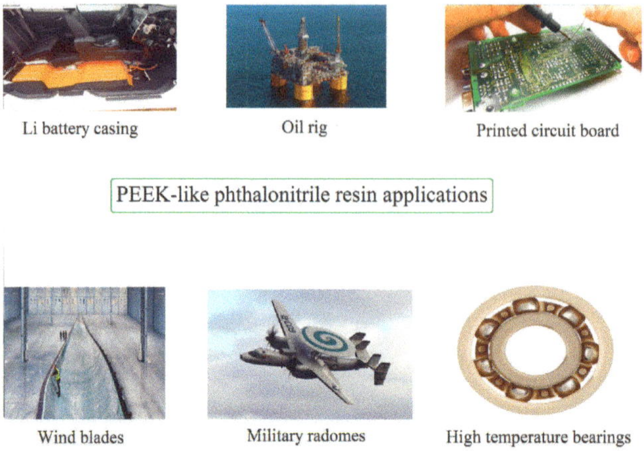

Figure 2.20 Some of the PEEK-like PN or 21-PN resin applications. *PN*, phthalonitrile.

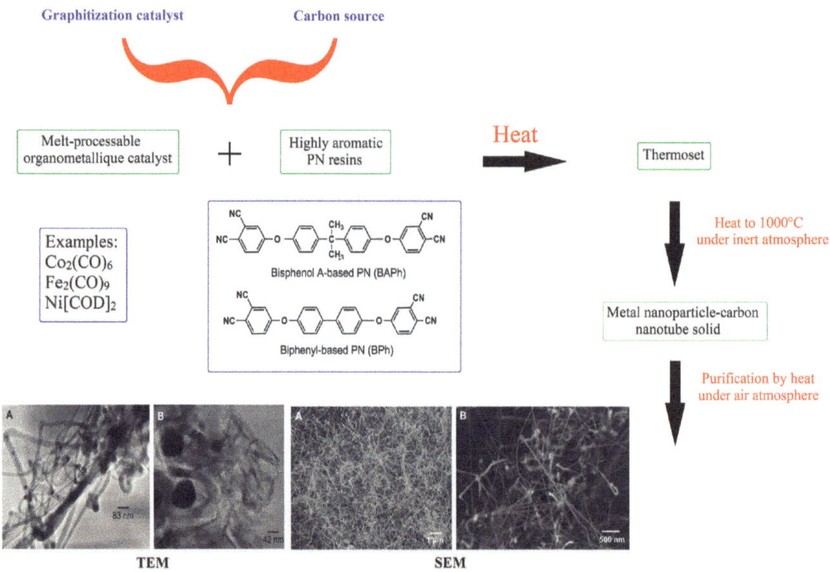

Scheme 2.1 Processes of making carbon nanotubes from PN precursors. *PN*, phthalonitrile.

for the preparation of the carbon nanotubes is shown in Scheme 2.1. Generally, the PN monomers/oligomers are mixed with a metal salt such as $Co_2(CO)_8$ and heated to form the thermoset. The obtained thermoset is then carbonized under argon or nitrogen atmospheres up to 1200°C. Finally, the purification step takes place by heating under an air atmosphere between 300 and 500°C. The formation of the carbon nanotubes starts with the decomposition of the metallic salts yielding for instance Co atoms, which react with the developing fused aromatic rings from the PN precursors, and the ultimate formation of the metallic nanoparticles. Further proceeding with the carbonization process (> 500°C) induces the formation of a reactive carbon–cobalt nanocompound, which can be designated as the carbon nanotubes precursors. As the mixture undergoes further heating at about 800°C, the carbon–cobalt nanocompound is transformed into multiwalled, carbon nanotubes and cobalt nanoparticles. The SEM and TEM observations, also presented in Scheme 2.1, show the carbon nanotubes prepared from PN precursors. Additionally, carbon fibers can be easily afforded by drawing from the melt of the PN and the graphitization catalyst at elevated temperatures or by the conventional spinning techniques followed by carbonization. Overall, the preparation of carbon nanotubes/fibers through this method allows a low cost and large-scale production of these materials. Moreover, this process also permits the simple incorporation of heteroatoms and multiple metal nanoparticles. Indeed, hybrid compositions of nanotubes and metal nanoparticles may have useful structural, catalytic, electric, and/or magnetic properties.

By being designated for extremely exigent applications, the PN resins may be subjected to highly ionizing radiations such as the gamma rays. A NASA report indicated that all the materials (metal, ceramic, and polymer) entering in the fabrication of space vehicles or any related aerospace machinery may be severely damaged by nuclear particles and electromagnetic radiations from both external or onboard sources [29]. The polymeric-based structures were found to be the most vulnerable, resulting in significant degradations in performances. Thus many specialists in the field are looking for high-performance polymers with enhanced nuclear resistance for aerospace applications. Preliminary studies, performed by Derradji et al. *[unpublished results]*, revealed that the highly cross-linked, polymeric structure of the PN resins was quite benefic to the radiation resistance and shielding efficiency. For instance, a 2 cm thick sample of the first-generation BAPh cured resin sample subjected to 1.17 MeV of highly ionizing gamma rays successfully stopped 17% of the incident energy. In contrast with the typical epoxy resin which only stopped 18% for a 3 cm thick sample, the PN resin showed better screening performances. In terms of radiation resistance, the preliminary results indicated that up to 200 kGy of absorbed dose, no significant damage was detected in the main polymeric structure of the BAPh cured polymer. In the meantime, some enhancements were noted in the thermomechanical properties upon exposition to doses lower than 200 kGy, which indicated that the highly ionizing, gamma rays activated some of the unreacted nitrile groups. On the other hand, the epoxy resin showed a significant increase in the chain scissions for absorbed doses superior to 5 kGy. Recently, another NASA report also stated that the PN-based materials could be a good alternative to the currently in-use phenolic resins for aerospace applications. However, the study did not report about the nuclear and space radiations effects of the PN resins [30]. Overall, all these promising results confirm the great potential of the PN-based materials in the field of nuclear radiation resistance and shielding.

2.6 Conclusions

The present chapter not only effectively reviewed the extensive works on the subject but also clearly revealed the exceptional properties of the common types of PN resins. The main advantages of the PN polymers, including their outstanding thermal resistance, low water uptake, and high thermomechanical properties were highlighted and thoroughly discussed. Meanwhile, by putting each result in its context and by comparing the influence of various parameters on the final properties, new comers as well as other interested researchers can easily catch up with the earliest and latest developments in the field.

References

[1] Griffith JR, O'Rear JG, Walton TR. Phthalonitrile resins, vol. 142, ACS publications, Washington. 1975. p. 458–64.
[2] Keller TM, Griffith JR. The synthesis of highly fluorinated phthalonitrile resins and cure studies. J Fluorine Chem 1979;13:315–24.
[3] Ting R, Nash H. The development of C-10 phthalocyanine for composite matrix applications. Polym Eng Sci 1981;21:441–8.
[4] Ting RY, Keller TM, Price TR, Poranski JCF. Characterization of the cure of diether-linked phthalonitrile resins. ACS Symp Ser 1982;195:337–50.
[5] Keller TM, Price TR. Amine-cured bisphenol-linked phthalonitrile resins. J Macromol Sci Chem 1982;18:931–7.
[6] Keller TM. Phthalonitrile-based high temperature resin. J Polym Sci, A: Polym Chem 1988;26:3199–212.
[7] Keller T.M. Amino phenyl containing curing agent for high performance phthalonitrile resin. United States Patent US5003039. 1991 Mar 26.

[8] Sastri SB, Keller TM. Phthalonitrile polymers: cure behavior and properties. J Polym Sci A: Polym Chem 1998;37:2105–11.
[9] Sastri SB, Keller TM. Phthalonitrile cure reaction with aromatic diamines. J Polym Sci A: Polym Chem 1998;36:1885–90.
[10] Keller TM, Dominguez DD. High temperature resorcinol-based phthalonitrile polymer. Polymer 2005;46:4614–18.
[11] Chen Z, Yang X, Xu M, Liu X. Preparation and properties of bisphenol A-based bisphthalonitrile polymers. High Perform Polym 2013;26:3–11.
[12] Guo H, Zou Y, Chen Z, Zhang J, Zhan Y, Yang J, et al. Effects of self-promoted curing behaviors on properties of phthalonitrile/epoxy copolymer. High Perform Polym 2012;24:571–9.
[13] Huang Y, Jiang M, Xu M, Zou X, Luo Y, Lei Y, et al. Curing behavior and processability of BMI/3-APN system for advanced glass fiber composite laminates. J Appl Polym Sci 2016;133:43640-7.
[14] Sheng H, Peng X, Guo H, Yu X, Naito K, Qu X, et al. Synthesis of high performance bisphthalonitrile resins cured with self-catalyzed 4-aminophenoxy phthalonitrile. Thermochim Acta 2014;577:17–24.
[15] Warzel M, Keller T. Tensile and fracture properties of a phthalonitrile polymer. Polymer 1993;34:663–6.
[16] Derradji M, Ramdani N, Zhang T, Wang J, Gong L-d, Xu X-d, et al. Effect of silane surface modified titania nanoparticles on the thermal, mechanical, and corrosion protective properties of a bisphenol-A based phthalonitrile resin. Prog Org Coat 2016;90:34–43.
[17] Keller TM. Synthesis and polymerization of multiple aromatic ether phthalonitriles. Chem Mater 1994;6:302–5.
[18] Laskoski M, Dominguez DD, Keller TM. Processable phthalonitrile resins with high-thermal and oxidative stability. ACS Symp Ser 2005;922:378–88.
[19] Laskoski M, Dominguez DD, Keller TM. Synthesis and properties of a bisphenol A based phthalonitrile resin. J Polym Sci A: Polym Chem 2005;43:4136–43.
[20] Dominguez DD, Keller TM. Low-melting phthalonitrile oligomers: preparation, polymerization and polymer properties. High Perform Polym 2006;18:283–304.
[21] Laskoski M, Dominguez DD, Keller TM. Synthesis and properties of aromatic ether phosphine oxide containing oligomeric phthalonitrile resins with improved oxidative stability. Polymer 2007;48:6234–40.
[22] Laskoski M, Schear MB, Neal A, Dominguez DD, Ricks-Laskoski HL, Hervey J, et al. Improved synthesis and properties of aryl ether-based oligomeric phthalonitrile resins and polymers. Polymer 2015;67:185–91.
[23] Laskoski M, Neal A, Schear MB, Keller TM, Ricks-Laskoski HL, Saab AP. Oligomeric aliphatic-aromatic ether containing phthalonitrile resins. J Polym Sci A: Polym Chem 2015;53:2186–91.
[24] Laskoski M, Neal A, Keller TM, Dominguez DD, Klug CA, Saab AP. Improved synthesis of oligomeric phthalonitriles and studies designed for low temperature cure. J Polym Sci, A: Polym Chem 2014;52:1662–8.

[25] Keller TM, Laskoski M, Manoj KV, Dominguez DD. High temperature polymers for extreme environmental applications. In: Proc. 2012 AFOSR/ Australian National Fabrication Facility (ANFF) program review; 2012.

[26] Keller TM, Laskoski M. Synthesis of and curing additives for phthalonitriles. United States Patent US8981063. 2015 Mar 17.

[27] Parry D. NRL licenses new polymer resin for commercial applications. NRL releases 2015.

[28] Laskoski M, Keller TM, Qadri SB. Direct conversion of highly aromatic phthalonitrile thermosetting resins into carbon nanotube containing solids. Polymer 2007;48:7484−9.

[29] Shulman H, Ginell WS. Nuclear and space radiation effects on materials. NASA technical reports. 1970.

[30] Boghozian T., Stackpoole M.M., Gasch M. Cyanate ester and phthalonitrile impregnated carbon ablative TPS. NASA technical reports; 2016.

3 X-Functional Phthalonitrile Monomers and Polymers

3.1 Introduction

During the last few years, the phthalonitrile (PN) resins have gained much attention owing to their exceptional properties. The PN resins possess outstanding thermal and thermo-oxidative stabilities with initial decompositions temperatures over 450°C for the typical PN resins. The enhanced thermal resistance of these resins is also accompanied with a myriad of other highly attractive performances, such as a superior flame-resistance, absence of a glass transition temperature (T_g) before the thermal decomposition, good mechanical properties at high temperature, low water uptake, excellent corrosion resistance, and advanced UV-shielding behavior [1–10]. Nonetheless, the PN materials suffer from the brittleness of their cured networks. Indeed, the brittleness, which is an intrinsic characteristic of the PN resins, derives from the rigidity of the monomeric precursors and the high crosslinking degree of their cured forms. In addition, typical PN monomers also need higher curing temperatures and extended time to achieve fully cured systems. Although the addition of multiple curing catalyzers has been found to improve the curing behavior to a certain extent, the curing temperatures and times still remained higher than those of other high-performance thermosets.

The previously cited PN shortcomings challenged researchers and many solutions were proposed. One of the solutions involved the preparation of physical blends and copolymers with various other thermosets and thermoplastics [11–15]. The prepared hybrids successfully circumvented some of the issues of the PN materials. In the meantime, phase separations were sometimes noted for these systems, and the overall improvements in the performances were, to a large extent, beneficial to the non-PN phase. Another solution focused on the preparation of micro- and nanocomposites based on PN resins and inorganic micro- or nanoparticles [16]. The obtained results showed improvements in the thermal and mechanical properties of the PN matrices. The efficiency of these combinations depended upon the properties of the reinforcing phases and the quality of their dispersion within the PN matrices. As far

Phthalonitrile Resins and Composites.
DOI: https://doi.org/10.1016/B978-0-12-812966-1.00003-2

as the curing behavior and processability were concerned, no major improvement was obtained by the addition of the fillers. Therefore, a third solution based upon the molecular design was proposed to address as many shortcomings as possible through the preparation of functional, PN monomers and polymers.

Recently, the development of functional materials has become one of the major branches of polymer chemistry, especially with the emergence of advanced characterization techniques that allow more efficiency in the confirmation of the molecular structures. Thus the preparation of X-functional, PN monomers and polymers, where X is the introduced functionality, were extensively studied and reported about. The introduced functionalities included alkene, alkyl, imide, bismaleimide, novolac, benzoxazine, benzimidazole, and so on. The choice of the functionality is generally driven by the targeted properties. For instance, structural rigidity can enhance the thermal stability, while the introduction of structure bearing reactive sites can improve the curing behavior and processability. Other functionalities can also provide both improvements. Moreover, the chain lengths of the moieties between the PN units can be easily adjusted to improve the flexibility of the chains while targeting reduced melting points, large processing windows, and toughness improvements. Clearly the molecular design allows many possibilities in tailoring the initial and final properties of the PN materials.

Similarly to the physical blends and copolymers, the preparation of the X-functional, PN materials was also beneficial to the functionality itself. Considering the excellent properties of the PN resins, other polymers, including phenolic resin, polyimide, polybenzoxazine, cyanate ester, poly(ether ether ketone) (PEEK), and so on took advantage from the cyano addition reaction of the PN units to improve some of their own properties. For instance, all the previously cited polymers had their thermal and thermomechanical properties exponentially increased by the addition of the PN units. Indeed, the X-functional PN can sometimes be seen as PN-functional X materials as well.

In the following sections, the major works regarding the preparation and characterization of the X-functional, PN monomers, oligomers, and polymers are reviewed. In each case, a full review is provided, starting from the synthetic route to the final properties of the cured polymers. Meanwhile, step-by-step, comparative discussions between the X-functional, PN materials are provided to assure a complete understanding of the differences between these materials.

3.2 X-Functional Phthalonitrile Monomers and Polymers

3.2.1 Alkene-Functional Phthalonitrile

Considering the sluggish curing character of the typical PN monomers, adding additional reactive functional groups in the PN monomers backbone to achieve a thermally accelerated polymerization mechanism has gained a lot of attention in the past few decades. In fact, when the self-catalyzed PN monomers started to emerge, the introduction of active hydroxyl groups or amino functionalities was the most famous way to initiate the cyano addition reaction [17,18]. Although many improvements in the overall properties were obtained, many nitriles were found to remain unreacted. Therefore, the addition of reactive crosslinking groups, including alkene functionalities, was expected to achieve better performance.

The first report about the preparation of an alkene-functional, PN monomer was attempted by Zou et al. in 2014 through the synthesis of an allyl-functional, PN monomer [19]. The targeted monomer was obtained via a classical substitution reaction between the diallyl-bisphenol-A (BPA) (DBPA) and 4-nitrophthalonitrile in the presence of potassium carbonate (K_2CO_3) in dimethyl sulfoxide solvent. A detailed thermal analysis of the polymerization reaction was carried out by dynamic scanning calorimetry (DSC) in the temperature range from 50 to 380°C. The resulting DSC curve of the allyl-functional, PN monomers is displayed in Fig. 3.1 along with that of the DBPA. The allyl-functional, PN monomers showed an endothermic transition at about 122°C corresponding to their melting transition. In contrast to the typical BPA-based PN (BAPh) monomers, the melting point was reduced by at least 72°C, which implied more processability for these newly developed monomers. The allyl-functional, PN monomers also exhibited two exothermic transitions confirming that the polymerization reaction took place in two stages. The first exotherm at 270°C was wide and short and was related to the polymerization of the allyl moieties. Meanwhile, the second exotherm at 332°C was intense and sharp and was ascribed to the cyano addition reaction of the PN units. If compared with the typical BAPh monomers, these new monomers clearly showed a self-curing behavior.

Fourier transform infrared (FTIR) was also used to monitor the structural changes occurring during the curing. The FTIR spectra of the allyl-functional, PN monomers and polymers at various curing

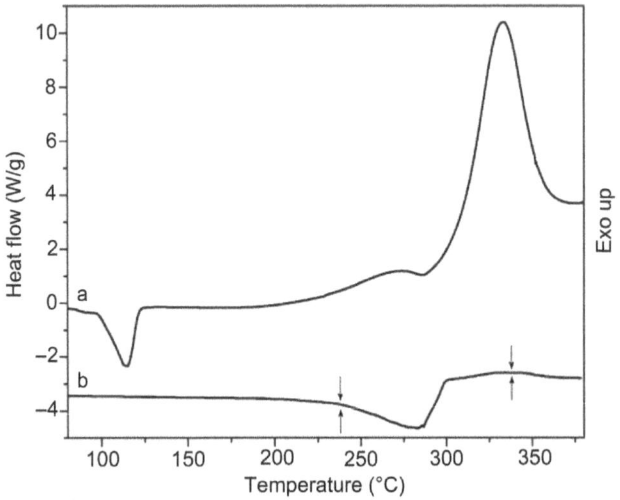

Figure 3.1 DSC curves of the allyl-functional, PN monomers and DBPA. *DBPA*, diallyl-bisphenol-A; *DSC*, dynamic scanning calorimetry; *PN*, phthalonitrile. Reprinted from Zou X, Xu M, Jia K, Liu X. Synthesis, polymerization, and properties of the allyl-functional phthalonitrile. J Appl Polym Sci 2014;131.

temperatures are showed in Fig. 3.2. By analyzing the data, it appeared that increasing the curing temperatures resulted in the intensity reduction of the allyl moiety characteristic bands. For instance, the band at 3082 cm^{-1} ascribed to the $=C-H$ absorption, as well as the bands at 1638 and 1010 cm^{-1} related to the vibrations of the $-CH=CH_2$, were all reduced by increasing the curing temperatures. Moreover, at high temperatures, evidences of the PN polymerization were clearly noticed through the appearance of the bands at 1010 and 1360 cm^{-1} corresponding to the formation of phthalocyanine and triazine structures, respectively. By combining the DSC and FTIR results, it was confirmed that the relatively low polymerization of the allyl moiety resulted in the formation of active hydroxyl groups which further initiated the cyano addition reaction.

Rheometric investigations revealed that the initial melt viscosity of the allyl-functional, PN monomers was relatively low and stable before the initiation of the polymerization process. In addition, the melt viscosity was found to be easily controlled by changing the processing temperature. Based on these rheometric observations, the allyl-functional, PN monomers were cured starting from 260 to 360°C by 20°C steps each 4 h. The as-obtained polymers were tested by dynamic mechanical analysis (DMA) to evaluate their thermomechanical properties. From

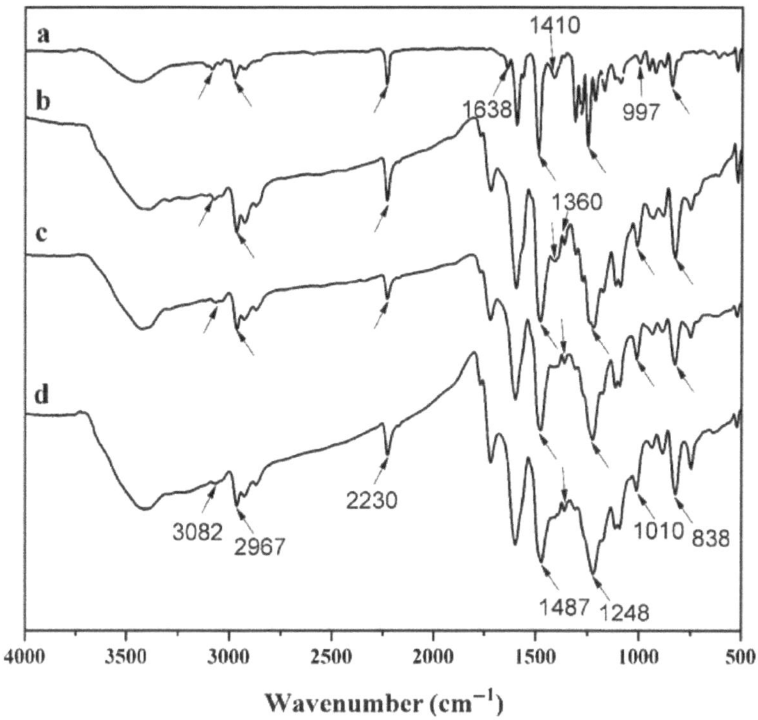

Figure 3.2 FTIR spectra of the allyl-functional, PN monomers and polymers at various curing temperatures: monomers (a), polymer cured at 280°C (b), polymer cured at 320°C (c), and polymer cured at 340°C (d). *FTIR*, Fourier transform infrared; *PN*, phthalonitrile. Reprinted from Zou X, Xu M, Jia K, Liu X. Synthesis, polymerization, and properties of the allyl-functional phthalonitrile. J Appl Polym Sci 2014;131.

the obtained curves shown in Fig. 3.3, the storage modulus reached 3.7 GPa at 50°C and conserved up to 2.2 GPa at 400°C. The T_g, obtained from the max of Tan Delta peak, was also confirmed to be above 350°C. These improved thermomechanical properties were mainly attributed to the rigidity of the polymeric structure and also, the high crosslinking degree afforded by the polymerization of the allyl moieties and their further positive impact on the PN cyano addition reaction.

Both the rigidity and the high, crosslinking degree were behind the improvement in the thermal stability. Hence, the allyl-functional, PN polymer cured up to 340°C presented a temperature of 5 and 10 wt.% weight loss ($T_{5\%}$ and $T_{10\%}$, respectively) of 478 and 552°C, respectively, with a char yield of 76% at 800°C, under nitrogen atmosphere.

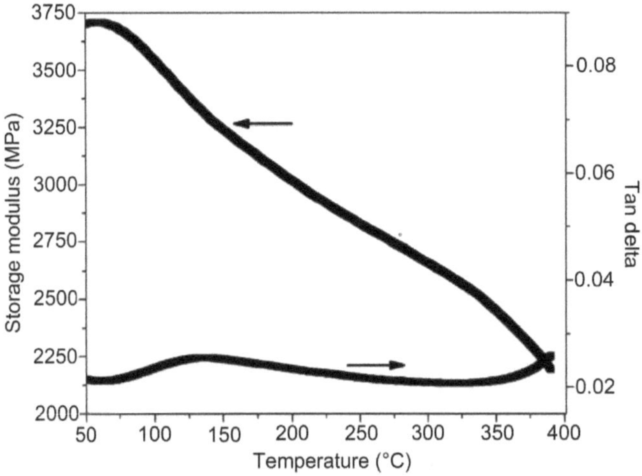

Figure 3.3 Storage modulus and loss factor (Tan Delta) variations for the allyl-functional, PN polymer. *PN*, phthalonitrile. Reprinted from Zou X, Xu M, Jia K, Liu X. Synthesis, polymerization, and properties of the allyl-functional phthalonitrile. J Appl Polym Sci 2014;131.

The same excellent thermal stability was observed under air atmosphere with a $T_{5\%}$ and $T_{10\%}$ of 464 and 521°C, respectively, and a char yield at 800°C of 46%. In summary, the allyl-functional, PN monomers showed an autocatalytic effect with an improved processability in contrast to the typical PN monomers. Moreover, the additional allyl polymerization markedly increased the crosslinking degree and resulted in an improvement in the thermal and thermomechanical properties.

3.2.2 Alkyne-Functional Phthalonitrile

The objective behind the preparation of alkyne-functional, PN monomers is the same as that of the alkene-functional PN. The introduction of alkyne functionalities can provide additional reactive crosslinking units allowing an improvement in the overall curing behavior and properties. Besides, the polymerization of both alkyne and PN units at different temperatures can also provide a better control over the final properties.

Laskoski et al. using the modified Ullmann reaction prepared PN oligomers containing alkyne and aromatic ether [20]. Basically, the synthesis route (Scheme 3.1) involved at first the reaction between BPA and 4,4-dibromotolane under basic environment in the presence $(PPh_3)_2CuBr$, then, 4-nitrophthalonitrile was added to achieve the targeted PN oligomers.

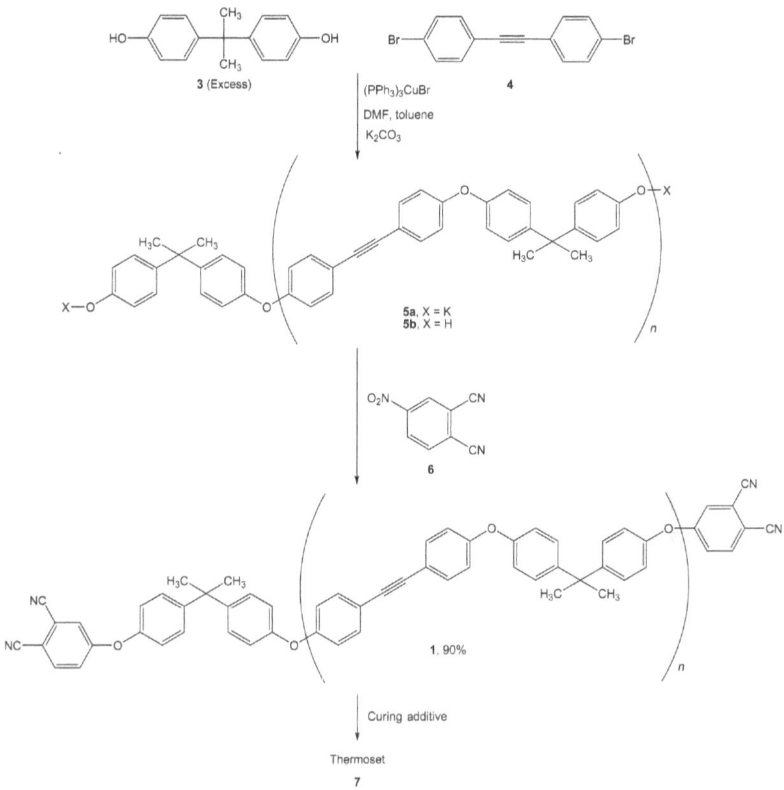

Scheme 3.1 Synthetic route of the alkyne-functional, PN oligomers and thermosets. *PN*, phthalonitrile. Reprinted from Laskoski M, Dominguez DD, Keller TM. Alkyne-containing phthalonitrile resins: controlling mechanical properties by selective curing. J Polym Sci: Part A: Polym Chem 2013;51:4774–8.

To the prepared alkyne-functional, PN oligomers and their nonalkyne counterparts (herein labeled as 22-PN oligomers see Fig. 2.9) were added 5 wt.% of *p*-BAPS to accelerate the cure reaction. The resulting blends were analyzed by DSC to study and compare their curing behavior. Both compositions exhibited a softening temperature at about 75°C. The alkyne-functional, PN oligomers also showed two exothermic transitions at 250 and 310°C related to cyano addition reaction of the PN units and to the polymerization of the alkyne units, respectively. The nonalkyne oligomers also showed two exothermic transitions ascribed to the polymerization of the PN units initiated by both the amine-curing agent and the high temperatures, respectively. Although the authors did not provide any DSC curve for the noncatalyzed alkyne-functional, PN oligomers, it is expected that the polymerization of the alkyne units would also initiate the cyano addition reaction at temperatures higher than 310°C even without the use of the *p*-BAPS.

The alkyne and nonalkyne, PN oligomers were also cured with 3 wt. % *p*-BAPS to compare their thermal and thermomechanical performances. The alkyne and nonalkyne, PN polymers were then afforded by curing the samples following the schedule of 270°C for 12 h, 300°C for 4 h, 350°C for 4 h, and 375°C for 4 h. Some samples were also further cured at 400°C for 8 h to study the postcure effect on the thermomechanical properties. Fig. 3.4A and B represent the storage modulus and the Tan Delta variations as a function of temperature, respectively. The storage modulus of the alkyne-functional, PN polymer ranged between 1850 MPa at room temperature and 330 MPa at 400°C. Meanwhile, the nonalkyne, PN polymer showed a reduced modulus values with 1340 MPa at room temperature and 190 MPa at 400°C. The improvement in the storage modulus values for the alkyne-functional, PN polymers was attributed to the higher crosslinking degree afforded by the polymerization of both PN and alkyne units. On the other hand, postcuring the alkyne-functional, PN polymer at 400°C for 8 h resulted in a decrease in the storage modulus values below 330°C and an increase in these values at temperatures higher than 330°C. For instance, the storage modulus at 400°C was 230 MPa higher at 400°C when compared with that of the non-post-cured sample. It was reported that the decrease in the storage modulus below 330°C was related to the inverse effect of the postcure. In fact, the postcure resulted in a highly crosslinked network where the internal damages and bond breakages could easily occur due to the enhancement in stress. Moreover, by analyzing the Tan Delta peaks, it appeared than the alkyne-functional, PN polymer had a T_g of about 395°C which was 95°C higher than that of the nonalkyne, PN polymer. Since the T_g is generally closely related to the rigidity and the crosslinking degree, it can be concluded that the additional polymerization of the alkyne moieties was also behind the improvement in the T_g. It should be also noted that the postcured sample showed a featureless Tan Delta curve with no T_g.

The thermal stability investigations revealed similar performances under nitrogen atmosphere with a $T_{5\%}$ of 495°C and a char yield at 1000°C of 70%. Under air atmosphere, the thermal stability performances were also similar before the starting of the thermal stability failures. Thereby, during the catastrophic decompositions, the alkyne-functional, PN polymer showed a slower decomposition rate. Once more, the high rigidity and additional crosslinking degree allowed by the polymerization of the alkyne units were behind the superior thermal properties of the alkyne-functional, PN polymer.

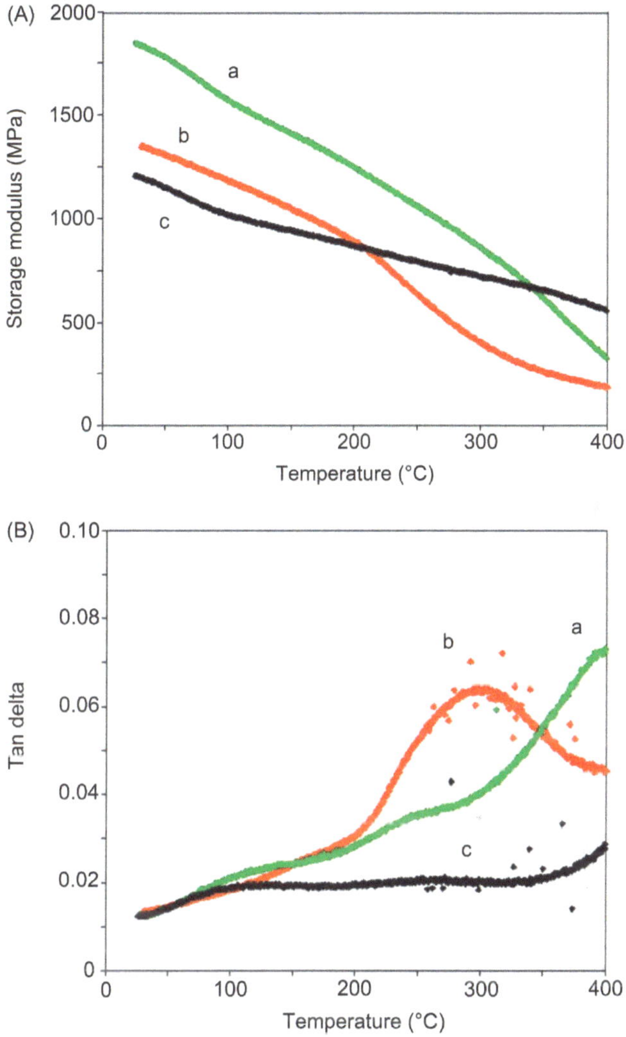

Figure 3.4 Storage modulus (A) and loss factor (Tan Delta) (B) variations for the alkyne-functional PN and 22-PN polymers cured with 3 wt.% of *p*-BAPS: alkyne-functional, PN polymer heated to 375°C for 4 h (a), 22-PN polymer heated to 375°C for 4 h (b), and alkyne-functional, PN polymer heated to 400°C for 8 h (c). *PN*, phthalonitrile. Reprinted from Laskoski M, Dominguez DD, Keller TM. Alkyne-containing phthalonitrile resins: controlling mechanical properties by selective curing. J Polym Sci: Part A: Polym Chem 2013;51:4774–8.

Augustine et al. also reported about a novel type of alkyne-functional, PN in which a new prepolymer containing PN and propargyl groups on the same backbone was designed and characterized [21]. The preparation process of these propargyl-functional, PN oligomers involved a two-step synthetic route. First, several novolac bearing different numbers of PN units were synthesized by a sample nucleophilic displacement of a nitro substituent from 4-nitrophthalonitrile. The amount of phenolic and PN functionalities were estimated from elemental analysis and hydroxyl values. The second step involved the propargylation of the remaining free phenol groups from the previous reaction. The global preparation process can be depicted in Scheme 3.2.

FTIR spectra for the novolac-containing, PN and propargyl-functional, PN before and after curing are displayed in Fig. 3.5. When both oligomers were compared, it was confirmed that the OH peak disappeared and new bands related to the propargyl appeared such as the absorption at 2120 cm^{-1} attributed to the C≡C stretching vibrations. After curing, the nitrile intensity was reduced for both compounds; however, the extent of decrease was greater for the propargyl-functional, PN. Therefore, the ability of the propargyl groups to involve more nitriles in the curing was clearly demonstrated. Concerning the activation mechanism, it should be first acknowledged that the polymerization of the propargyl into the chromene structure is preceded by the formation of hydroxyl groups, as seen in the spectrum of the cured propargyl-functional, PN. Both intermediate hydroxyls and chromene can initiate the cyano addition reaction of the PN units as further explained in Scheme 3.3.

DSC was also employed to further understand the curing process. The experimental tests were not only performed for oligomeric PN containing various proportions of propargyl units but also for the novolac-containing propargyl and novolac-containing PN. The obtained DSC curves are presented in Fig. 3.6. The novolac-containing propargyl showed a large exotherm peaking at 261°C attributed to the polymerization of the propargyl units. The novolac-containing PN also showed only one exothermic transition around 360°C ascribed to the self-catalyzed, cyano addition reaction of the PN units. As expected from the FTIR results, the DSC curves of the propargyl-functional PN revealed the presence of a large exotherm composed of two almost overlapped peaks. The first peak corresponded to the polymerization of the propargyl units, while the second one was related to the nitriles polymerization activated by the active hydroxyls and the chromene structure resulting from the previous polymerization of the propargyl units. Moreover, the position of these peaks depended on the proportion of the propargyl units in the PN oligomers. For example, increasing the number of propargyl units ensued reduced curing temperatures and when the proportion of propargyl to PN reached 2.3:1 molar ratio

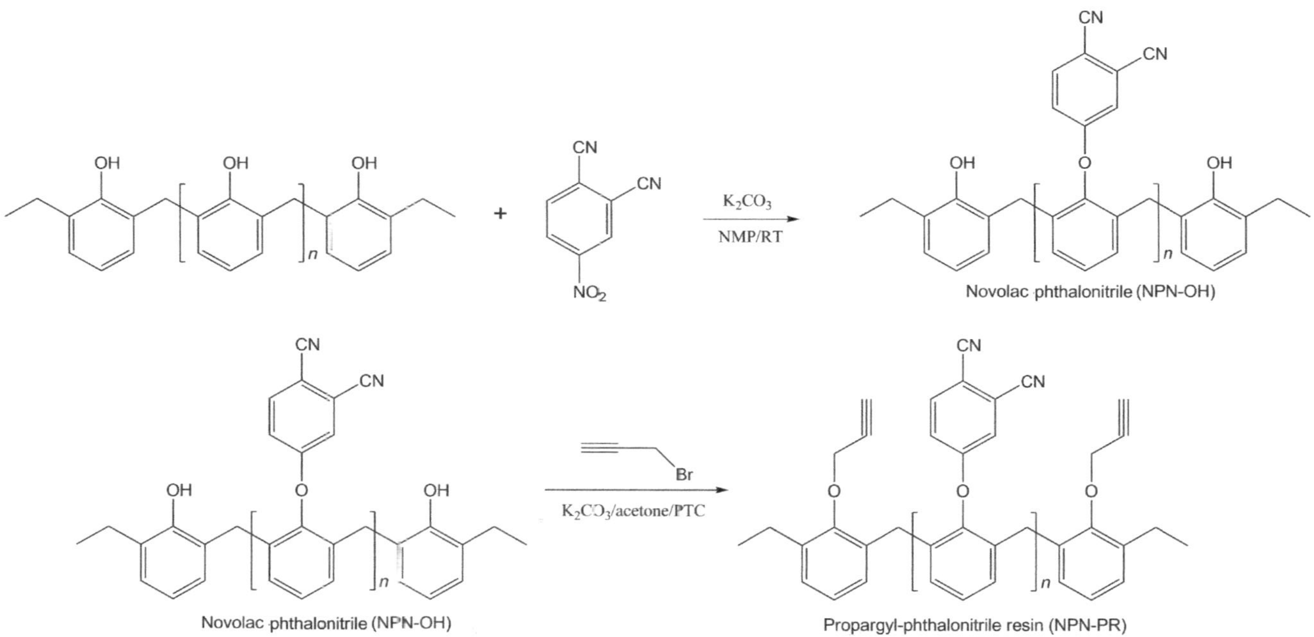

Scheme 3.2 Synthetic route of the propargyl-functional, PN oligomers. *PN*, phthalonitrile. Reprinted from Augustine D, Mathew D, Nair CR. One component propargyl phthalonitrile novolac: synthesis and characterization. Eur Polym J 2015;71:389–400.

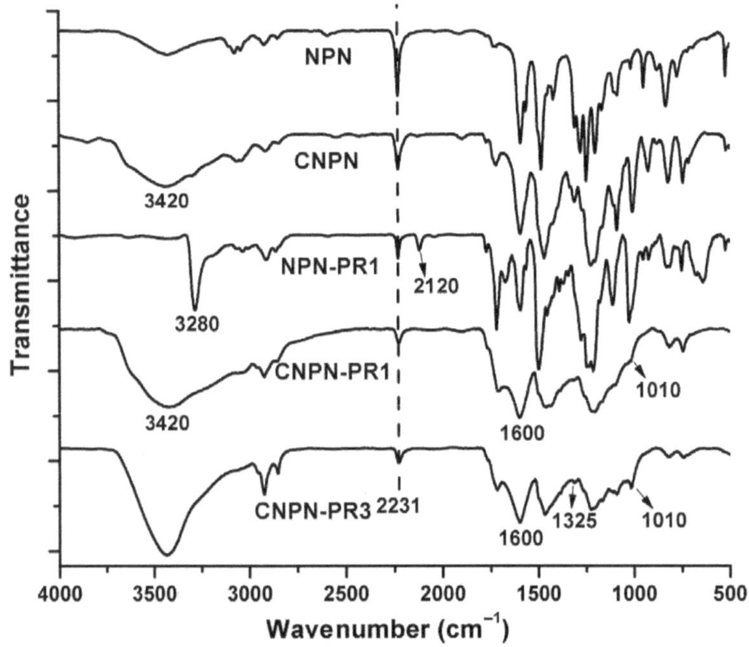

Figure 3.5 FTIR spectra for the novolac-containing PN and propargyl-functional PN before and after curing. *FTIR*, Fourier transform infrared; *PN*, phthalonitrile. Reprinted from Augustine D, Mathew D, Nair CR. One component propargyl phthalonitrile novolac: synthesis and characterization. Eur Polym J 2015;71:389–400.

(NPN−PR3), the two exotherms were completely overlapped. Thus, the DSC results corroborated those of the FTIR and also revealed some other interesting features of the propargyl-functional, PN oligomers.

The evaluation of the thermal stability by thermogravimetric analysis (TGA) revealed that the polymers containing elevated ratios of PN units were thermally more stable, while, higher contents of propargyl negatively affected the thermal stability. In summary, the propargyl-functional, PN oligomers were demonstrated to exhibit a self-curing behavior and good processability which were traded for slightly lower thermal stability.

3.2.3 Imide-Functional Phthalonitrile

Polyimides are another class of high-performance polymers exhibiting many interesting features including high thermal stability, good mechanical properties and improved resistance to chemicals and irradiations. The common types of imide monomers are prepared in a two-step

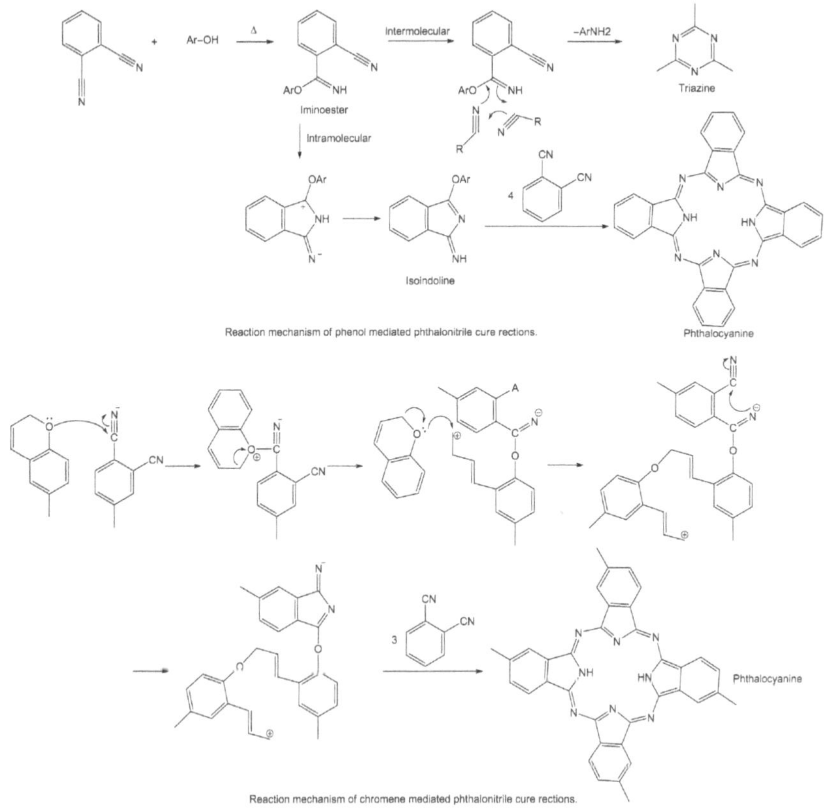

Scheme 3.3 Hydroxyls and chromene initiated cyano addition reaction. Reprinted from Augustine D, Mathew D, Nair CR. One component propargyl phthalonitrile novolac: synthesis and characterization. Eur Polym J 2015;71:389–400.

reaction involving the reaction of a dianhydride with an aromatic or aliphatic diamine. The initial reaction results in the formation of the poly (amic-acid) intermediate, which is soluble in many solvents. Further pursuing the reaction, either by chemical or heat treatments, results in the formation of the imide linkages. The as-such prepared polyimides present some insolubility issues restricting the fabrication of finished products. In addition, the evolution of volatile by-products during the polymerization causes void problems negatively affecting the mechanical performances of the resulting polyimides [22–25]. One of the proposed solutions focuses on developing fully-imidized monomers by end-capping with reactive groups able to increase the crosslink degree. As such, the volatile by-products are eliminated before the starting of the polymerization reaction.

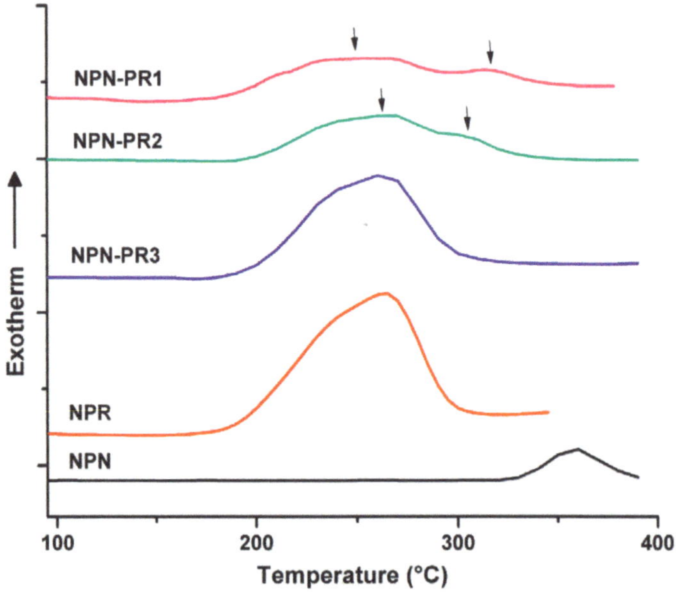

Figure 3.6 DSC curves of the PN containing various proportions of propargyl units (NPN-PR1–3), NPR, and NPN. *DSC*, dynamic scanning calorimetry; *NPR*, novolac-containing propargyl; *NPN*, novolac-containing PN; *PN*, phthalonitrile. Reprinted from Augustine D, Mathew D, Nair CR. One component propargyl phthalonitrile novolac: synthesis and characterization. Eur Polym J 2015;71:389–400.

The PN monomers/oligomers are known for their ability to form a highly crosslinked network afforded by their specific polymerization process which takes place through the cyano groups by an addition mechanism. Therefore, end-capping imide monomers with PN ends not only ensures a high crosslinking degree, but also the resulting monomers can take advantage from the superior thermal and thermomechanical of the PN. On the same time, the curing behavior and the processability of the PN monomers can also be improved and easily tailored by incorporating different imide linkages within the PN backbones. Indeed, the imide-containing, PN monomers and polymers have been the most extensively studied type of functional, PN materials.

The first imide-containing, PN resin was reported by Keller [26]. The synthesis procedure was a two-step reaction involving a displacement reaction to prepare the ether linkage and a condensation reaction to form the imide ring. The aforementioned synthetic route can be resumed as seen in Scheme 3.4. The prepared imide-containing, PN monomers and their cured polymers were studied for their curing

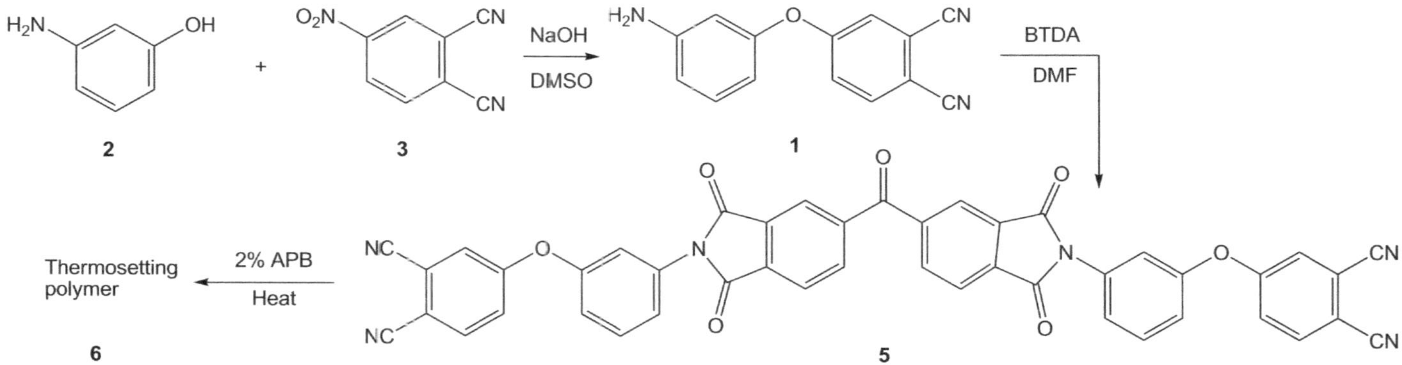

Scheme 3.4 Synthetic route of the first reported imide-containing PN monomer. *PN*, phthalonitrile. Reprinted from Keller TM. Imide-containing phthalonitrile resin. Polymer 1993;34:952–5.

behavior and thermal properties. Globally, it was found that, similarly to any other PN monomers, the polymerization reaction was difficult to initiate without the use of external curing additives. Thus, the author opted for the addition of 2 wt.% of 1,3-bis(3-aminophenoxy)benzene (*m*-APB) to activate the addition reaction of the cyano groups. The curing behavior was also found to be closely related to the amount of *m*-APB and curing temperatures. For instance, increasing the amount of the catalyst resulted in a shorter reaction time to gelation.

The prepared imide-containing, PN monomers were cured at 210°C for 4 h, 260°C for 8 h, and 316°C for 16 h in the presence of 2 wt.% of *m*-APB. Moreover, samples were also postcured at 350°C for 4 h and 375°C for 4, 12, and 24 h, respectively. All the obtained polymers were tested by TGA to investigate their thermal resistances under nitrogen and air atmospheres. The obtained data revealed that, under nitrogen atmosphere, increasing the curing temperature from 316 to 350°C allowed an improvement in the thermal stability. On the other hand, further postcuring the samples to 375°C only showed minor amelioration in the thermal stability with similar performances even by extending the postcure up to 24 h. Under air atmosphere, the samples cured up to 316°C and postcured at 350°C exhibited similar behavior with crucial failure above 550°C. Meanwhile, increasing the postcure temperature and time did not achieve any further improvement in the thermo-oxidative stability. Globally, it was confirmed that the thermal and thermo-oxidative stabilities are a function of the curing temperatures and time and the catastrophic degradations seen under air atmosphere were probably due to the breakdown of the linking structures connecting the terminal PN units. Indeed, for temperatures in excess of 350°C, the reaction of the free radical with the oxygen became more pronounced.

If compared with the typical, bisphenols-based PN resins, these imide-functional, PN showed reduced thermo-oxidative stability under the same curing conditions. Relatively, the imide-linked, PN resin had better thermal stability than its corresponding neat polyimide. Also, during the catastrophic breakdown occurring at high temperatures, the imide-functional, PN resin was more oxidatively stable than the typical PN resins prepared from various bisphenols under the same curing conditions. The presence of both ether and imide linkages within the imide-functional polymer is behind this slow down during the thermal failure. Furthermore, it was also confirmed that the initial breakdown corresponded to the failure of the imide moieties. It is important to point that this first report about the imide-functional, PN set the basics for further researches on the field and the following studies reported on advanced imide-functional, PN with better performances.

In attempt to prepare high-performance, imide-functional, PN resins, Selvakumar et al. prepared PN oligomers containing both aromatic ether and imide linkages [27,28]. The objective of the study was to obtain polymeric materials with superior thermal stability and good processability by combining highly stable structural units with flexible linkages within the same backbone. The oligomeric imide-containing, PN were obtained after a two-step process involving the reaction of pyromellitic dianhydride with various aromatic diamines in an appropriate mixture of solvents to form an oligomeric anhydride intermediate which was further reacted with 4-aminophenoxy PN (4-APN) to finally obtain the targeted oligomers. The detailed synthesis process of the imide-functional, PN is presented in Scheme 3.5. The oligomeric, imide-linked, PN, labeled Ia−c and IIa−c, were cured in the presence of 4,4′-diaminodiphenyl sulfone (DDS) at high temperature to afford fully cured polymers.

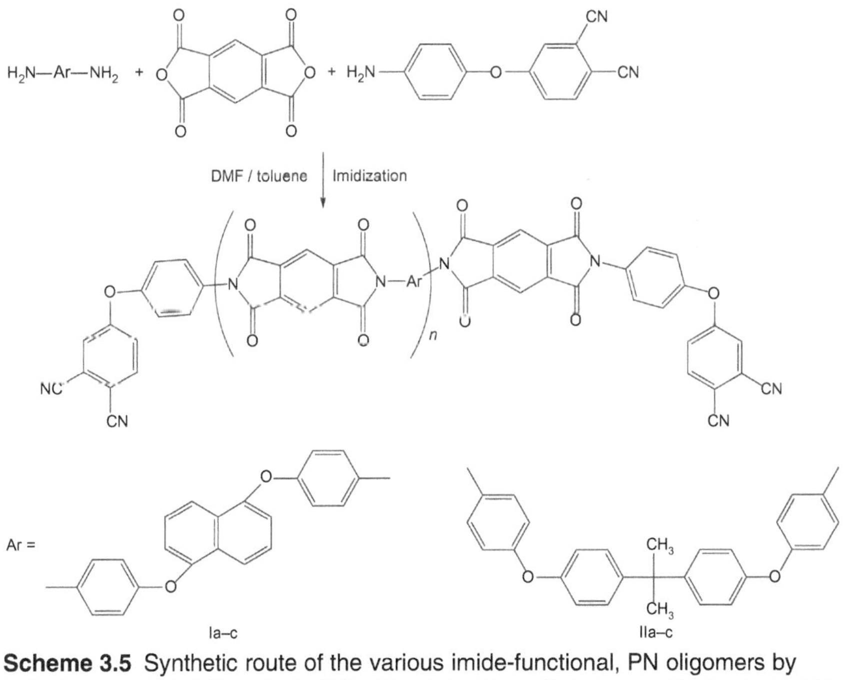

Scheme 3.5 Synthetic route of the various imide-functional, PN oligomers by Selvakumar et al. *PN*, phthalonitrile. Reprinted from Selvakumar P, Sarojadevi M. Development of oligomeric phthalonitrile resins for advanced composite applications. Macromol Symp 2009;277:190−200.

The appropriate amount of the amine-curing agent was determined in accordance with the rheological measurements which were performed on the imide-functional, PN prepolymers containing various amounts of DDS (4, 4.5, and 5 wt.%). The viscosity changes were evaluated at various times at 290°C. The obtained results revealed that the increase in the viscosity was gradual over 1 h for the prepolymers containing 4 and 4.5 wt.% DDS, while a faster increase rate was depicted for the prepolymers of 5 wt.% DDS. Therefore, the authors opted for the last choice to enhance the curing process. The curing behavior of the imide-functional, PN prepolymers with 5 wt.% DDS was investigated by DSC analysis. The obtained curves revealed the presence of one melting transition around 168–176°C and one exothermic transition ranging between 274 and 307°C. The presence of one endothermic transition implied that the curing agent and the oligomeric PN already copolymerized during the prepolymers preparations. Subsequently, the exothermic transition was ascribed to the cyano addition reaction on the PN units triggered by the DDS curing agent. Interestingly, the processing window was measured to be up to 100°C which was markedly larger than that of the first imide-functional PN reported by Keller and previously discussed. Taking into account the rheometric and DSC results, the imide-functional, PN prepolymers were cured according to the cure schedule of 260°C for 8 h, 280°C for 5 h, and 325°C for 10 h. The obtained samples were also postcured at 325°C for 5 h and 380°C for 10 h.

The thermal stability of the studied imide-functional, PN polymers was investigated by TGA under nitrogen and air atmospheres and the obtained data are listed in Table 3.1. Under nitrogen atmosphere, all the compositions showed no weight loss up to 450°C and the $T_{10\%}$ ranged between 498 and 511°C according to the molecular weight of the oligomer precursors. The char yield at 800°C also reached up to 83% revealing the exceptional thermal stability of these imide-containing, PN polymers. Postcuring the samples at 380°C for 10 h resulted in further improvements in the thermal stability. Under air atmosphere, all the polymers also showed excellent weight retention up to 430°C and the $T_{10\%}$ ranged between 443 and 461°C. Postcuring the samples at 380°C for 10 h also improved the thermal oxidative stability of the imide-functional PN. For instance, the onset degradation temperature reached 454°C for the aforementioned postcuring conditions. Conversely, it was found that the rate of thermal degradation at higher temperatures was faster for the postcured samples. Globally, the imide-functional, PN oligomers exhibited ameliorated curing behavior and processability and

Table 3.1 Thermal Properties of the Imide-Functional, PN Polymers Prepared by Selvakumar et al. under Nitrogen and Air Atmospheres

Imide-Functional PN	Nitrogen		Air	
	$T_{10\%}$ (°C)	Char Yield at 800°C (%)	$T_{10\%}$ (°C)	Char Yield at 800°C (%)
Ia	498	74.5	448	44.2
Ib	504	80.0	456	47.4
Ic	511	83.1	461	50.2
IIa	492	70.6	443	41.7
IIb	501	75.4	452	45.3
IIc	509	81.3	459	49.1

Reprinted from Selvakumar P, Sarojadevi M. Development of oligomeric phthalonitrile resins for advanced composite applications. Macromol Symp 2009;277:190–200.

their resulting polymers showed enhanced thermal stability. Thus, the primary objective of combining the flexibility with the thermal stability was successfully achieved.

The imide-functional, PN monomers previously discussed were all cured in the presence of small amounts of curing agents to achieve fully polymerized polymers in a timely manner. Thankfully, with the emergence of the self-catalyzed, PN monomers, based on simple molecular modifications, many new possibilities appeared to overcome the need of catalyst to cure the PN monomers/oligomers. Taking advantage from this smart chemistry, Badshah et al. synthetized some PN monomers bearing imide moieties and containing an amino function at different positions (*ortho*, *meta*, and *para*) [29]. The authors investigated the self-promoting behavior of these compounds and studied the influence of the amino position on the processability and the curing behavior. The obtained results concerning the monomers preparation, curing behavior, and processability have been already discussed in Chapter 1, Introduction to the Phthalonitrile Resins and their Curing Behavior, and herein we report about the thermal and thermomechanical properties of their resulting polymers.

The thermal stability of the cured imide-linked, PN polymers was investigated by TGA under nitrogen and air atmospheres and the obtained results are shown in Fig. 3.7A and B, respectively. The measurement of the $T_{5\%}$ indicated that all the self-catalyzed, imide-functional,

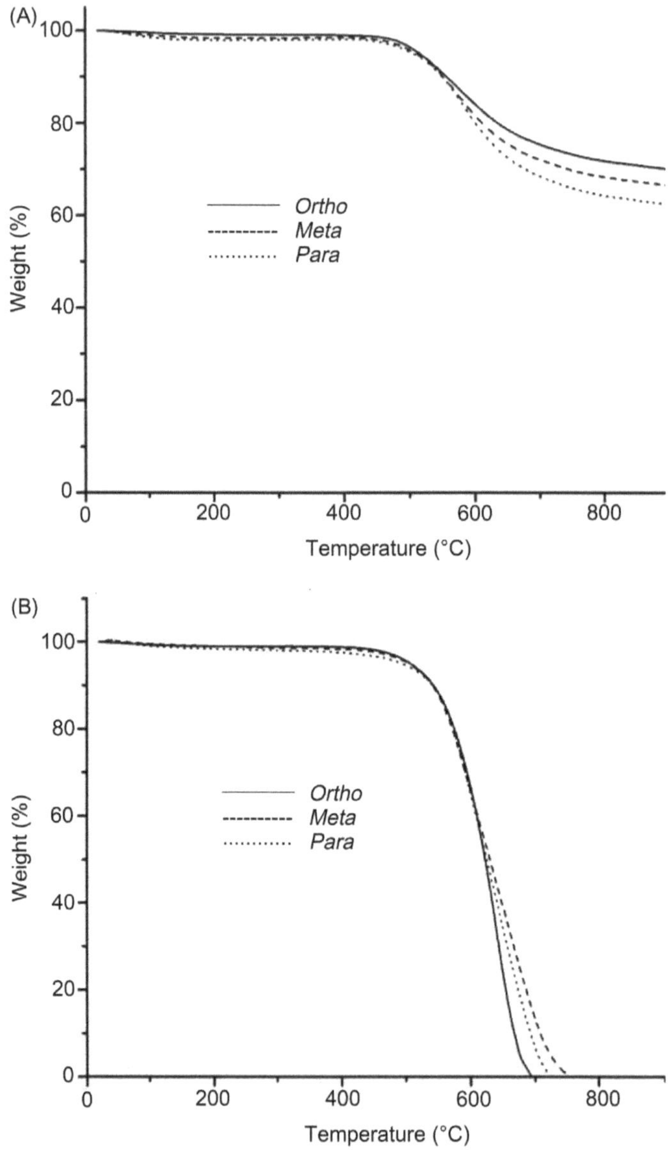

Figure 3.7 TGA curves in nitrogen (A) and air (B) atmospheres of the autocatalytic imide-functional, PN polymers at various position of the amino group. *PN*, phthalonitrile. Reprinted from Amir B, Zhou H, Liu F, Aurangzeb H. Synthesis and characterization of self-catalyzed imide-containing pthalonitrile resins. J Polym Sci: Part A: Polym Chem 2010;48:5916–20.

PN polymers were extremely thermally resistant. In fact, according to the position of the amine, the $T_{5\%}$ value under nitrogen atmosphere for the *ortho*, *meta*, and *para* positions was 529, 524, and 528°C, respectively. The char yield at 800°C also reached up to 70% for the *ortho*-amine-based polymer. Similarly, the evaluation of the thermo-oxidative stability showed $T_{5\%}$ values in the range of 514–520°C with catastrophic degradation between 550 and 650°C. In all instances, the self-catalyzed imide-containing PN polymers were thermally more stable than their noncatalyzed counterparts. Thus, the preparation of self-cured monomers not only allowed improvements in the curing behavior and processability, but also the resulting polymers showed an enhanced thermal stability.

The variations of the storage modulus and Tan Delta for the samples cured up to 350°C for 5 h and those further postcured at 375°C for an additional 5 h are shown in Fig 3.8A–C. The evaluation of the T_g from the maximum of Tan Delta peak revealed that all the polymers had high T_g values ranging from 435°C for the *meta*-amine-based compound and around 465°C for both *ortho*- and *para*-amine-based polymers. In all cases, postcuring the samples at high temperature resulted in an improvement in the T_g. For instance, the imide-functional, PN polymer containing an amine at the *ortho* position showed no T_g when postcured at 375°C for 5 h. The storage modulus was also affected by the postcuring conditions. In fact, the *meta*- and *para*-amine-based polymers showed an enhancement in the storage modulus values upon postcuring, while the *ortho*-amine based compound showed an opposite behavior. In summary, the introduction of amine groups in the imide-functional, PN monomers resulted in multiple improvements in the curing behavior and processability, and the cured polymers showed better thermal and thermomechanical properties than their noncatalyzed counterparts.

Another interesting work relating about the introduction of PN units as pendent substituents into two different polyimide systems, was performed by Zeng et al. [30]. This study differs from the previously discussed one since the PN units were added to a whole polyimide structure to balance the so-called contradiction between the processability and the thermal properties. The synthetic route first involved the preparation of a PN-containing, diamines monomer in a three-step reaction process. The reaction was continued to obtain the polyimides by reacting the PN-containing diamines with the appropriate dianhydride via the typical two-step polymerization approach. The overall preparation process can be depicted in Scheme 3.6. The prepared

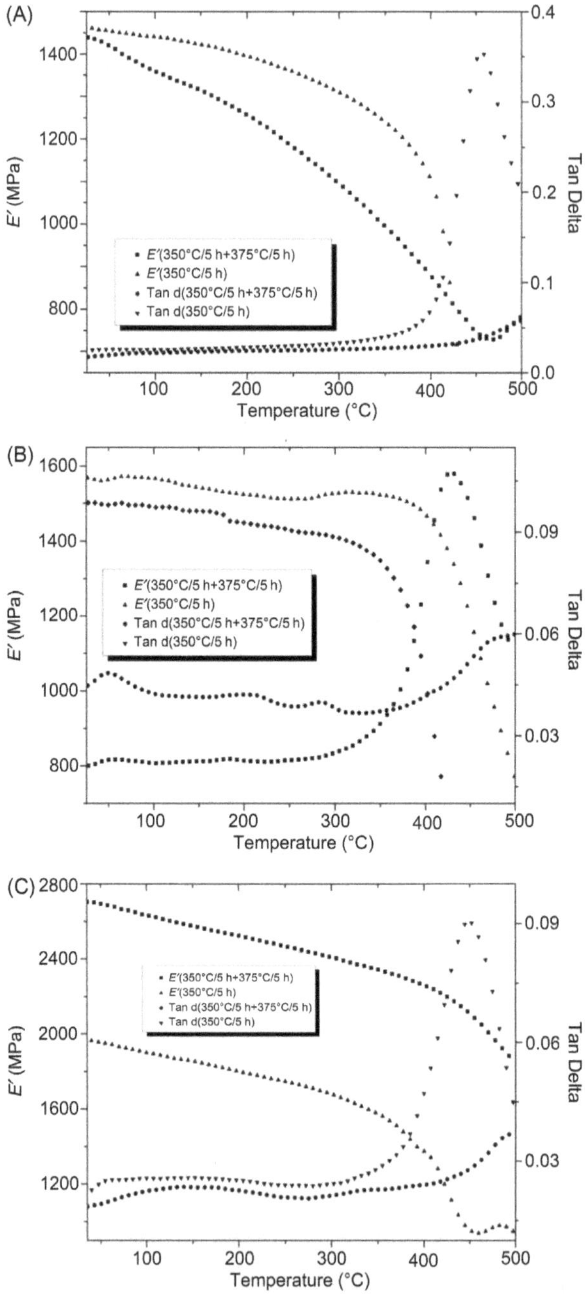

Figure 3.8 Storage modulus (E') and loss factor (Tan Delta) variations for the autocatalytic imide-functional, PN polymers at various position of the amino group: *ortho* (A), *meta* (B), and *para* (C). Reprinted from Amir B, Zhou H, Liu F, Aurangzeb H. Synthesis and characterization of self-catalyzed imide-containing pthalonitrile resins. J Polym Sci: Part A: Polym Chem 2010;48:5916–20.

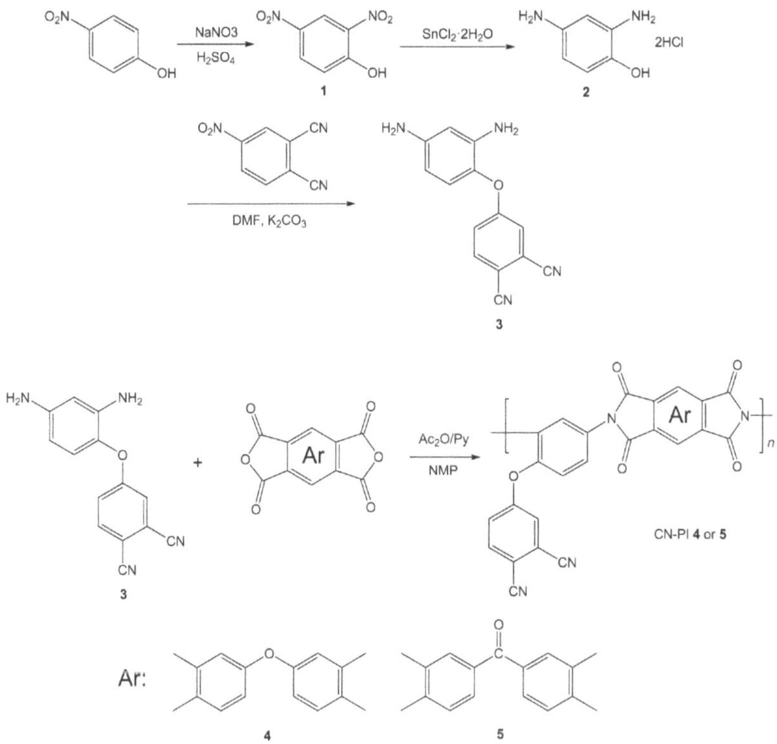

Scheme 3.6 Synthetic route of the polyimides with pendent PN units. *PN*, phthalonitrile. Reprinted from Zeng K, Li L, Xiang S, Zhou Y, Yang G. Synthesis and thermal polymerization of new polyimides with pendant phthalonitrile units. Polym Bull 2011;68:1879–88.

polyimide-containing, PN materials were referred as CN-PI 4 and 5 according their aryl groups.

The thermal stability investigations were carried out by TGA under nitrogen atmosphere for both uncured and cured (at 300°C for 10 h) polyimide-containing PN. The thermal stability parameters obtained from the TGA tests can be depicted in Table 3.2. The uncured CN-PI 4 exhibited a $T_{5\%}$ and $T_{10\%}$ of 438 and 462°C with a char yield at 800°C of 51%. Meanwhile, the cured samples showed an improvement of 110 and 113°C in $T_{5\%}$ and $T_{10\%}$, respectively, with a char yield at 800°C of 61%. The same observations were noted for the CN-PI 5 and their cured materials. For example, the cured CN-PI 5 had the outstanding $T_{5\%}$ and $T_{10\%}$ of 588 and 609°C, respectively. These improvements in the thermal stability are due to the originally good thermal resistance of the polyimide and also to the additional thermal stability allowed by the crosslink reaction of the reactive PN units. It should be also pointed out

Table 3.2 Thermal Stability Parameters of the Uncured and Cured Polyimide-Functional, PN Materials

Imide-Functional PN code	$T_{5\%}$ (°C)	$T_{10\%}$ (°C)	Char Yield at 800°C (%)
CN-PI 4	438	462	51
CN-PI 4 (cured)	548	575	61
CN-PI 5	504	542	55
CN-PI 5 (cured)	588	609	68

Reprinted from Zeng K, Li L, Xiang S, Zhou Y, Yang G. Synthesis and thermal polymerization of new polyimides with pendant phthalonitrile units. Polym Bull 2011;68:1879–88.

Table 3.3 Solvent Resistance of the Uncured and Cured Polyimide-Functional, PN Materials

Imide-Functional PN Code	H_2SO_4	NMP	DMAc	DMSO
CN-PI 4	++	++	++	++
CN-PI 4 (cured)	+−	−−	−−	−−
CN-PI 5	++	++	++	++
CN-PI 5 (cured)	+−	−−	−−	−−

Solubility: ++soluble at room temperature; +− partly soluble; − insoluble even on heating. *DMSO*, dimethyl sulfoxide; *NMP*, *n*-methyl-2-pyrrolidone.
Reprinted from Zeng K, Li L, Xiang S, Zhou Y, Yang G. Synthesis and thermal polymerization of new polyimides with pendant phthalonitrile units. Polym Bull 2011;68:1879–88.

that the $T_{5\%}$ and $T_{10\%}$ of the cured CN-PI 5 are one of the highest thermal stability parameters obtained with polymeric materials.

The solvent resistance of these prepared polyimide-containing, PN materials was evaluated and the obtained results are summarized in Table 3.3. All the uncured samples were found to be easily soluble in many polar solvents as well as in concentrated sulfuric acid. These findings confirmed that the prepared materials possess good solubility and processability which is highly appreciated for the preparation of composite materials via the conventional injection molding techniques. As expected, the cured, polyimide-containing PN showed a reduced solubility obviously attributed to the formation of a highly crosslinked network after the cyano addition reaction of the PN units. Globally, the polyimide-containing PN showed many interesting properties with thermal performances not seen in any other imide-functional, PN polymers. Thus, further works need to be performed to acknowledge about their curing behavior and their resulting mechanical and thermomechanical properties.

Yuan et al. reported about a different approach to achieve an improved curing behavior and processability for the PN monomers/oligomers, while maintaining the excellent thermal and thermomechanical properties of their resulting polymers. The authors ingeniously prepared imide-containing PN and phenylethynyl end-groups [31]. The reason behind the introduction of the phenylethynyl groups was based upon previous investigations which revealed that the phenylethynyl groups markedly improved the processing window of some polyimides. In addition, the presence of the triple bond on the phenylethynyl was also expected to catalyze the cure addition reaction of the PN units. On the other hand, the high curing enthalpy of the phenylethynyl-based monomers/oligomers could be reduced by the presence of the PN units. Therefore, the imide-containing PN and phenylethynyl were designed to combine the advantages of each compound to achieve self-catalyzed monomers with improved curing behavior and processability along with many appreciable thermal and mechanical properties. The synthetic route, shown in Scheme 3.7, was a three-step reaction during which 4,4′-oxydiphthalic dianhydride (ODPA) was added first then 3-APN was slowly introduced to achieve an initial dominant concentration of the ODPA in order to promote as much as possible the mono-end-capped imide intermediates. Most importantly, the volume ratio of ethanol to *n*-methyl-2-pyrrolidone was also critical to the separation of the mono- and di-end-capped imide products. Globally, the reaction was difficult to achieve considering that there were much di-end-capped imides that reduced the reaction yield to 54%.

The rheological measurements were evaluated for the imide-containing PN and phenylethynyl (3PN3PEODPA) and also for the imides separately containing PN and phenylethynyl (3PNODPA and 3PEODPA, respectively). The evolution of the melt viscosity was monitored as a function of time at 330 and 350°C for all the samples and the obtained results are displayed in Fig. 3.9A and B. All the studied samples showed a reduced initial melt viscosity less than 0.1 Pa s. Meanwhile, increasing the temperature from 330 to 350°C resulted in a faster increase in the viscosity. The 3PEODPA had the highest increase rate while the 3PNODPA showed slight changes in viscosity over time. Interestingly, the 3PN3PEODPA exhibited a middle behavior with large

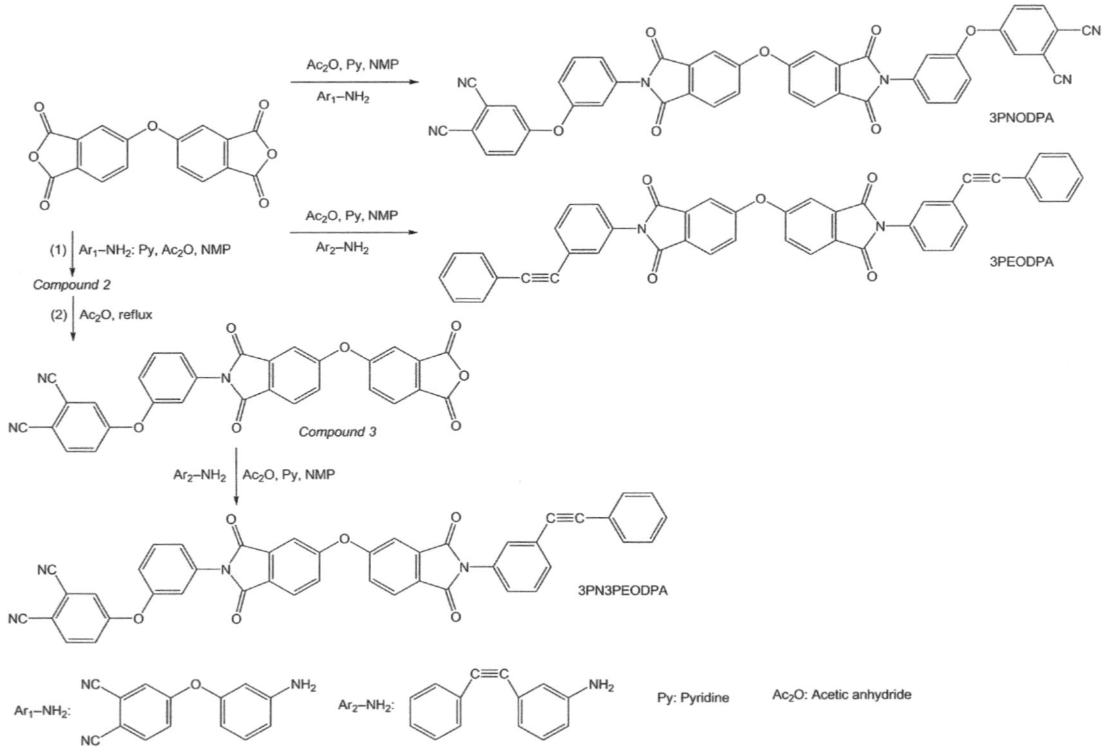

Scheme 3.7 Synthetic route and designation of the imide-containing PN and phenylethynyl end-groups. *PN*, phthalonitrile. Reprinted from Yuan P, Liu Y, Zeng K, Yang G. Synthesis and characterization of a new imide compound containing phthalonitrile and phenylethynyl end-groups. Des Monomers Polym 2015;18:343–9.

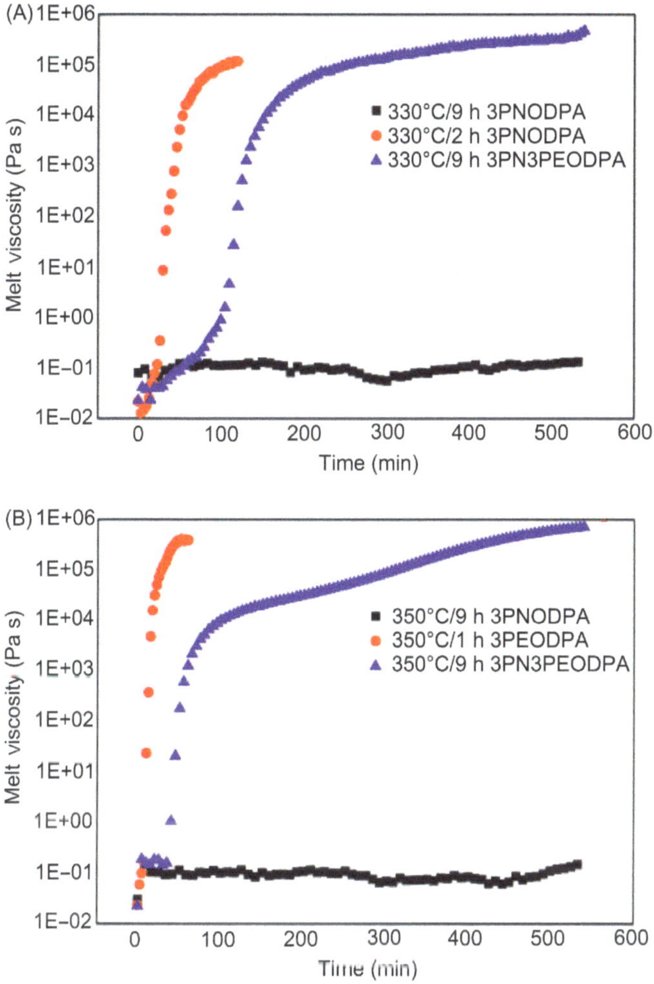

Figure 3.9 Melt viscosity as a function of time for the 3PNODPA, 3PEODPA, and 3PN3PEODPA compounds at 330 (A) and 350°C (B). Reprinted from Yuan P, Liu Y, Zeng K, Yang G. Synthesis and characterization of a new imide compound containing phthalonitrile and phenylethynyl end-groups. Des Monomers Polym 2015;18:343–9.

and adjustable processing window. Thus, the primary objective reducing the phenylethynyl reactivity and on the same time increasing that of the PN was clearly achieved.

The ability of the prepared imide-containing PN and phenylethynyl to cure without the addition of any catalyst was confirmed by DSC analyzes. Obviously, it was confirmed that the 3PNODPA monomers had no exothermic transition confirming their sluggish curing character.

Although the 3PEODPA exhibited a large curing exotherm, the high curing enthalpy of 438 J/g is not appealing for the preparation of thick composite sections in the industry. On the other hand, the 3PN3PEODPA oligomers displayed an exothermic peak at around 245°C with a processing window of about 140°C, confirming the ability of the triple bond from the phenylethynyl to effectively initiate the addition reaction of the PN units. In addition, the curing enthalpy of the 3PN3PEODPA was highly reduced, in contrast to that of the 3PNODPA. Once more, the prepared, imide-containing PN and phenylethynyl showed an interesting behavior combining the benefits of the PN and phenylethynyl units.

The thermal stability investigations were carried out by TGA under nitrogen atmosphere for all the samples cured at 350°C for 3 and 12 h. The resulting TGA curves are presented in Fig. 3.10. The data revealed that all the compositions showed good thermal stability with $T_{5\%}$ temperatures ranging between 474–532 and 491–516°C for samples cured at 350°C for 3 and 12 h, respectively. The char yield at 800°C was also at least higher than 69%. It was also found that increasing the curing time for the 3PN3PEODPA slightly affected the thermal resistance, which may be due to the lower curing rate of these materials.

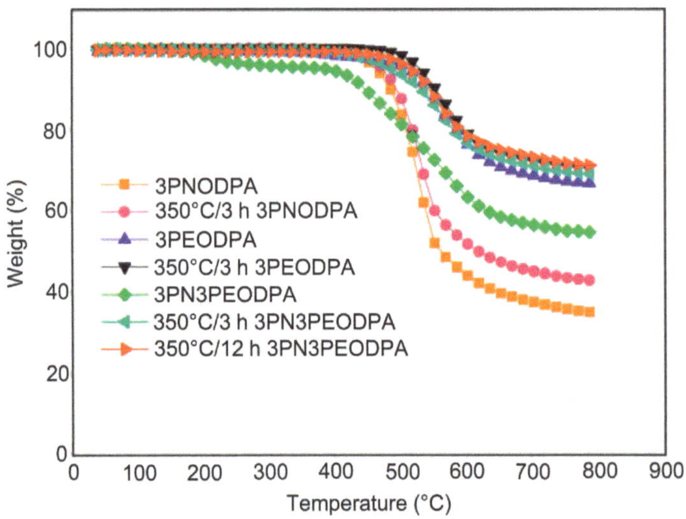

Figure 3.10 TGA curves under nitrogen atmosphere of the 3PNODPA, 3PEODPA, and 3PN3PEODPA compounds and their related polymers. Reprinted from Yuan P, Liu Y, Zeng K, Yang G. Synthesis and characterization of a new imide compound containing phthalonitrile and phenylethynyl end-groups. Des Monomers Polym 2015;18:343–9.

Fortunately, the 3PN3PEODPA showed equivalent or even higher temperature than the polymers from 3PEODPA, indicating that the reaction of the triple bond from the phenylethynyl with the PN units positively affected the thermal stability. Generally, the prepared imide-containing PN and phenylethynyl monomers exhibited one of the best-known processing window (about 140°C) and their resulting polymers were in the range of other imide-functional, PN polymers.

In summary, the imide-functional, PN monomers and resins have reached a high degree of maturity since the preparation of the first monomer by Keller. These materials are now able to be used as advanced materials for composite manufacturing designated to meet the requirements of multiple industries.

3.2.4 Bismaleimide-Functional Phthalonitrile

Bismaleimide resins are another class of high-performance polymers with some physical properties quite similar to that of the PN resins. Therefore, the preparation of monomers, combining both bismaleimide and PN units, was investigated to see whether a synergistic effect can result from these combinations. The first bismaleimide-functional, PN monomer was reported by Zhang et al. [32]. The monomer was obtained by reacting 4,4′-bismaleimidodiphenylmethane and 4-APN in DMA solvent, as seen in Scheme 3.8. The resulting monomers mixed with 4,4′-diaminodiphenylmethane as a curing agent were cured up to 240°C for 10 h and postcured at 300°C for 1 h.

The thermal stability of the bismaleimide-functional, PN polymer was investigated by TGA under both nitrogen and air atmospheres. Considering the relatively low curing temperatures chosen by the authors, the obtained polymers showed excellent thermal stability with

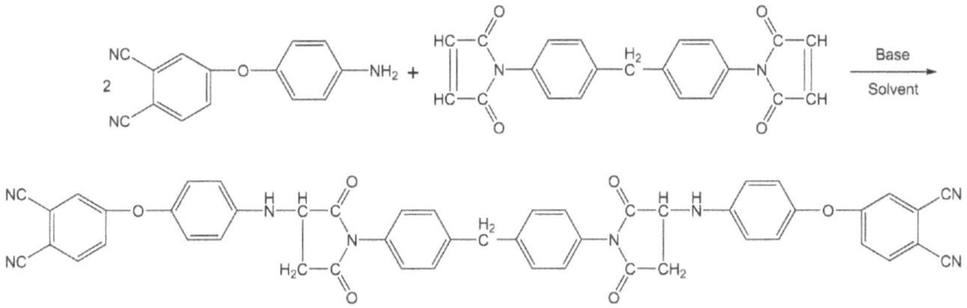

Scheme 3.8 Synthetic route of the bismaleimide-functional, PN monomers. *PN*, phthalonitrile. Reprinted from Zhang J, Liu X, Wen H, Xie M, Cai X. Investigation of the properties of phthalocyanine resin containing bismaleimide groups. Polym Inter 1997;42:363–6.

starting decomposition temperatures of 483 and 368°C in nitrogen and air atmospheres, respectively. Moreover, the char yield at 778°C reached up to 83% under nitrogen atmosphere. Obviously, the improvement in the thermal stability in contrast with some other typical PN polymers was mainly attributed to the enhanced, crosslinking degree afforded by the polymerization of both the bismaleimide and PN units.

The work performed by Zhang et al. revealed that the benefits obtained from the preparation of bismaleimide-functional, PN are limited, mainly because of the similarities in the physical properties of both precursors. Moreover, it was confirmed that the polymerization of the bismaleimide and the PN units were happening separately with no inter-catalytic effect such as the one seen with other X-functional, PN monomers/oligomers.

3.2.5 Novolac-Functional Phthalonitrile

Belonging to the family of phenolic resins, novolac-based materials are still attracting much interest considering their combination of many useful properties. Indeed, novolac resins are well-known for their good mechanical properties and good flame resistance, which are attributed to their highly aromatic structures and to the inhibiting effect of their phenolic compounds. Nonetheless, the evolution of volatiles during the novolac condensation reaction results in void problems and limited crosslinking degrees, which negatively affect their overall performances [33,34]. Therefore the introduction of additional curable compounds within the novolac backbone can be a good alternative to the phenolic condensation reaction to achieve good processability, along with a higher crosslinking degree. During the last few decades, many addition curable groups were added to the novolac backbone to achieve enhanced performances. PN resins are well-known for their superior properties mainly attributed to their effective cyano addition reaction, which can result in rigid and highly crosslinked networks. However, one of the limitations of the PN resins is the need of higher temperatures to initiate the cure reaction. Meanwhile, as discussed in Chapter 1, Introduction to the Phthalonitrile Resins and their Curing Behavior, the PN monomers/oligomers can be easily cured in the presence of a variety of curing agents including phenols, amines, metals and their salts, and so on. In fact, from the emergence of the PN materials, the combinations of novolac and PN in forms of physical blends or as single compounds was studied and reported in the open literature. The synergistic effect between the phenols of novolac and the cyano addition reaction of the PN units was always easily obtained and the resulting resins exhibited many interesting properties. Herein, we focus on the novolac-functional, PN monomers and polymers, while the physical blends are discussed in another chapter.

Yang et al. were the first to report about various novolac-containing multiple percentages of PN units anchored to benzene rings through phenyl azo linkages [35]. The synthetic route was a coupling reaction between novolac and PN diazonium salt as seen in Scheme 3.9. The curing behavior of these novolac-functional, PN oligomers was studied

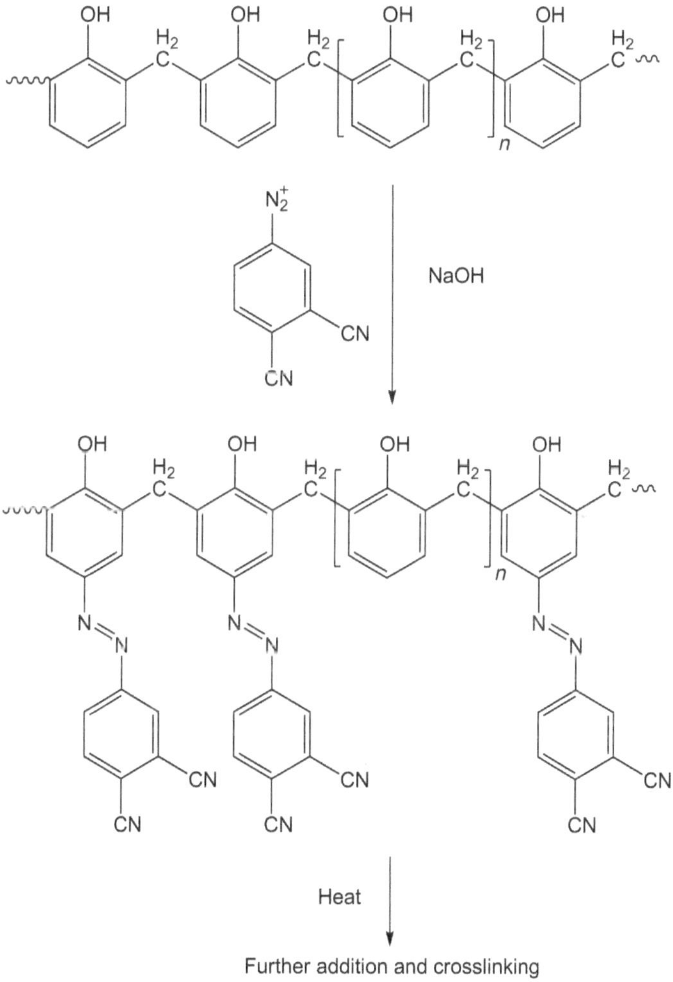

Scheme 3.9 Synthetic route of the novolac-containing multiple percentages of PN units anchored to the benzene rings through phenyl azo linkages. *PN*, phthalonitrile. Reprinted from Yang Y, Min Z, Yi L. A novel addition curable novolac bearing phthalonitrile groups: synthesis, characterization and thermal properties. Polym Bull 2007;59:185–94.

by DSC technique. In all instances, two exothermic transitions were depicted for each degree of substitution. The first exotherm ranged between 200 and 270°C and was confirmed to be due to the partially decomposition of the azo bonds. The second exotherm ranged between 310 and 340°C and was related to the PN addition reaction. Although at that time the curing was reported to be obscure, we can now advance that the addition reaction of the PN units was activated by the free phenolic groups attacking the nitrile group to produce isoindoline structure, which further reacted with other nitrile groups to propagate the reaction. Importantly, the partial decomposition of the azo bonds resulted in catastrophic decreases in the thermal stability. For instance, the starting decomposition temperatures were all below 250°C and extremely far from the outstanding thermal resistances generally afforded by the polymerization of the PN units. Obviously the novolac-functional, PN oligomers with azo phenyl linkages were not suitable to be further investigated.

The second work concerned about the novolac-functional, PN resins was published by Zhang et al. [36]. The researchers used the typical nucleophilic substitution reaction between the hydroxyl groups of novolac and 4-nitrophthalonitrile to prepare novolac-functional, PN oligomers containing an equal molar ratio of hydroxyl to PN groups. The polymerization reaction was tracked by DSC analyze. The novolac-functional, PN oligomers showed a large exothermic transition ranging between 150 and 350°C which was ascribed to the cyano addition reaction of the PN units triggered by the free hydroxyls from the novolac backbone. In contrast with the typical PN monomers, the heat of polymerization was excessively high, with up to 138 J/g, implying that these newly developed, PN oligomers were highly reactive due to the presence of large amounts of free hydroxyl groups. As seen in Chapter 1, Introduction to the Phthalonitrile Resins and their Curing Behavior, few amounts of phenolic curing agents can easily cure the PN precursors; however, large amounts can increase the curing rate.

The curing behavior study was followed by rheological measurements to get a clear picture of the processability of these novolac-functional, PN oligomers. The complex viscosity changes were monitored as a function of temperature. The obtained data showed a sharp decrease in viscosity at 110°C where the viscosity reached values below 4 Pa s. Starting from 140°C and up to 225°C the viscosity remained at its lowest values. Interestingly, the viscosity at 185°C was as low as 0.44 Pa s and the processing window reached about 85°C. In summary, although the novolac-functional, PN oligomers showed enhanced reactivity, their processability remained as good as that of other PN monomers.

The novolac-functional, PN oligomers were cured following the curing schedule of 150°C for 4 h, 180°C for 4 h, 220°C for 4 h, and 250°C for 4 h. The DMA revealed that the resulting polymer had a highly crosslinked structure that allowed a T_g value of almost 372°C. Considering the low curing temperatures, such a high T_g value cannot be achieved by other typical bisphenols-based PN polymers if cured following the same curing program. The thermal stability of the novolac-functional, PN polymer was also inspected by TGA. By always taking into account the chosen curing procedure, the thermal stability parameters under both nitrogen and air atmospheres confirmed the superior thermal resistance of this polymer. For instance, the $T_{5\%}$ reached about 449 and 455°C under nitrogen and air atmospheres, respectively. In contrast with the unmodified novolac resin, improvements of more than 100°C were obtained by the introduction of 50 mol.% of PN units.

As an extension of the work of Zhang et al., Augustine et al. realized multiple novolac-functional, PN oligomers by varying the substitution degree of the PN groups [37]. Thus, 44%, 71%, 87%, and 99% of the hydroxyl groups were replaced by PN units to understand how the overall performances could be altered by these changes in the novolac backbone. The synthetic route was similar to that of the previously discussed novolac-functional, PN oligomers with minor changes in the molar ratios of the precursors to achieve the targeted substitution degrees. The influence of the hydroxyl fractions was revealed by DSC analyzes. By analyzing the data, it appeared that larger fractions of hydroxyls resulted in lower curing temperatures, while increasing the PN fraction, shifted the curing exotherm to higher temperatures. For example, the initial cure temperature moved from 143 to 265°C when the percentage of hydroxyls changed from 56% to 1%. FTIR investigations corroborated the DSC results and also revealed that the resulting polymeric structures were affected by the relative proportions of each phase. However, in all cases, the isoindoline, triazine, and phthalocyanine structures were all depicted in the FTIR spectra.

Polymeric samples, obtained from curing these novolac-functional, PN oligomers following the curing schedule of 100, 150, and 200°C for 30 min each, 250 and 300°C for 1 h each, and 350°C for 3 h, were analyzed by TGA under nitrogen atmosphere, and the results are displayed in Fig. 3.11. First, a noticeable improvement in the thermal stability parameters was depicted by increasing the PN fraction in the novolac backbone. Indeed, the char yield at 1000°C reached up to 78% for the

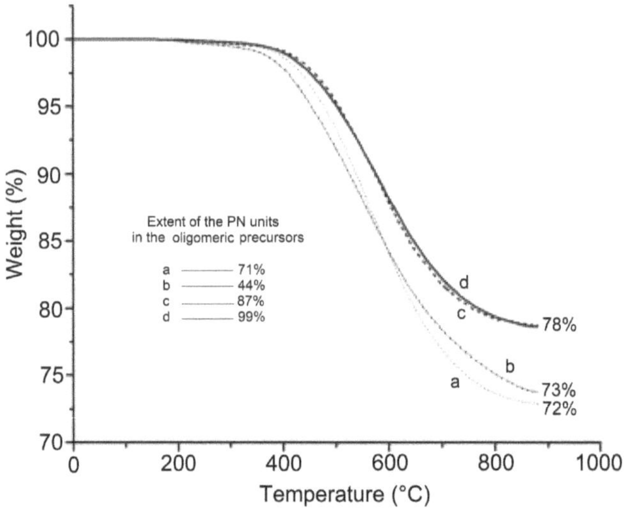

Figure 3.11 TGA curves under nitrogen atmosphere of the novolac-functional, PN polymers from oligomeric precursors with various extents of PN units. *PN*, phthalonitrile. Reprinted from Augustine D, Mathew D, Nair CPR. Phenol-containing phthalonitrile polymers-synthesis, cure characteristics and laminate properties. Polym Inter 2012;62:1068–76.

PN fraction of 99%. The improvement in the thermal stability was mainly attributed to the enhanced crosslinking degree achieved by the polymerization of the PN units.

In summary, all the discussed novolac-functional, PN oligomers showed a self-curing behavior and were cured at temperatures largely lower than those of the typical PN monomers/oligomers. Except for the novolac-functional, PN containing azo linkages, all the other novolac-functional, PN polymers showed remarkable thermal and thermomechanical properties considering their appealing low curing temperatures. Therefore, such materials can easily be used in many exigent military and civilian applications.

3.2.6 Benzoxazine-Functional Phthalonitrile

Polybenzoxazines, as a new family of thermosetting polymer, exhibit various attractive properties including high thermal stability, high char yields, high T_g, near-zero volumetric change upon curing, good mechanical and dielectric properties, low water absorption, and low flammability. These characteristics promote polybenzoxazines over epoxies and traditional phenolic resins in many industries like the electronic encapsulation, composites production, coatings, adhesives, and flame retardant materials [38–40]. The main advantage of the benzoxazine monomers is their ability to be cured without the addition of any external catalyst. Moreover, the thermally activated, ring-opening polymerization of the benzoxazine units results in the formation of phenolic hydroxyl groups, which can be used to react with other polymeric structures to achieve interpenetrating networks.

By understanding the shortcomings of benzoxazines and PN, the importance of their combination will appear as an inevitable choice. In fact, the polybenzoxazine resins suffer from a limited thermal and thermomechanical properties due to low crosslinking degree, while the typical PN precursors need activated hydroxyl groups to initiate their cyano addition reaction. Thus, the preparation of benzoxazine functional, PN monomers/oligomers allows a self-curing behavior while their resulting polymers can take advantage of the high crosslinking degree of the PN.

Brunovska and Ishida first reported about several benzoxazine-functional, PN monomers prepared according to the solventless method from monofunctional phenols or bisphenols and APN [41]. The studied monomers are shown in Fig. 3.12.

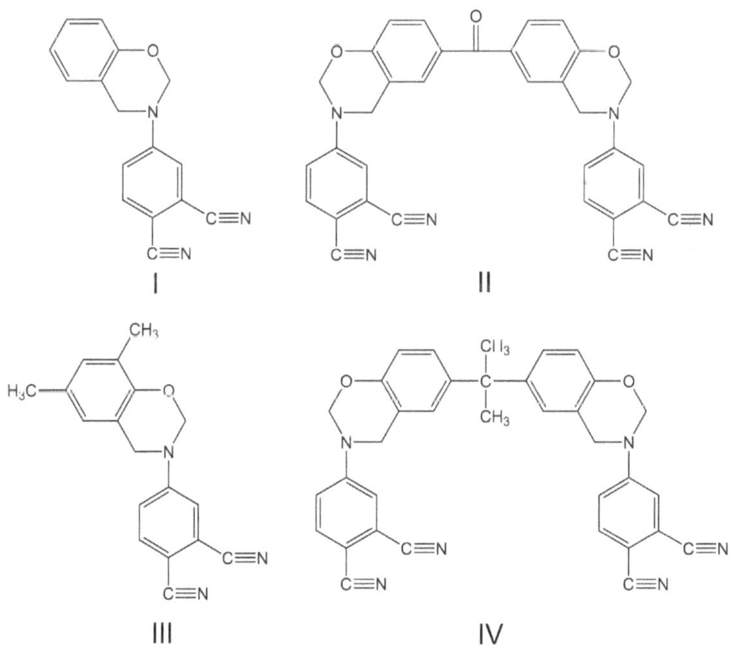

Figure 3.12 Chemical structures of the benzoxazine-functional, PN monomers synthetized by Brunovska et al. Reprinted from Brunovska Z, Lyon R, Ishida H. Thermal properties of phthalonitrile functional polybenzoxazines. Thermochim Acta 2000;357:195–203.

Table 3.4 DSC Results and Curing Conditions of the Benzoxazine-Functional, PN Monomers Prepared by Brunovska et al.

Monomer	$T_{exo(benzoxazine)}$ (°C)	$T_{exo(nitrile)}$ (°C)	T_{cure} (°C)	T_{cure} (h)
I	268	285	240	6
II	314	330	280	6
III	234	262	250	6
IV	227	268	240	6

Reprinted from Brunovska Z, Lyon R, Ishida H. Thermal properties of phthalonitrile functional polybenzoxazines. Thermochim Acta 2000;357:195–203.

Investigations of the curing behavior were performed by DSC and the obtained results are resumed in Table 3.4. All the studied samples showed two exothermic transitions related to the oxazine ring-opening and to the PN cyano addition reaction, respectively. Thus, the primary objective of the study was clearly achieved since the prepared monomers showed a self-curing behavior. The addition reaction of the PN was initiated after the formation of active hydroxyl groups resulting from the oxazine ring-opening. If the monofunctional and bifunctional monomers are compared, the DSC analyzes revealed that the bifunctional monomers had exothermic transition at lower temperatures. Indeed, the bifunctional monomers possess a higher concentration of active hydroxyl groups, which means that the polymerization reaction can proceed more readily for these compounds at lower temperatures.

TGA measurements were performed under nitrogen and air atmospheres on samples cured up to 220°C and postcured at 250°C for 6 h. The thermal stability parameters including the $T_{5\%}$, $T_{10\%}$, and char yields at 600 or 800°C are listed in Table 3.5. By analyzing the data, it appeared that the best match was achieved from the cured monomers of III with a $T_{5\%}$ and $T_{10\%}$ of 544 and 596°C, respectively, and a char yield at 800°C of 80% under nitrogen atmosphere. Interestingly, the cured monomers from I presented better thermal resistance than that of the polymer from IV. For instance, its $T_{5\%}$, $T_{10\%}$ reached up to 450 and 560°C with a char yield at 800°C of 76%. Considering the relatively low curing temperatures, the obtained thermal stability performances of the polymers from I and III can be compared with those from high-performance, PN polymers cured at temperatures in excess of 300°C and for an extended period of time. This excellent thermal stability is due the large increase in the crosslinking degree resulting from the

Table 3.5 Thermal Properties of the Benzoxazine-Functional, PN Polymers, Prepared by Brunovska et al., under Nitrogen and Air Atmospheres

	Nitrogen			Air		
Benzoxazine-Functional, PN Polymer Code	$T_{5\%}$ (°C)	$T_{10\%}$ (°C)	Char Yield at 800°C (%)	$T_{5\%}$ (°C)	$T_{10\%}$ (°C)	Char Yield at 600°C (%)
I	450	560	76	420	440	50
II	414	505	73	378	420	0
III	544	596	80	390	450	75
IV	423	468	68	–	–	–

Reprinted from Brunovska Z, Lyon R, Ishida H. Thermal properties of phthalonitrile functional polybenzoxazines. Thermochim Acta 2000;357:195–203.

Table 3.6 Flammability Properties of the Benzoxazine-Functional, PN Polymer (I) and Other Polymeric Systems

Polymer Designation	Peak μHRR (W/g)	Total μHRR (kJ/g)
Benzoxazine-functional PN (I)	100.9	4.97
Polyetherimide Ultem	187.4	6.72
Phenolic resin	80	6.5
PPS	400	11.2
Nylon	700	19.0

Reprinted from Brunovska Z, Lyon R, Ishida H. Thermal properties of phthalonitrile functional polybenzoxazines. Thermochim Acta 2000;357:195–203.

boosted reaction of the nitriles groups. It should be also pointed out that the reduced thermal stability of the polymer from II was due to the low, crosslinked, Mannich-based networks because the precursor blocked the *ortho* position of the benzene rings. As a result, the oxazine ring-opening reaction through the *meta* position only proceeded to a low extent.

Flammability tests were also carried out for the polymer from I using a microcone calorimeter. The maximum peak heat release (PHR) and total heat release (THR) were measured and compared with other known, flame resistant polymers. From the obtained results, shown in

Table 3.6, the polymer from I had the PHR of 100.9 W/g and a THR of 4.97 kJ/g which are almost better than those of the phenolic resins. Hence, the preparation of benzoxazine functional. PN monomers not only allowed ameliorations in the curing temperature and the overall curing behavior, but also the resulting resins exhibited outstanding thermal and flame resistances.

The benzoxazine-functional, PN materials were further studied by Cao et al. [42]. The authors reported on a new type of PN monomer containing both aromatic ether and oxazine linkages. The as-described monomer was synthetized from BPA, 4-APN, and paraformaldehyde in a mixture of toluene and 1,4-dioxane as a solvent. The synthetic route is resumed in Scheme 3.10. The prepared benzoxazine-functional, PN monomers were noted as benzoxazine-functional PN (BZPN).

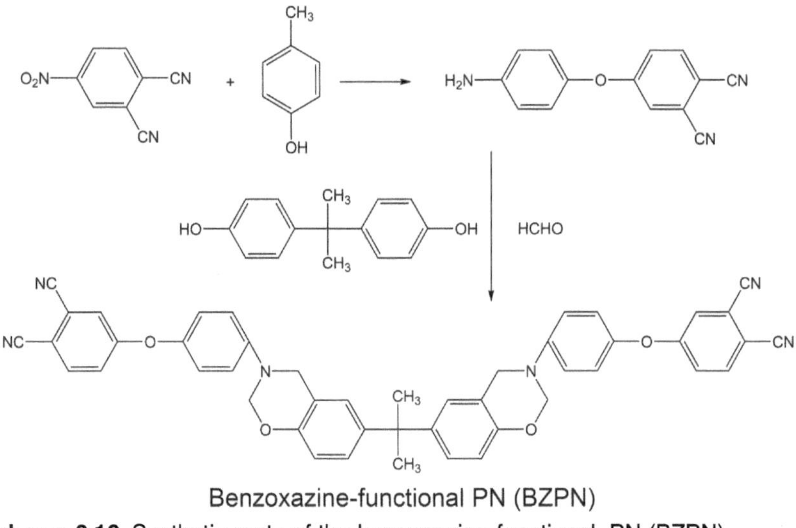

Scheme 3.10 Synthetic route of the benzoxazine-functional, PN (BZPN) monomers. *PN*, phthalonitrile. Reprinted from Cao GP. Synthesis and characterization of a novel bisphthalonitrile containing benzoxazine. Exp Polym Lett 2007;1:512–8.

The DSC thermograms of the newly developed, benzoxazine-functional, PN monomers and their related precured resins at different temperatures are displayed in Fig. 3.13. The self-curing behavior of the prepared monomer was evidenced by the presence of two distinct exothermic transitions. The first exotherm ranged between 170 and 265°C and was ascribed to the oxazine ring-opening polymerization. The second exotherm peaked at about 265°C and corresponded to the cyano

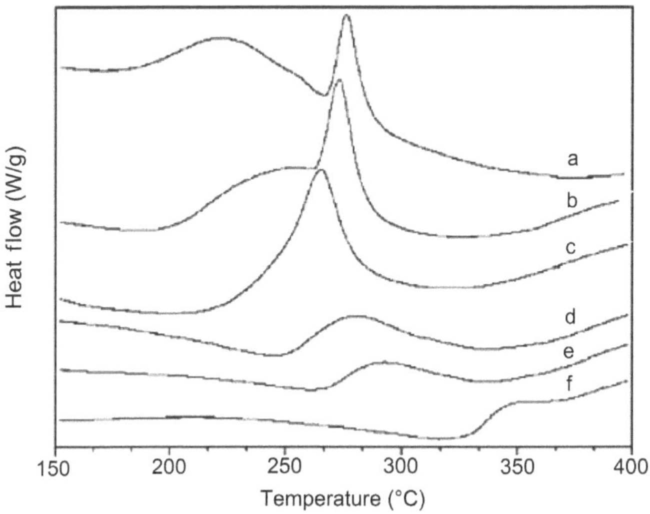

Figure 3.13 DSC thermograms of the BZPN monomers (a) and their related precured resins at different temperatures: 160 (b), 180 (c), 200 (d), 220 (e), and 250°C (f). *DSC*, dynamic scanning calorimetry. Reprinted from Cao GP. Synthesis and characterization of a novel bisphthalonitrile containing benzoxazine. Exp Polym Lett 2007;1:512–8.

addition reaction of the nitriles groups afforded by the presence of the active hydroxyl groups, generated from the oxazine ring-opening reaction. It was also found that both exothermic peaks were reduced and later eliminated by increasing the curing time and temperature. For instance, the first exotherm disappeared after curing at 200°C for 8 h and the second one was completely eliminated after at 220°C for 6 h and 250°C for 5 h These observations indicated that the two steps polymerization reaction can be easily controlled by varying the curing conditions and further confirmed the ability of these PN monomers to be cured at lower temperatures and times. Moreover, curing the monomers up to 250°C for 5 h resulted in the formation of a large peak between 325 and 360°C. The 325°C temperature can be used as the ultimate postcuring temperature and the 360°C temperature was later confirmed by TGA to be the initial decomposition temperature.

The thermal and thermo-oxidative stabilities of the benzoxazine-functional, PN polymer were investigated by TGA and the obtained curves are presented in Fig. 3.14. As expected from the DSC analyzes, the initial decomposition temperature (T_{id}) was depicted at 343 and 354°C under nitrogen and air atmospheres, respectively. The $T_{5\%}$ and $T_{10\%}$ reached 438 and 505°C under nitrogen atmosphere and 435

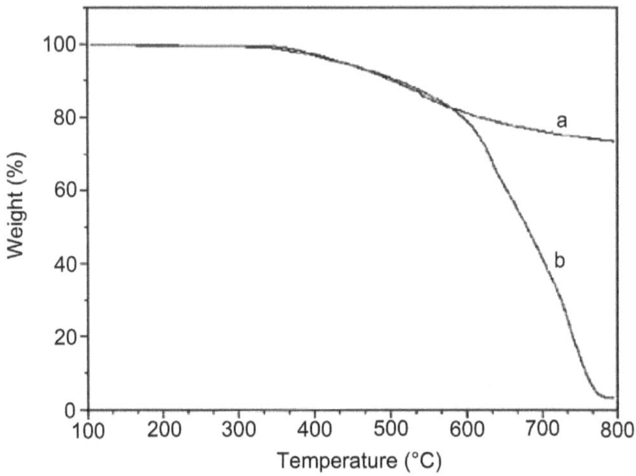

Figure 3.14 TGA curves of the BZPN polymers under nitrogen (a) and air (b) atmospheres. Reprinted from Cao GP. Synthesis and characterization of a novel bisphthalonitrile containing benzoxazine. Exp Polym Lett 2007;1:512–8.

and 504°C under air atmosphere, respectively. The char yield measured at 800°C under nitrogen atmosphere was 74% and at 600°C under air atmosphere was 79%. Globally, the thermal resistance of the studied polymer was comparable to that of other, typical PN resins that were not postcured. Moreover, when compared with the benzoxazine-functional, PN polymers prepared by Brunovska and Ishida, the BZPN polymer showed slightly inferior thermal stability under nitrogen atmosphere; nonetheless, under air atmosphere significant improvements were achieved in the thermo-oxidative stability.

As a further extension of the work of Cao et al., Xu et al. studied the effect of different amounts of BPA on the curing behavior and processability of the BZPN monomers [43]. Considering that a large number of nitrile groups were found to remain unreacted, the authors proposed the introduction of additional hydroxyl groups to achieve a higher degree of polymerization. Meanwhile, these hydroxyl groups could also reduce the curing temperature, enlarge the processing window, and improve the overall processability of these monomers. Hence, the authors prepared series of blends with various molar ratios of BZPN monomers to BPA (2:1, 2:2, 2:3) and studied their properties. The proposed reaction process in the presence of BPA is shown in Scheme 3.11. Basically, the BPA started by catalyzing the oxazine ring-opening, and linear polybenzoxazine was formed. Subsequently, the cyano addition reaction of the nitriles was activated by the hydroxyl groups from

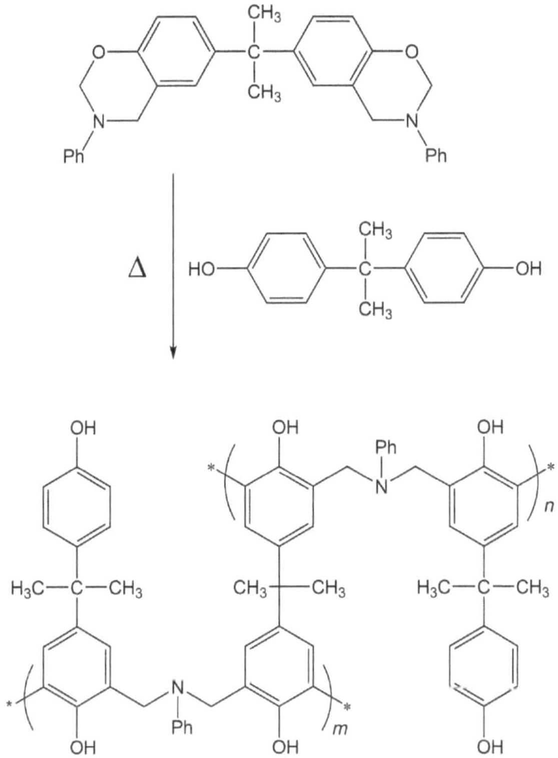

Scheme 3.11 Reaction mechanism of BZPN in the presence of BPA. *BPA*, bisphenol-A. Reprinted from Xu M, Jia K, Liu X. Effect of bisphenol-A on the structures and properties of phthalonitrile-based resin containing benzoxazine. Exp Polym Lett 2015;9.

both the remaining BPA and oxazine rings. Then, with higher curing temperatures and an extended curing time, the phthalocyanine structures self-aggregated and paralleled to the linear linked network of polybenzoxazine.

To get more insights about the phenomenon of phase separation, the polymerization reaction of the 2:2 BZPN to BPA blends was followed by an scanning electron microscopy (SEM) visualization of the polymer surfaces, as seen in Fig. 3.15. After moving forward with the curing process up to 160°C, the main reaction was the oxazine ring-opening with the formation of the polybenzoxazine homopolymer. Indeed, the SEM image (Fig. 3.15A) showed a homogenous state of surface. Further increasing the curing temperature resulted in the formation of the phthalocyanine from the reaction of the nitrile groups. At this step, the SEM images (Fig. 3.15B and C) revealed the presence of minor

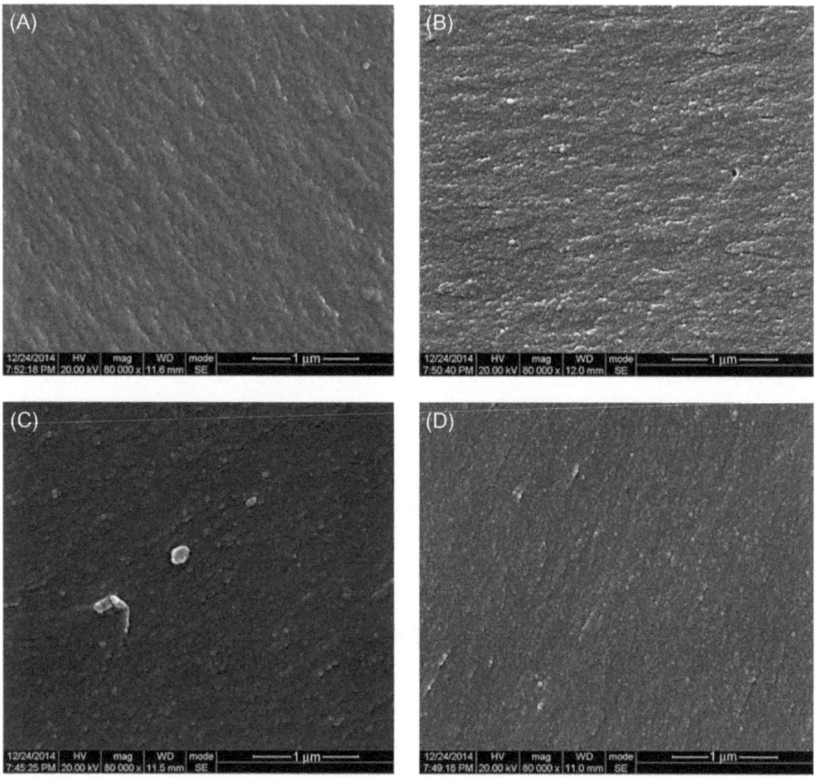

Figure 3.15 SEM images of BZPN/BPA 2:2 blend cured at various temperatures: (A) 160, (B) 200, (C) 240, and (D) 280°C. *BPA*, bisphenol-A. Reprinted from Xu M, Jia K, Liu X. Effect of bisphenol-A on the structures and properties of phthalonitrile-based resin containing benzoxazine. Exp Polym Lett 2015;9.

tendency of phase separation because the polymeric network contained a large proportion of nonconnected polybenzoxazine and phthalocyanine structures. After higher curing temperatures, the formation of an interlinked network reduced the severity of phase separation and the state of surface was more homogeneous.

The impact of the BPA on the processability of the BZPN was studied by rheology. The evolution of the complex viscosity as a function of temperature for the neat BZPN monomers and their BPA-based blends are displayed in Fig. 3.16. In contrast with the neat sample, all the blends showed reduced viscosity values with minimum melt viscosities of 1–6 Pa s. This minimum melt viscosity remained stable from 80 to 180°C implying an improved processability for the blends. Between 180 and 200°C, a rapid increase in the viscosity was noticed for the samples. However, increasing the amount of the BPA ensued an

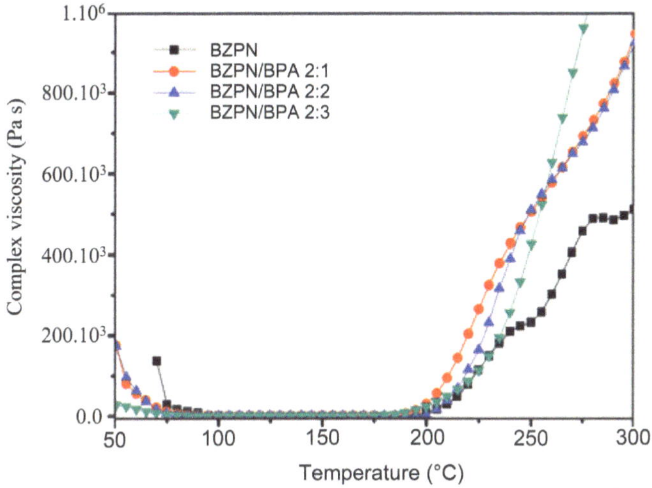

Figure 3.16 Complex viscosity as a function of temperature for BZPN/BPA system with various content of BPA. *BPA*, bisphenol-A. Reprinted from Xu M, Jia K, Liu X. Effect of bisphenol-A on the structures and properties of phthalonitrile-based resin containing benzoxazine. Exp Polym Lett 2015;9.

increase in the viscosity at lower temperatures. In addition, the isothermal viscosity measurements at 160°C (Fig. 3.17) revealed that the initial melt viscosity decreased by increasing the proportion of the BPA in the blend.

In summary, the introduction of the BPA as an additional source of hydroxyl groups allowed improvements in the curing behavior and processability of the neat benzoxazine-functional PN. The as-such prepared blends were then used as matrix for fiber-reinforced composites, a proper discussion about those findings can be found on Chapter 5, Fiber-Reinforced Phthalonitrile Composites.

Xu et al., also reported about an interesting PN monomer containing both benzoxazine and allyl-functional groups [44]. The aim was to prepare self-catalyzed monomers by taking advantage from the allyl and benzoxazine-mediated cyano addition reaction. The resulting PN monomers were expected to be cured at lower temperatures, while their overall thermal properties could be improved by the additional crosslinking degree allowed by the participation of a much larger number of nitrile groups in the polymerization reaction. The as-described monomer was synthetized from DBPA, 3-APN, and paraformaldehyde in a mixture of toluene and 1,4-dioxane as a solvent.

The self-catalyzed, curing behavior of the allyl and benzoxazine functional, PN monomers was studied by DSC. For a comparison

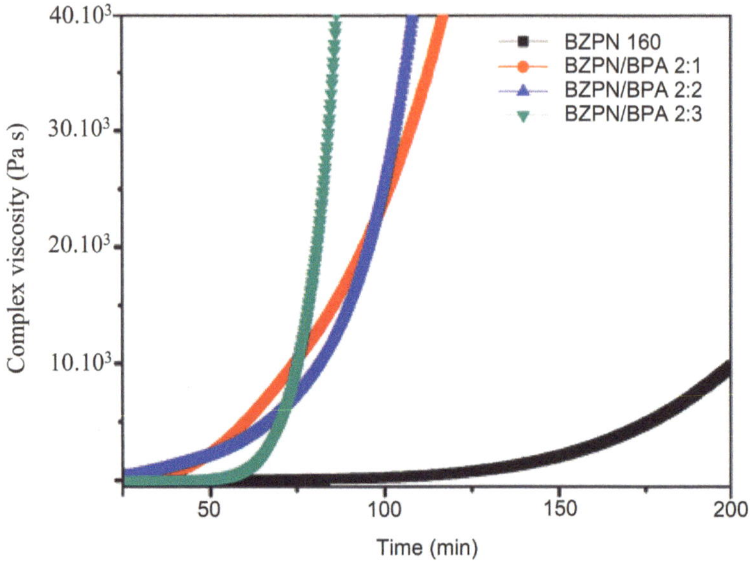

Figure 3.17 Complex viscosity as a function of time at 160°C for the BZPN/BPA system with various content of BPA. *BPA*, bisphenol-A. Reprinted from Xu M, Jia K, Liu X. Effect of bisphenol-A on the structures and properties of phthalonitrile-based resin containing benzoxazine. Exp Polym Lett 2015;9.

purpose, the DSC results of the DBPA and benzoxazine functional, PN monomers were also introduced. The DSC curve of the allyl and benzoxazine functional, PN monomers showed one weak and large exothermic transition between 150 and 270°C ascribed to the overlapped polymerization reactions of the allyl and benzoxazine moieties. Another sharp exothermic transition peaking at 285°C corresponded to the cyano addition reaction of the nitrile groups, activated by both the allyl and hydroxyl groups from the oxazine rings. If compared with the benzoxazine functional, PN monomers, the ring-opening of the oxazine rings of these newly developed monomers had a reduced exothermicity and the whole curing exotherms shifted to higher temperatures. Furthermore, the exothermicity of the cyano addition reaction was markedly increased, confirming the triple curing behavior of these monomers.

The suitable initial curing temperature and time were also determined by a rheological test. The evolution of the viscosity as a function of time at various temperatures revealed that the initial viscosity values were relatively low; ranging between 0.96 and 2.56 Pa s. A low melt viscosity is important for the preparation of composites by the conventional industrial processes. Moreover, the 180°C was chosen as the most suitable initial curing temperature considering that the polymerization reaction occurred at an appropriate rate at this temperature.

The DMA and TGA analyzes were employed to ascertain about the thermomechanical and thermal properties of the allyl and benzoxazine functional, PN polymer cured following the optimized curing schedule of 180°C for 2 h, 220°C for 2 h, 260°C for 2 h, and 280°C for 3 h. The storage modulus at 50°C was measured to be about 3 GPa and remained about 500 MPa at 400°C. In contrast with the BZPN polymer described by Xu et al., which only showed 10 MPa of storage modulus at 400°C, this newly developed polymer showed a thermal relaxation behavior. Indeed, the increased rigidity of this polymer afforded by the additional polymerization processes of the allyl moieties positively affected the stiffness of the polymer. The T_g was also improved and reached 380°C, confirming that the crosslinking density was enhanced by the triple polymerization processes. The thermal stability of the allyl and benzoxazine functional, PN polymer was also markedly improved. For instance, the $T_{5\%}$ was 497°C and the char yield at 800°C was 77%. The same explanations about the increase in rigidity and crosslinking degree are also valid for the thermal stability improvements.

Generally, the benzoxazine functional, PN materials can be classified as advanced materials owing to their combination of many interesting features from benzoxazine and PN. The preparation of composites from these resins can be advantageous in many ways considering their relatively low melt viscosity, large processing window, and long storage time at room temperature. The benzoxazine functional, PN materials can be easily applied to the process of vacuum-assisted transfer molding (VARTM) and to any other conventional industrial process. Therefore, the benzoxazine functional, PN materials are expected to gain in influence in the next few years in both academia and industrial fields.

3.2.7 Benzimidazole-Functional Phthalonitrile

Developing PN monomers/oligomers able to self-catalyze while exhibiting good processability with excellent thermal and mechanical properties of the resulting polymers was and remains until now the ultimate goal pursued by many researchers. As discussed in the previous sections, the first autocatalytic, PN monomers were prepared by the introduction of an amino or hydroxyl groups in the monomer backbone. The as-such prepared, PN resins were found to exhibit slightly lower global performances than their noncatalytic counterparts. Hence, the challenge is to improve the properties of the self-catalyzed, PN resins by the preparation of X-functional, PN monomers, where the "X" functionalities are preferably able to simultaneously polymerize and activate the cyano addition reaction of the PN units. It should also be pointed out that self-catalytic, PN monomers suffer from the increased reactivity generated from the presence of large amounts of active hydroxyls which affect their processing by the common industrials processes such as RTM, VARTM, and filament winding.

One of the rare PN monomer able to satisfy the majority of these conditions was reported by Wu et al. [45]. The authors synthetized benzimidazole-functional, PN (BIPN) monomers and studied their curing behavior, processability, and thermal properties of their cured polymer. The unique characteristics of the benzimidazole include its excellent thermal stability and the high acidity of the N–H proton on the ring. Interestingly, the prepared BIPN monomers showed an unexpected mild curing process which opened the way for the use of higher initial processing temperature crucial for industrialization purposes. Prior to more explanations, it should be noted that the new monomer was synthetized via a three-step reaction process as seen in Scheme 3.12. Basically, the first step consisted on reacting 3,5-dihydroxybenzoic acid and 1,2-phenylenediamine in the presence of dicyclohexylcarbodiimide

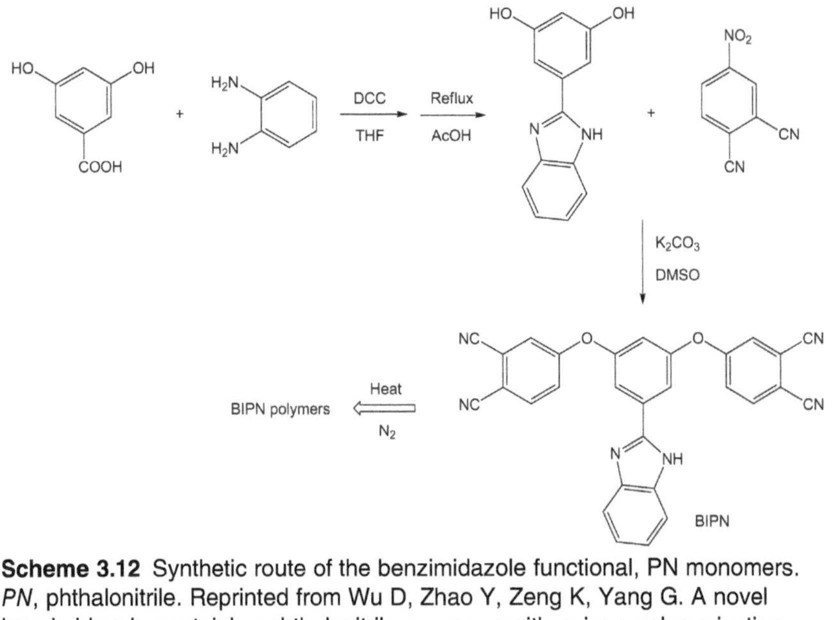

Scheme 3.12 Synthetic route of the benzimidazole functional, PN monomers. *PN*, phthalonitrile. Reprinted from Wu D, Zhao Y, Zeng K, Yang G. A novel benzimidazole-containing phthalonitrile monomer with unique polymerization behavior. J Polym Sci: Part A: Polym Chem 2012;50:4977–82.

as the dehydration catalysis. The second step was the cyclodehydration in acetic acid, which was forwarded by a nucleophilic displacement reaction in the presence of 4-nitrophthalonitrile to afford the targeted BIPN monomers.

The unique curing behavior of the BIPN monomers was studied by DSC and the obtained curves are displayed in Fig. 3.18. The first DSC scan revealed the presence of two endothermic transitions at 175 and 250°C attributed to the monomers loss of crystalline water and to the melting transition, respectively. In addition, an exothermic peak was depicted at 185°C and was corresponded to the recrystallization of the BIPN monomers/oligomers after the water loss. Another DSC scan was also performed immediately after the first one and the results were quite similar, except for the appearance of a T_g at 120°C. Interestingly, both DSC scans did not show any exothermic transition at high temperatures,

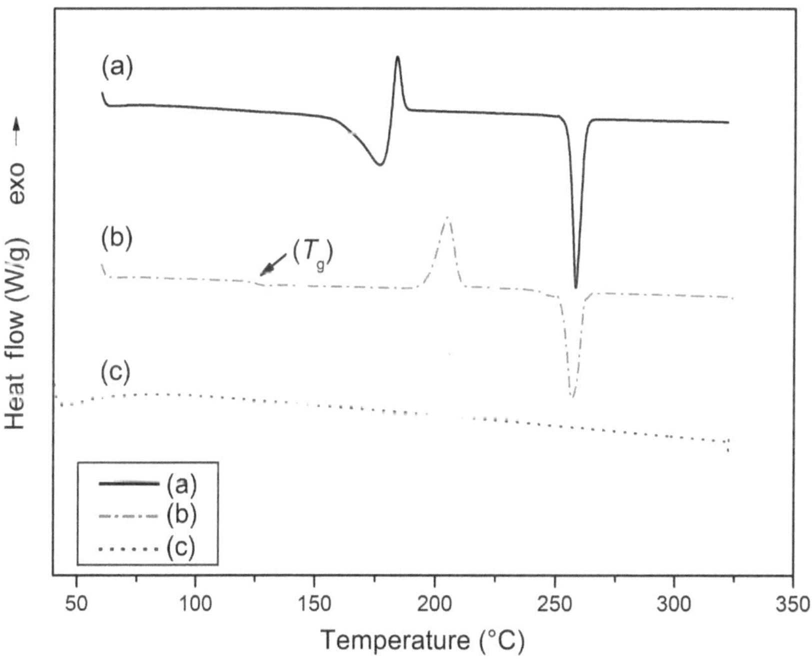

Figure 3.18 DSC curves of BIPN under nitrogen atmosphere: (a) first scan at a heating rate of 10°C/min; (b) second scan at a heating rate of 10°C/min immediately after first scan; (c) second scan at a heating rate of 20°C/min immediately after isothermal scan at 310°C for 8 h in the DSC. *DSC*, dynamic scanning calorimetry. Reprinted from Wu D, Zhao Y, Zeng K, Yang G. A novel benzimidazole-containing phthalonitrile monomer with unique polymerization behavior. J Polym Sci: Part A: Polym Chem 2012;50:4977–82.

meaning that either the polymerization did not happen or the polymerization rate was slow and the enthalpy of cure was weak. Since such behavior was not expected, considering the high acidity of the benzimidazole hydroxyl groups, the researchers decided to perform an additional heat treatment of 310°C for 8 h in the DSC chamber on the BIPN monomers. The heat treatment was followed by a DSC scan which revealed that the sample did not exhibit any T_g. The absence of a T_g implied that the crosslinking degree was higher as the mobility of the chains was obviously restricted. The authors further tested the solubility of the resulting materials by several common solvents and the results confirmed that the BIPN monomers were effectively transformed into a polymeric structure by an uncommonly mild, polymerization reaction.

Encouraged by the obtained results, rheological measurements on the BIPN monomers were also performed to assess their processability. Investigations of the storage modulus variations as a function of the temperature revealed that the monomers showed a rapid decrease in the storage modulus upon melting at 250°C and that low modulus value was maintained up to 400°C. Thus, the authors decided to further investigate the stability of the melt modulus over time at various temperatures. As seen from the obtained results presented in Fig. 3.19, a stable and low modulus was surprisingly kept for longer times. For instance, the modulus was low and stable at 300°C for 4 h, at 310°C for 2 h, and at 330°C for 50 min. Meanwhile, after 5 h at 300°C the modulus showed a rapid increase, implying the starting of the polymerization reaction. Such a low initial modulus can only be seen for some typical PN monomers with no self-curing ability. Therefore, the BIPN monomers have been proven to combine all the benefits required for easy industrialization purposes.

Although the authors did not report the exact curing schedule for the preparation of the BIPN polymer, it was confirmed that curing up to temperatures in excess of 350°C had a slight effect on the overall thermal stability. In all instances, the BIPN polymer showed starting decomposition temperatures similar to that of the typical resorcinol-based PN (RPh) polymer (in excess of 500°C for the $T_{5\%}$ under nitrogen atmosphere), while the char yield at 800°C of 80% was, in this case, higher than that of the RPh polymer with respect to the curing schedule and the type and amount of the curing agent. Thus the work performed by Wu et al., came across a novel and unique new PN resin combining properties not seen in many other thermosetting resins including the PN. The obtained results can also be seen as a breakthrough in the field of PN monomers design and a myriad of opportunities are still available for future researchers.

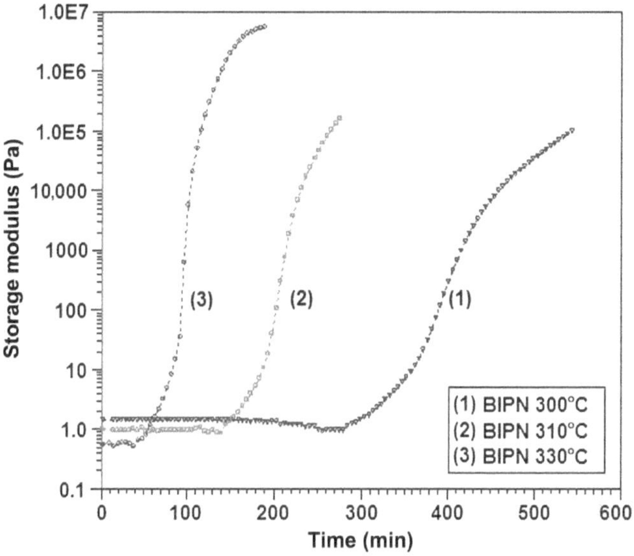

Figure 3.19 Storage modulus vs time curves for the BIPN monomers at 300, 310, and 330°C. *BIPN*, benzimidazole-functional, PN. Reprinted from Wu D, Zhao Y, Zeng K, Yang G. A novel benzimidazole-containing phthalonitrile monomer with unique polymerization behavior. J Polym Sci: Part A: Polym Chem 2012;50:4977–82.

3.2.8 Cyanate Ester-Functional Phthalonitrile

Considering the similarities in the curing behavior between the OCN groups of the cyanate esters and the CN groups in the PNs, it was interesting to study the possible compatibility between these groups when anchored on the same backbone. Thus, the objective of the preparation of cyanate ester-functional, PN oligomers and polymers was to see whether the cyanate ester groups can provide any improvements in the curing behavior and thermal properties of the PN resins.

The only report about the synthesis of cyanate ester-functional, PN oligomers was performed by Augustine et al. [46]. The synthetic route was quite similar to that adopted for the preparation of the propargyl-functional, PN oligomers. First, novolac prepolymers were chosen as the precursor since they contain several free phenolic functions. Then, various novolac-containing PN compounds were prepared by varying the initial molar ratio of novolac and 4-nitrophthalonitrile. Finally, the remaining phenolic functionalities were further reacted with cyanogen bromide to afford the cyanate ester-functional, PN oligomers. The ratio of PN to cyanate ester groups was measured by means of elemental

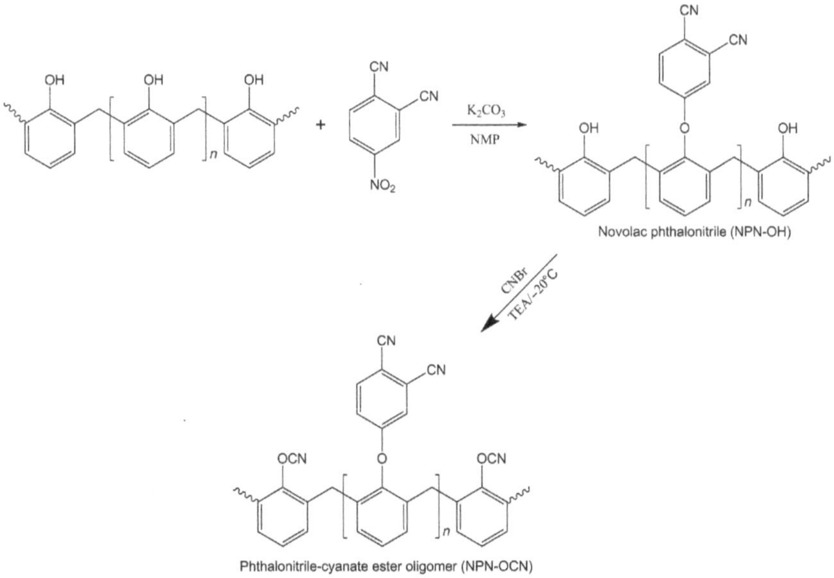

Scheme 3.13 Synthetic route of the cyanate ester-functional, PN oligomers. *PN*, phthalonitrile. Reprinted from Augustine D, Mathew D, Nair CR. Phthalonitrile resin bearing cyanate ester groups: synthesis and characterization. RSC Adva 2015;5:91254–61.

analysis and expressed in terms of molar ratios. The global synthetic route is presented in Scheme 3.13.

The curing behavior of the cyanate ester-functional, PN oligomers was studied by DSC and compared with that of several other neat and functional, PN and cyanate ester oligomers as seen in Fig. 3.20. The neat novolac-containing PN and cyanate esters oligomers with a complete phenols transformation into functionalities only showed one exothermic transition, ascribed in each case to the polymerization of the CN and OCN groups, respectively. For example, the novolac-functional, PN oligomers exothermic transition peaked at 320°C and that of the novolac-functional cyanate ester oligomers was depicted at 210°C. On the other hand, the cyanate ester-functional, PN oligomers exhibited two well resolved exothermic transitions at 120 and 335°C. Obviously, these transitions were seen as the combination of that of their self-precursors and it appeared that the polymerization of the cyanate ester and PN groups were separately happening with no effect on each other. These observations can be understood by taking a closer look on the curing behavior of each phase alone. In fact, the cyanate ester and PN units do not show any self-curing behavior, while their

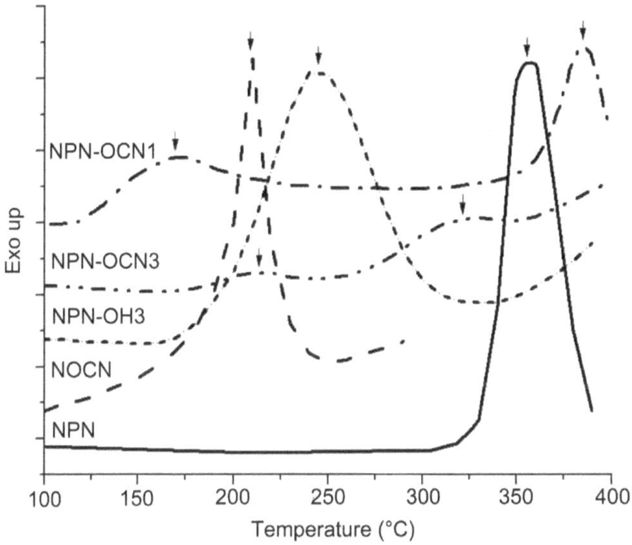

Figure 3.20 DSC curves of the cyanate ester-functional PN (NPN-OCN), novolac-functional PN with free hydroxyl groups (NPN-OH), and novolac-functional cyanate ester (NOCN) and novolac-functional PN (NPN) with no free hydroxyl groups. *DSC*, dynamic scanning calorimetry. Reprinted from Augustine D, Mathew D, Nair CR. Phthalonitrile resin bearing cyanate ester groups: synthesis and characterization. RSC Adv 2015;5:91254–61.

resulting networks are also quite similar. As a result, the polymeric structure resulting from the polymerization of the cyanate esters had no effect on the curing behavior of the PN units.

As both PN and cyanate ester polymers are well known for their reduced water absorbance, it was interesting to explore the water uptake properties of the cyanate ester-functional, PN polymers. The aforementioned polymers were obtained by curing their oligomeric precursors according to the curing schedule of 150, 200, and 250°C for 1 h each followed by 300°C for 2 h and 330°C for 3 h. The water uptake measurements were performed in accordance with the ASTM standard D 570-98. The percentage of water absorption was measured as a function of the molar fraction of the OCN groups. As seen in Fig. 3.21, the water uptake increased by more OCN groups contents. The reason behind these reduced performances is related to the polymeric structure resulting from the polymerization of the cyanate ester groups. Indeed, during the formation of the triazine structure, moisture can be accommodated in the free volume formed on curing. On the other hand, the PN monomers polymerize into different structures including the phthalocyanine, triazine, and isoindoline which give them superior water repellent performances.

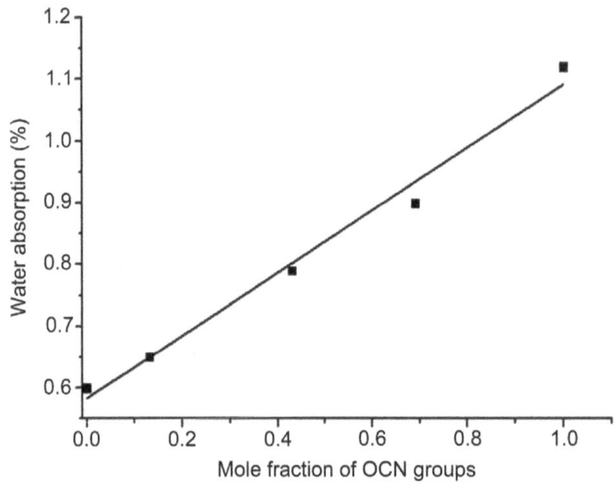

Figure 3.21 Water uptake as a function of the OCN groups molar fraction. Reprinted from Augustine D, Mathew D, Nair CR. Phthalonitrile resin bearing cyanate ester groups: synthesis and characterization. RSC Adv 2015;5:91254–61.

As seen in our previous discussions about the X-functional, PN materials when the "X" functionality was able to provide additional reactive sites, the crosslinking density was generally enhanced, resulting in ameliorations in the thermal and thermomechanical properties. These observations were not confirmed for the cyanate, ester-functional, PN polymers, since slight reductions in the starting decomposition temperatures were noticed upon increasing the proportion of the OCN groups. Therefore, it was advanced that the triazine structure may be the weakest thermally stable structure in the PN polymeric network. In summary, although the cyanate, ester-functional PN did not show major improvements in the overall performances, they remain a class of high-performance polymers with many interesting features.

3.2.9 Naphthyl-Functional Phthalonitrile

The naphthyl structure has for a long time inspired many works especially when excellent thermal and thermomechanical properties are needed. The naphthyl-functional, PN resins were designed to achieve additional improvements in these properties by targeting the hydrogen free polymeric structures.

The research group led by professor Zhang reported about the first known naphthyl-functional, PN monomers and revealed their remarkable

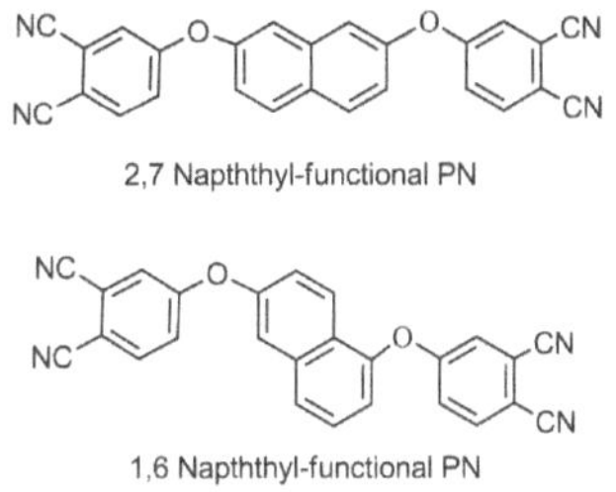

Figure 3.22 Chemical structure of the studied naphthyl-functional, PN monomers.

behavior [47,48]. The studied naphthyl-functional, PN materials were prepared following the typical nucleophilic substitution reaction between 4-nitrophthalonitrile and 2,7 and 1,6-dihydroxynaphthalene, respectively. The prepared naphthyl-functional, PN monomers are drawn in Fig. 3.22.

Since the prepared PN monomers were not able to self-catalyze due to the absence of any known initiator in their backbone, the authors decided to study the curing behavior of these monomers by adding amine-curing agents. Thus the 2,7 and 1,6 naphthyl-functional, PN monomers were tested by DSC in the presence of 5.8 mol.% *m*-APB and 2 mol.% 4,4′-diaminodiphenyl ether, respectively. The obtained DSC curves are presented in Fig. 3.23. As seen in Fig. 3.23A, the 2,7 naphthyl-functional, PN monomers showed two endothermic transitions at about 114 and 195°C ascribed to the melting of the *m*-APB and PN monomers, respectively. In addition, a large exothermic transition was also depicted between 216 and 313°C, which was related to the cyano addition reaction of the PN units activated by the amino-curing agent. The processability of these new PN monomers was estimated through the measurement of the processing window. The 2,7 naphthyl-functional, PN monomers had a large processing window of about 97°C. A similar behavior was also depicted for the 1,6 naphthyl-functional, PN monomers as seen in Fig. 3.23B. However, the choice of the authors to change the curing agent rendered the comparison between both monomers difficult to achieve. Regardless of the choice of the curing agent, the 1,6 naphthyl-functional, PN monomers had also a large processing window of almost 106°C. In all instance, the naphthyl-functional, PN monomers showed a similar curing behavior with the typical PN monomers with some improvements in the processability.

With respect to the choice of the type and amount of the curing agent as well as the curing procedure, both new naphthyl-functional, PN polymers showed excellent thermal and thermo-oxidative stabilities. By

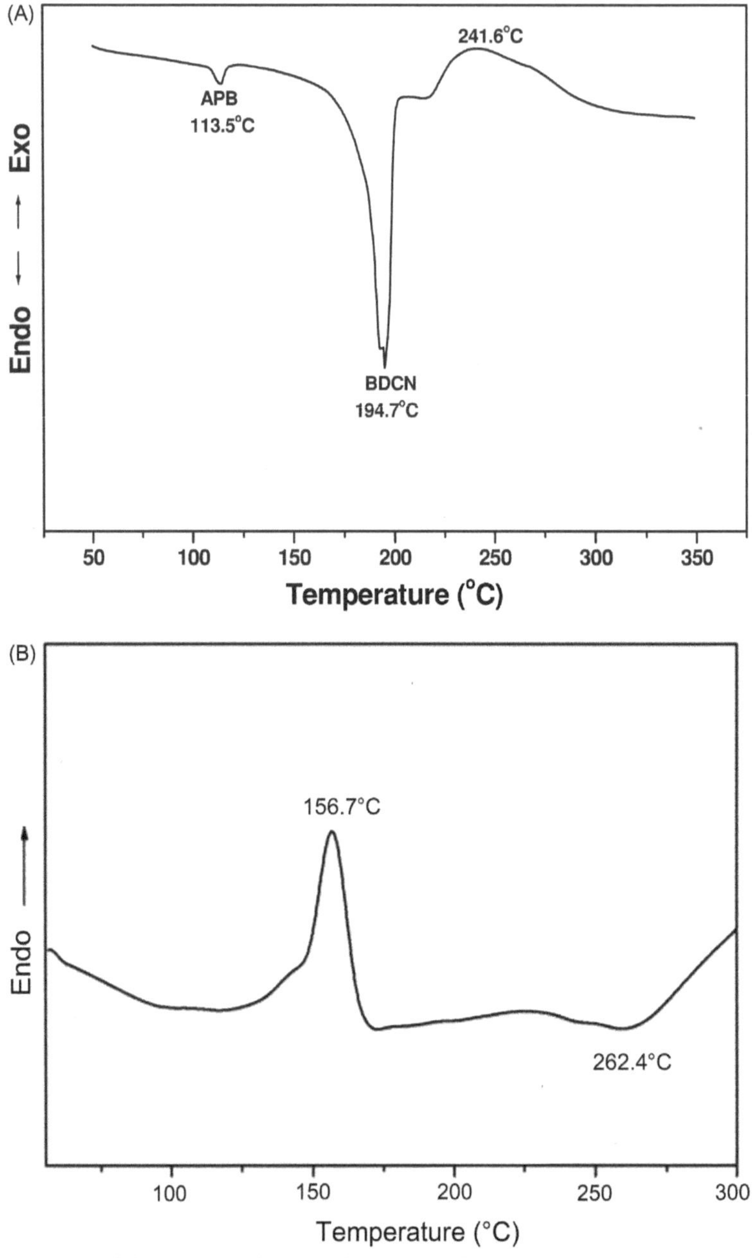

Figure 3.23 DSC curves of the 2,7 (A) and 1,6 (B) naphthyl-functional, PN monomers. *DSC*, dynamic scanning calorimetry. (A) Reprinted from Yu X-Y, Naito K, Kang C, Qu X-W, Zhang Q-X. Synthesis and properties of a high-temperature naphthyl-based phthalonitrile polymer. Macromol Chem Phys 2013;214:361–9 and (B) Reprinted from Zhao F, Liu R, Kang C, Yu X, Naito K, Qu X, et al. A novel high-temperature naphthyl-based phthalonitrile polymer: synthesis and properties. RSC Adv 2014;4:8383–90.

focusing on the thermo-oxidative stabilities, it was found that the relatively low free hydrogens in the polymeric structures allowed one of the highest thermo-oxidative stability. For instance, the $T_{5\%}$ reached up to 506 and 549°C for the 2,7 and 1,6 naphthyl-functional, PN polymers, respectively. Therefore the naphthyl structure filled the gap usually seen between the nitrogen and air atmospheres stabilities. Once more, the different type and amount of the amino-curing agent made it difficult to advance any explanation regarding the differences in performances between the 2,7 and 1,6 naphthyl-functional, PN resins. Moreover, the thermomechanical properties investigated by DMA also revealed consequent ameliorations in the storage modulus and T_g in contrast with some typical PN monomers. For example, the 1,6 naphthyl-functional, PN polymer showed a storage modulus value at 38 of about 3.8 GPa which was as high as that obtained by Derradji et al. with ceramic-based, typical, BAPh nanocomposites with an elevated amount of reinforcements [49,50]. Therefore, the naphthyl-functional, PN monomers can be a good substitute to some typical, PN monomers for the preparation of advanced PN composites. The T_g of the 1,6 naphthyl-functional, PN polymer was also confirmed to be superior to 465°C implying that this polymer had a high crosslinking degree with very limited segmental motions. Overall, the naphthyl-functional, PN materials brought new insights and improvements in the PN and these materials can be further improved by the addition of self-curing units within the naphthyl-functional, PN monomers.

3.2.10 Pyridine-Functional Phthalonitrile

Pyridine is a heteroaromatic structure generally used to design polymers with improved thermal stability, owing to the excellent stability derived from corresponding molecular symmetry and aromaticity. Pyridine has for a long time inspired many researchers targeting advanced polymeric materials. The pyridine-based polymers exhibited superior thermal and chemical stabilities, good mechanical properties even at higher temperatures and good solubility in organic solvents [51]. Thus, the introduction of the pyridine within the PN backbone was expected to bring some improvements in the thermal stability and mechanical properties in contrast with other ether- or poly(aryl ether)-based PN resins. It should be reminded that a proper comparison between the PN polymers is always a difficult task considering the multitude of parameters influencing the final properties.

The only report about the pyridine-functional, PN materials was performed by Xi et al. [52]. The synthetic route was the typical nucleophilic displacement reaction between 4-nitrophthalonitrile and 2,6-bis(4-hydroxybenzoyl)pyridine. The latter was afforded through the Friedel–Crafts acylation of phenetole with 2,6-pyridinedicarbonyl chloride using anhydrous $AlCl_3$ as a catalyst as seen in Scheme 3.14. Since the prepared monomers were not able to self-catalyze, 10 wt.% 4-APN were added to the monomers. As discussed in previous sections, 4-APN is kind of monofunctional, self-catalytic, PN monomer, which is used as an advanced curing agent, since it not only allows the initiation of the cyano addition reaction but also 4-APN fully integrates into the polymeric structure.

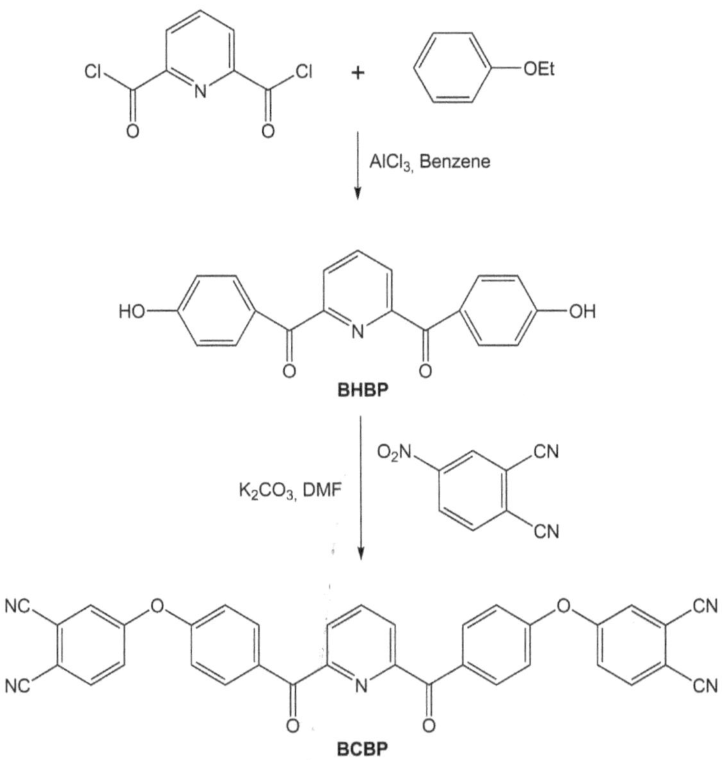

Scheme 3.14 Synthetic route of the pyridine-functional, PN monomers. Reprinted from Xi Z, Chen X, Yu X, Ma Y, Ji P, Naito K, et al. Synthesis and properties of a novel high temperature pyridine-containing phthalonitrile polymer. J Polym Sci Part A: Polym Chem 2016;54:3819–25.

Upon heating the pyridine-functional, PN monomers by DSC in the presence of 10 wt.% 4-APN, two endothermic transitions appeared at 124 and 180°C. These transitions were ascribed to the melting points of the pyridine-functional, PN and 4-APN monomers, respectively. In addition, a sharp exothermic transition was depicted at 244°C corresponding to the cyano addition reaction of the PN units from both pyridine-functional, PN and 4-APN monomers. The processing window was measured to be 64°C which was, for instance, larger than that of the typical, BAPh monomers (about 55°C in the presence of 5 wt.% *p*-BAPS) [53].

Polymers of pyridine-functional, PN were obtained by heating the monomeric precursors in the presence of 5 wt.% 4-APN following the cure schedule of 220°C for 2 h, 250°C for 4 h, 280°C for 4 h, 320°C for 4 h, 350°C for 4 h, and 380°C for 4 h. The thermal stability of the cured polymers was analyzed by TGA under nitrogen and air atmospheres and the obtained curves are displayed in Fig. 3.24. Under nitrogen atmosphere, the $T_{5\%}$ and $T_{10\%}$ reached up to 521 and 537°C, respectively, and the char yield at 1000°C was about 70%. Under air atmosphere, $T_{5\%}$ and $T_{10\%}$ were depicted at 505 and 530°C. With respect to the type and amount of the curing agent, the curing schedule and the postcure conditions if any, the pyridine-functional, PN showed thermal and thermo-oxidative stabilities similar or even better than those of the ether- and poly(aryl ether)-based PN polymers [47,54–56]. Specifying these conditions is important because increasing the 4-APN to 10 wt.% resulted in an improvement of about 17 and 25°C in the $T_{5\%}$ and $T_{10\%}$, respectively. Globally, the pyridine structures confirmed their ability to improve the thermal resistance of the PN resins to a certain extent.

The thermomechanical properties, namely the storage modulus and T_g, of the pyridine-functional, PN polymers cured in the presence of 5 and 10 wt.% 4-APN were analyzed by DMA and the obtained results are shown in Fig. 3.25. The storage modulus of both polymers at 50°C was about 2.2 GPa and by increasing the temperature to 400°C, about 50% of the initial values were maintained. However, the polymer containing 10 wt.% of 4-APN showed a better retention of modulus at 400°C. Subsequently, the Tan Delta graph revealed that the T_g was superior to 400°C for both cases. These excellent thermomechanical properties were attributed to the high-stiff, pyridine moiety and the orderly, linear, *para*-oriented linkages to the rigidity of the polymer backbone. In summary, the pyridine-functional, PN resins were confirmed to exhibit superior thermal and thermomechanical properties and can also be candidate for extremely exigent applications.

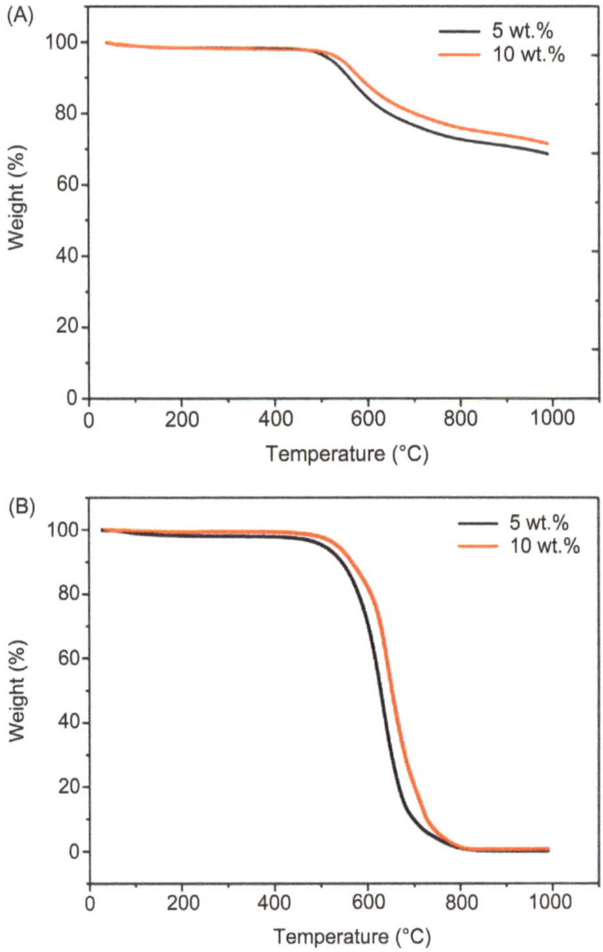

Figure 3.24 TGA curves of the pyridine-functional, PN polymers under nitrogen (A) and air (B) atmospheres. Reprinted from Xi Z, Chen X, Yu X, Ma Y, Ji P, Naito K, et al. Synthesis and properties of a novel high temperature pyridine-containing phthalonitrile polymer. J Polym Sci Part A: Polym Chem 2016;54:3819–25.

3.2.11 Poly(Ether Ether Ketone)-Functional Phthalonitrile

PEEK is an engineering, high-performance thermoplastic useful in many applications. The preparation of PEEK-functional, PN oligomers was driven by the need to reduce the curing temperatures and improve the processability of the PN monomers. The introduction of flexible PEEK linkage could also reduce the brittleness of the PN resins.

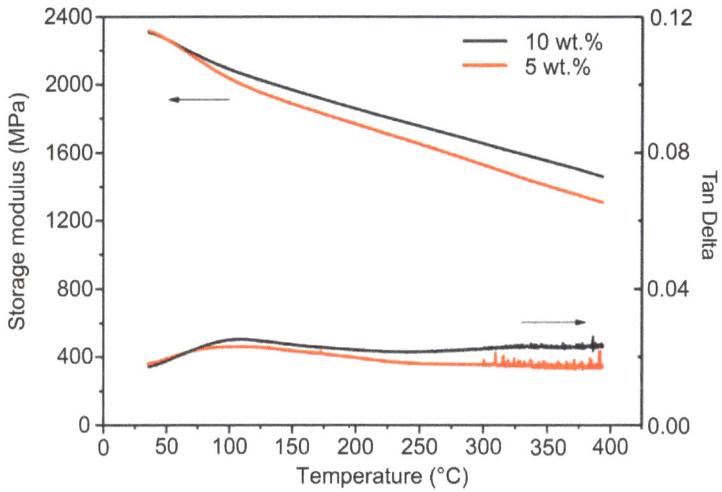

Figure 3.25 Storage modulus and loss factor (Tan Delta) variations for the pyridine-functional, PN polymers cured in the presence of 5 and 10 wt.% 4-APN. *4-APN*, 4-aminophenoxy PN. Reprinted from Xi Z, Chen X, Yu X, Ma Y, Ji P, Naito K, et al. Synthesis and properties of a novel high temperature pyridine-containing phthalonitrile polymer. J Polym Sci Part A: Polym Chem 2016;54:3819–25.

Laskoski et al., synthetized multiple, oligomeric, BPA-based, PEEK-functional, PN oligomers and studied their curing behavior and processability, as well as the thermal and mechanical properties of their cured polymers [53]. These PEEK-functional, PN oligomers were prepared in a two-step, one-pot reaction involving the reaction of an excess amount of BPA with 4,4′-difluorobenzophenone in a basic environment. The resulting product was further reacted with 4-nitrophthalonitrile to afford the targeted oligomers. It should be also pointed out that the length of the spacers between the terminal PN units can be easily extended or reduced by changes in the molar ratios of the precursors. Based on the synthetic route, these PN oligomers can also be classified as second generation PN resins.

The PEEK-functional, PN oligomers similarly to other typical PN monomers exhibited a sluggish curing behavior in the absence of a catalyzer. Thus, 5 wt.% of *p*-BAPS were added to these oligomers to achieve the polymerization in a timely manner. For a comparison purpose, the well-known BAPh monomers were cured with the same type and amount of curing agent. DSC measurements revealed that the PEEK-functional, PN oligomers had a softening point at about 75°C and completely flowed at 145°C. The cure reaction initiated by the amino-curing agent resulted in an exothermic peak at 315°C. On the other hand, the BAPh monomers showed a sharp endothermic transition at 195°C related to their melting point, while an exothermic peak appeared only 55°C after the latter. Obviously, the PEEK-functional, PN oligomers showed an improved curing behavior with a processing window of 240°C extensively larger than that of the BAPh monomers. Viscosity studies also confirmed that the presence of these flexible linkages allowed easy fabrication of shaped components above 125°C. Additionally, the rheological investigations revealed that the melt viscosity was very low, about 1.2 Pa s at 150°C and 0.05 Pa s at 250°C. Such a reduced melt viscosity allows these resins to be processed into thick and complex shapes by cost-effective processing methods.

Thermal stability measurements for both PEEK-functional, PN and BAPh polymers were performed by TGA under nitrogen and air atmosphere to see whether the improvements in the curing behavior and processability were followed by acceptable thermal resistance. Under nitrogen atmosphere, The $T_{10\%}$ was measured to be 505 and 520°C and the char yield at 1000°C was 70% and 72% for the PEEK-functional, PN and BAPh polymers, respectively. Postcuring these resins at elevated temperatures revealed improvements up to 60°C in the $T_{10\%}$. Under air atmosphere, both polymers showed quite similar thermo-oxidative stability with a $T_{10\%}$ of 530 and 550°C for the PEEK-functional, PN and BAPh polymers, respectively. Therefore it was confirmed that the PEEK linkages greatly improved the curing behavior and processability of the PN, while the thermal stability of the resulting polymers was slightly affected.

In another recently published study, the same authors further investigated the properties of these PEEK-functional, PN resins [57]. For instance, the cure mechanism was explored under the presence of various type and amount of curing agents. Moreover, different cure and postcure conditions were applied to study their effect on the thermal, mechanical, and thermomechanical properties. The obtained results corroborated the earliest one and it was advanced that the PEEK-functional, PN resins are emerging as an excellent replacement to the currently in use high-performance polymers especially in composites fabrication.

3.2.12 Poly(Arylene Ether Nitrile)-Functional Phthalonitrile

Poly(arylene ether nitrile) polymers, known as poly(arylene ether nitrile) (PEN), are a kind of advanced engineering thermoplastics. These materials have gained in importance during the last few decades owing to their exceptional properties. The most noteworthy advantage of the PEN is the presence of nitrile groups in their backbone. These nitriles have a strong polarizability which allows improved adhesion performances. In addition, the nitriles can undergo a cyclization reaction resulting in an increased crosslinking degree. As a result, PEN resins possess good mechanical and dielectrical properties. On the other hand, PEN polymers suffer from a limited thermal stability which restricted their development and application in many fields [58–60]. One way to overcome these shortcomings is to provide the PEN with addition reactive groups able to bring additional crosslinking sites. Since the PN resins are the leading thermosets in terms of thermal stability and thermomechanical properties, the preparation of PEN-containing, PN ends was found to give the best match in terms of overall performances. Although the PEN-functional, PN are closer to the thermoplastics, some of their properties surpass those of advanced thermosets. The PEN-functional, PN materials were first synthetized and further developed by the research group led by Prof. Liu. Indeed, thanks to the dedication of these researchers the PEN-functional, PN polymers are now becoming one of the leading polymers combining the ease of fabrication of the thermoplastics and the properties of the thermosets. Herein are summarized the main achievement regarding the PEN-functional, PN oligomers and polymers.

The PEN-functional, PN oligomers are prepared according to the following process: first PENs are end-capped with hydroxyl groups via extra biphenyl contents, then the PEN-functional PN are obtained by reacting the previously prepared compounds with 4-nitropthalonitrile as per the typical nucleophilic substitution reaction [61,62]. An example of the synthetic route is presented in Scheme 3.15. Depending on the excess amount of biphenyl, various molecular weights of the PEN-functional, PN oligomers can be easily obtained.

The thermal behavior of the PEN-functional, PN polymers under DSC (Fig. 3.26) showed the appearance of a T_g which position was independent of the molecular weight. Globally, the T_g value ranged between 210 and 214°C. The reason behind the limiting values is obviously related to the chain length and extent of curing. Indeed, longer chains increase the free volume and limit the interactions between the reactive sites. Globally, the PEN-functional, PN polymers presented excellent T_g values owing the enhanced crosslinking degree afforded by the PN cyano addition reaction.

The PEN-functional, PN polymers were also found to be as thermally resistant as the typical PN polymers cured at higher temperatures. The TGA investigations, shown in Fig. 3.27, showed that all the

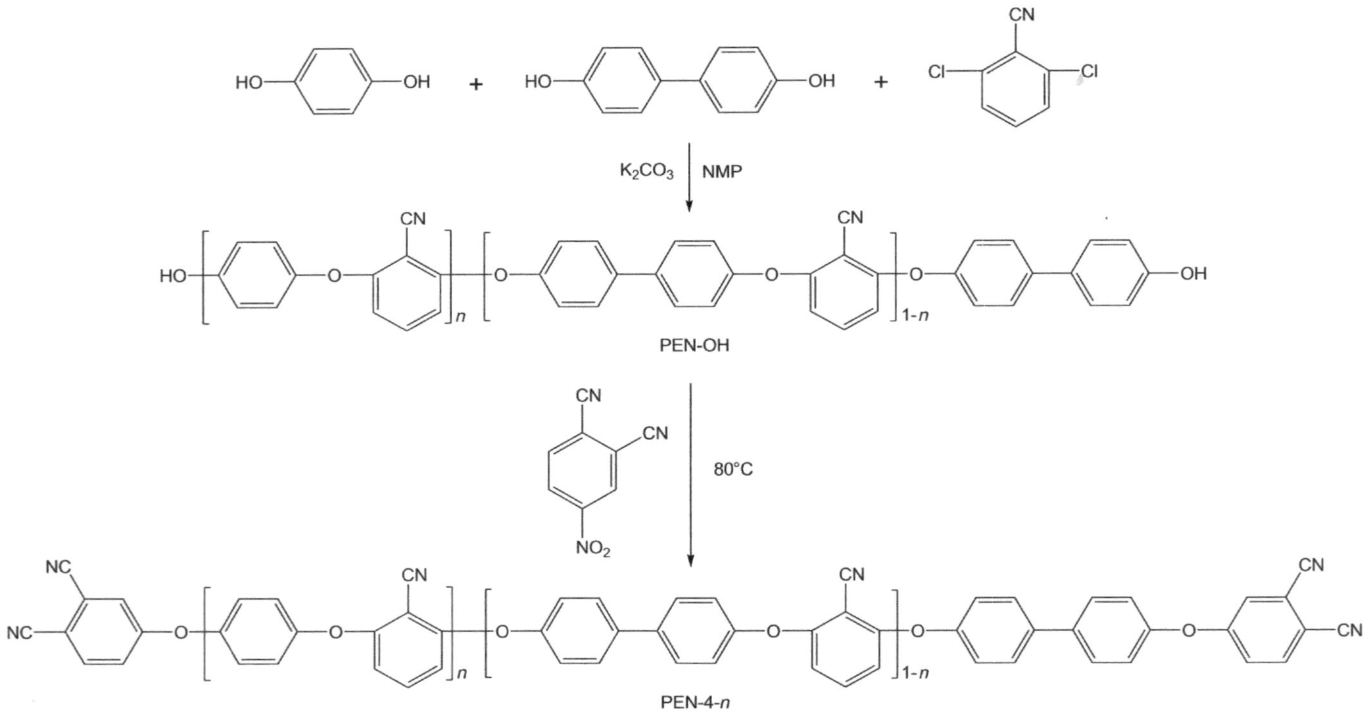

Scheme 3.15 Synthetic route of the PEN-functional, PN oligomers. *PN*, phthalonitrile. Reprinted from Zou Y, Yang J, Zhan Y, Yang X, Zhong J, Zhao R, et al. Effect of curing behaviors on the properties of poly(arylene ether nitrile) end-capped with phthalonitrile. J Appl Polym Sci 2012;125:3829–35.

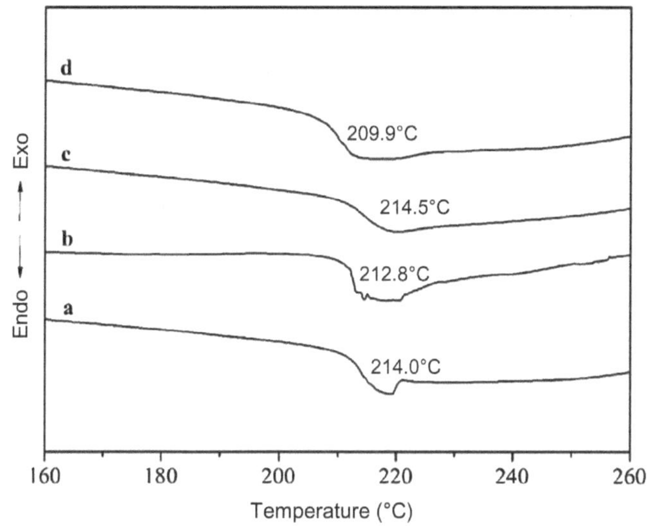

Figure 3.26 DSC curves of the PEN-functional, PN polymers issued from oligomeric precursors of various average molecular weights: 119.4 (a), 78.6 (b), 65.6 (c), and 22.5 (d). *DSC*, dynamic scanning calorimetry. Reprinted from Tong L, Jia K, Zhan Y, Liu XB. Novel phthalonitrile-terminated polyarylene ether nitrile with high glass transition temperature and enhanced thermal stability. Mater Lett 2014;128:267–70.

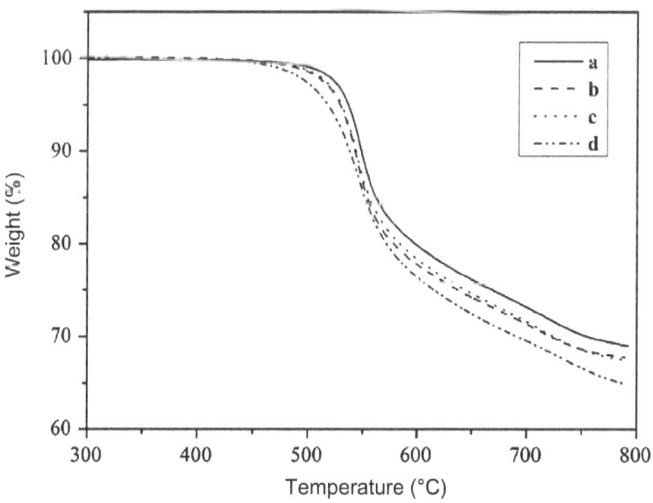

Figure 3.27 TGA curves of the PEN-functional, PN polymers issued from oligomeric precursors of various average molecular weights: 119.4 (a), 78.6 (b), 65.6 (c), and 22.5 (d). Reprinted from Tong L, Jia K, Zhan Y, Liu XB. Novel phthalonitrile-terminated polyarylene ether nitrile with high glass transition temperature and enhanced thermal stability. Mater Lett 2014;128:267–70.

PEN-functional, PN polymers had extremely high starting decomposition temperatures (over 518°C for the $T_{5\%}$). On the same time, it was also noticed that increasing the molecular weight reduced the thermal stability. As seen in the DSC section, the crosslinking degree was closely related to the number of PN units participating in the reaction. In fact, with higher molecular weight the crosslinking became less intense. These observations should not be seen as shortcomings, since according to the desired application, the properties of the PEN-functional, PN polymers could be easily tailored. Moreover, the tensile tests of the PEN-functional, PN polymers showed that all the samples had at least 90 MPa of strength and 1.7 GPa of elastic modulus. Meanwhile, varying the molecular weight affected the tensile properties in different ways.

The previous discussions gave a small introduction to the exceptional properties of the PEN-functional, PN materials and several reports concerned with many modifications can also be found in the open literature. In fact, during the last decade, the number of reports on the PEN-functional, PN polymers has exponentially increased including the preparation of blends and composites with other systems.

3.3 Conclusions

The present chapter reviewed the X-functional, PN materials and revealed their particularly appealing performances. Additionally, through an effective classification and comparison, readers and new comers to the field of PN can merely catch up with the first and latest works concerned with the X-functional, PN materials. Hopefully, this chapter will contribute to the future development of the PN resins.

References

[1] Ting RY, Keller TM, Price TR, Poranski JCF. Characterization of the cure of diether-linked phthalonitrile resins. ACS Sym Ser 1982;195:337–50.

[2] Achar BN, Fohlen GM, Parker JA. Synthesis and characterization of the novel type of polymerizable bisphthalonitrile monomers. J Polym Sci: Part A: Polym Chem 1986;24:1997–2010.

[3] Keller TM. Phthalonitrile-based high temperature resin. J Polym Sci: Part A: Polym Chem 1988;26:3199–212.

[4] Warzel M, Keller TM. Tensile and fracture properties of a phthalonitrile polymer. Polymer 1993;34:663–6.

[5] Sastri SB, Armistead JP, Keller TM. Phthalonitrile-carbon fiber composites. Polym Compos 1996;17:816–22.
[6] Sastri SB, Armistead JP, Keller TM. Phthalonitrile-glass fabric composites. Polym Compos 1997;18:48–54.
[7] Keller TM, Dominguez DD. High temperature resorcinol-based phthalonitrile polymer. Polymer 2005;46:4614–18.
[8] Dominguez DD, Jones HN, Keller TM. The effect of curing additive on the mechanical properties of phthalonitrile-carbon fiber composites. Polym Compos 2004;25:554–61.
[9] Derradji M, Ramdani N, Zhang T, Wang J, Gong LD, Xu XD, et al. Effect of silane surface modified titania nanoparticles on the thermal, mechanical, and corrosion protective properties of a bisphenol-A based phthalonitrile resin. Prog Org Coat 2016;90:34–43.
[10] Derradji M, Ramdani N, Gong LD, Wang J, Xu XD, Lin ZW, et al. Mechanical, thermal, and UV-shielding behavior of silane surface modified ZnO-reinforced phthalonitrile nanocomposites. Polym Adv Technol 2016;27:882–6.
[11] Keller T.M. High temperature epoxy-phthalonitrile blends. United States Patent US 5939508. 1999 Aug 17.
[12] Dominguez DD, Keller TM. Properties of phthalonitrile monomer blends and thermosetting phthalonitrile copolymers. Polymer 2007;48:91–7.
[13] Dominguez DD, Keller TM. Phthalonitrile-epoxy blends: cure behavior and copolymer properties. J Appl Polym Sci 2008;110:2504–15.
[14] Yang X, Lei Y, Zhong J, Zhao R, Liu X. Preparation and thermal properties of novel phthalonitrile oligomer containing biphenyl ethernitrile/bisphthalonitrile blends. J Appl Polym Sci 2011;119:882–7.
[15] Augustine D, Vijayalakshmi KP, Sadhana R, Mathew D, Reghunadhan Nair CP. Hydroxyl terminated PEEK-toughened epoxy–amino novolac phthalonitrile blends—synthesis, cure studies and adhesive properties. Polymer 2014;55:6006–16.
[16] Derradji M, Wang J, Liu WB. High performance ceramic-based phthalonitrile micro and nanocomposites. Mater Lett 2016;182:380–5.
[17] Zeng K, Zhou K, Tang WR, Tang Y, Zhou HF, Liu T, et al. Synthesis and curing of a novel amino-containing phthalonitrile derivative. Chin Chem Lett 2007;18:523–6.
[18] Zeng K, Zhou K, Zhou S, Hong H, Zhou H, Wang Y, et al. Studies on self-promoted cure behaviors of hydroxy-containing phthalonitrile model compounds. Eur Polym J 2009;45:1328–35.
[19] Zou X, Xu M, Jia K, Liu X. Synthesis, polymerization, and properties of the allyl-functional phthalonitrile. J Appl Polym Sci 2014;131:41203–7.
[20] Laskoski M, Dominguez DD, Keller TM. Alkyne-containing phthalonitrile resins: controlling mechanical properties by selective curing. J Polym Sci: Part A: Polym Chem 2013;51:4774–8.
[21] Augustine D, Mathew D, Nair CR. One component propargyl phthalonitrile novolac: synthesis and characterization. Eur Polym J 2015;71:389–400.

[22] Keller T.M. Synthesis of phthalonitrile resins containing ether and imide linkages. United States Patent US 5159054. 1992 Oct 27.
[23] Keller T.M. Phthalonitrile monomers containing imide and/or phenoxy linkages, and polymers thereof. United States Patent US 5132396. 1992 Jul 21.
[24] Sasaki S, Nishi S, Ghosh M, Mittal K. Polyimides: fundamentals and applications. New York: Marcel Dekker; 1996.
[25] Kim JH, Moore JA. A low-temperature route to polyimides. Macromolecules 1993;26:3510−13.
[26] Keller TM. Imide-containing phthalonitrile resin. Polymer 1993;34:952−5.
[27] Selvakumar P, Sarojadevi M. Development of oligomeric phthalonitrile resins for advanced composite applications. Macromol Symp 2009;277:190−200.
[28] Selvakumar P, Padmini K, Sarojadevi M, Leelavathy M. Synthesis, characterization and thermal properties of imide-containing phthalonitrile polymers. J Macromol Sci Part A 2009;47:76−88.
[29] Amir B, Zhou H, Liu F, Aurangzeb H. Synthesis and characterization of self-catalyzed imide-containing pthalonitrile resins. J Polym Sci: Part A: Polym Chem 2010;48:5916−20.
[30] Zeng K, Li L, Xiang S, Zhou Y, Yang G. Synthesis and thermal polymerization of new polyimides with pendant phthalonitrile units. Polym Bull 2011;68:1879−88.
[31] Yuan P, Liu Y, Zeng K, Yang G. Synthesis and characterization of a new imide compound containing phthalonitrile and phenylethynyl end-groups. Des Monomers Polym 2015;18:343−9.
[32] Zhang J, Liu X, Wen H, Xie M, Cai X. Investigation of the properties of phthalocyanine resin containing bismaleimide groups. Polym Inter 1997;42:363−6.
[33] Sumnera MJ, Sankarapandiana M, McGratha JE, Rifflea JS, Sorathiab U. Flame retardant novolac−bisphthalonitrile structural thermosets. Polymer 2002;43:5069−76.
[34] Guo H, Lei Y, Zhao X, Yang X, Zhao R, Liu X. Curing behaviors and properties of novolac/bisphthalonitrile blends. J Appl Polym Sci 2012;125:649−56.
[35] Yang Y, Min Z, Yi L. A novel addition curable novolac bearing phthalonitrile groups: synthesis, characterization and thermal properties. Polym Bull 2007;59:185−94.
[36] Zhang B, Luo Z, Zhou H, Liu F, Yu R, Pan Y, et al. Addition-curable phthalonitrile-functionalized novolac resin. High Perform Polym 2012;24:398−404.
[37] Augustine D, Mathew D, Nair CPR. Phenol-containing phthalonitrile polymers-synthesis, cure characteristics and laminate properties. Polym Inter 2012;62:1068−76.

[38] Ghosh NN, Kiskan B. Polybenzoxazines—new high performance thermosetting resins: synthesis and properties. Prog Polym Sci 2007;32:1344–91.

[39] Ishida H, Agag T. Handbook of benzoxazine resins. Amsterdam: Elsevier; 2011.

[40] Rimdusit S, Jubsilp C, Tiptipakorn S. Alloys and composites of polybenzoxazines. Singapore: Springer; 2013.

[41] Brunovska Z, Lyon R, Ishida H. Thermal properties of phthalonitrile functional polybenzoxazines. Thermochim Acta 2000;357:195–203.

[42] Cao GP. Synthesis and characterization of a novel bisphthalonitrile containing benzoxazine. Exp Polym Lett 2007;1:512–18.

[43] Xu M, Jia K, Liu X. Effect of bisphenol-A on the structures and properties of phthalonitrile-based resin containing benzoxazine. Exp Polym Lett 2015;9:567–81.

[44] Xu M, Jia K, Liu X. Self-cured phthalonitrile resin via multistage polymerization mediated by allyl and benzoxazine functional groups. High Perform Polym 2016;28:1161–71.

[45] Wu D, Zhao Y, Zeng K, Yang G. A novel benzimidazole-containing phthalonitrile monomer with unique polymerization behavior. J Polym Sci: Part A: Polym Chem 2012;50:4977–82.

[46] Augustine D, Mathew D, Nair CR. Phthalonitrile resin bearing cyanate ester groups: synthesis and characterization. RSC Adv 2015;5:91254–61.

[47] Yu X-Y, Naito K, Kang C, Qu X-W, Zhang Q-X. Synthesis and properties of a high-temperature naphthyl-based phthalonitrile polymer. Macromol Chem Phys 2013;214:361–9.

[48] Zhao F, Liu R, Kang C, Yu X, Naito K, Qu X, et al. A novel high-temperature naphthyl-based phthalonitrile polymer: synthesis and properties. RSC Adv 2014;4:8383–90.

[49] Derradji M, Song X, Dayo AQ, Wang J, Liu WB. Highly filled boron nitride-phthalonitrile nanocomposites for exigent thermally conductive applications. Appl Therm Eng 2017;115:630–6.

[50] Derradji M, Ramdani N, Zhang T, Wang J, Lin Z-w, Yang M, et al. High thermal and thermomechanical properties obtained by reinforcing a bisphenol-A based phthalonitrile resin with silicon nitride nanoparticles. Mater Lett 2015;149:81–4.

[51] Ataei S, Heidari H. Preparation of novel pyridine-based, thermally stable poly (ether imide) s. J Appl Polym Sci 2004;91:22–6.

[52] Xi Z, Chen X, Yu X, Ma Y, Ji P, Naito K, et al. Synthesis and properties of a novel high temperature pyridine-containing phthalonitrile polymer. J Polym Sci Part A: Polym Chem 2016;54:3819–25.

[53] Laskoski M, Dominguez DD, Keller TM. Synthesis and properties of a bisphenol A based phthalonitrile resin. J Polym Sci: Part A: Polym Chem 2005;43:4136–43.

[54] Sastri SB, Keller TM. Phthalonitrile polymers: cure behavior and properties. J Polym Sci: Part A: Polym Chem 1999;37:2105–11.
[55] Peng X, Sheng H, Guo H, Naito K, Yu X, Ding H, et al. Synthesis and properties of a novel high-temperature diphenyl sulfone-based phthalonitrile polymer. High Perform Polym 2014;26:837–45.
[56] Liu T, Yang Y, Wang T, Wang H, Zhang H, Su Y, et al. Synthesis and properties of poly(aryl ether ketone)-based phthalonitrile resins. Polym Eng Sci 2014;54:1695–703.
[57] Keller TM, Dominguez DD, Laskoski M. Oligomeric bisphenol A-based PEEK-like phthalonitrile-cure and polymer properties. J Polym Sci: Part A: Polym Chem 2016;54:3769–77.
[58] Liu XB. Synthesis, characterization and rheological property of biphenyl-based polyarylene ether nitrile copolymers. Exp Polym Lett 2007;1:499–505.
[59] Zhao X, Guo H, Lei Y, Zhao R, Zhong J, Liu X. Effect of polyarylene ether nitriles on processing and mechanical behaviors of phthalonitrile-epoxy copolymers and glass fiber laminated composites. J Appl Polym Sci 2013;127:4873–8.
[60] Tong L, Jia K, Liu X. Novel phthalonitrile-terminated polyarylene ether nitrile with high glass transition temperature and enhanced thermal stability. Mater Lett 2014;128:267–70.
[61] Zou Y, Yang J, Zhan Y, Yang X, Zhong J, Zhao R, et al. Effect of curing behaviors on the properties of poly(arylene ether nitrile) end-capped with phthalonitrile. J Appl Polym Sci 2012;125:3829–35.
[62] Tong L, Jia K, Zhan Y, Liu XB. Novel phthalonitrile-terminated polyarylene ether nitrile with high glass transition temperature and enhanced thermal stability. Mater Lett 2014;128:267–70.

4 Phthalonitrile-Based Blends and Copolymers

4.1 Introduction

Phthalonitrile (PN) resins are one of the leading, high-performance, polymeric materials applicable in many civilian and military applications requiring light-weight structures with metal or ceramic equivalent properties. The most noteworthy advantage of PN resins over other classes of polymers is their extremely high thermal stability with starting decomposition temperatures over 450°C for the most common type of PN resins. Other interesting properties of these resins include good processability, superior flame resistance, absence of a glass transition temperature (T_g) before their thermal decomposition, good mechanical properties, low-water absorption, advanced UV shielding, and excellent corrosion protective properties [1−11]. Thus the PN-based materials are particularly appealing for the aerospace, marine, and electronic applications.

The polymerization reaction of the PN monomers is an addition reaction through the cyano groups present at the end of the PN structure. The aforementioned reaction affords a void-free, highly crosslinked network mainly responsible for the outstanding thermal resistance of the PN resins. However, triggering the cyano addition reaction without any catalyst requires elevated temperatures and extended curing time. Therefore many efforts were dedicated to overcome the sluggish curing character of the typical PN monomers. For instance, various curing agents based on metals and metallic salts [12−14], phenols [15−17], organic amines [18−20], and strong organic acid/amine salts [21,22] were developed to address this challenge. Despite the positive impact of these curing additives of the overall curing process, some challenges related the low curing efficiency, high curing temperatures, and lack of toughness of the resulting cured polymers still needed to be addressed.

The recent advances in the polymer field, with a growing focus on the molecular design and the development of blends/copolymers with many interesting functionalities, were beneficial to the PN upgrowth. Indeed, the past 15 years showed an exponential increase in the number of studies related to the improvement of the PN curing behavior and

Phthalonitrile Resins and Composites.
DOI: https://doi.org/10.1016/B978-0-12-812966-1.00004-4

properties. Following this path, many functional, PN monomers containing various structures within the same molecule were developed and reported in the literature. For instance, functional, PN monomers containing imide [23], bismaleimide [24], benzoxazine [25], novolac [26], vinyl ester [27], benzimidazole [28], and so on were found to exhibit interesting properties not obtained with any phase alone. However, the design and preparation of such monomers is quite time consuming and expensive. Another cost-effective method to improve the PN resins properties is through the preparation of blends and copolymers with other polymeric materials. It should also be pointed out that the preparation of blends and copolymers from the PN materials can be benefic to other polymers considering the superior thermal and thermomechanical properties of the PN resins. The formation of blends and copolymers is in fact the easiest way to modify the polymer properties when an individual material does not quite meet all the requirements for a specific application. The resulting materials generally have a useful combination of properties derived from each phase which can be tuned by modifications in the blend composition.

In this chapter, all the PN-based alloys prepared from either typical, self-catalyzed or functional, PN monomers/oligomers blended with other polymeric systems including thermosets, thermoplastics, and rubbers are discussed in terms of their resulting blends, curing behavior and processability, as well as the properties of their cured copolymers. Therefore, this chapter aims to review the considerable work performed by many PN researchers in order to reveal the potential of the PN-based materials and build a solid base for further investigations.

4.2 Phthalonitrile Blends and Copolymers

4.2.1 Phthalonitrile/Phthalonitrile

When an individual PN monomer does not meet all the requirements for a specific application, the formation of blends and copolymers with another PN monomer is the first alternative, prior to considering blending with other type of monomers. The variety of existing PN monomers including typical, functional, and self-catalyzed, PN monomers and oligomers with various properties can address some of the limitations of an individual PN monomer. In this section, only the blends and copolymers formed from the typical PN monomers are discussed. The main reason behind this restriction is due to a classification purpose and book arrangement. For instance, the self-catalyzed, PN monomers blended with typical PN monomers were previously discussed in Chapter 1, Introduction to the Phthalonitrile Resins and Their Curing Behavior and presented as curing additives for the PN. Herein, the functional, PN monomers blended with typical PN monomers were categorized according to their functionalities.

In 2006, Dominguez and Keller reported on the first PN/PN blends and copolymers from the biphenyl PN (BPh) and oligomeric $n = 4$ PN ($n = 4$ PN) (see Fig. 4.1) [29]. The BPh resin is one of the first, ever-reported, typical, PN monomers. Some of its shortcomings include a high melting point (about 233°C) and narrow processing windows even when cured with various curing agents. On the other hand, the $n = 4$ PN is an oligomeric mixture of components with a lower melting point (around 40°C) and a lower degree of crystallinity. However, the $n = 4$ PN monomers need higher curing temperatures, up to 425°C. Thus, it was intended that the resulting blends could exhibit an improved processability while their cured copolymers could still maintain the necessary, high-temperature properties. The blends of 75:25, 50:50, and 25:75 weight ratios of BPh to $n = 4$ PN were prepared by powder mixing the monomers, then 5 mol.% of bis[4-(4-aminophenoxy)phenyl]sulfone (p-BAPS) curing agent was added to the mixtures at their melting stage. The chosen curing schedule to obtain the copolymers was 250°C for 1 h, 325°C for 3 h, 350°C for 2 h, and 375°C for 8 h under argon atmosphere. Some of the obtained samples were further cured at 425°C under argon atmosphere to evaluate the postcure effect.

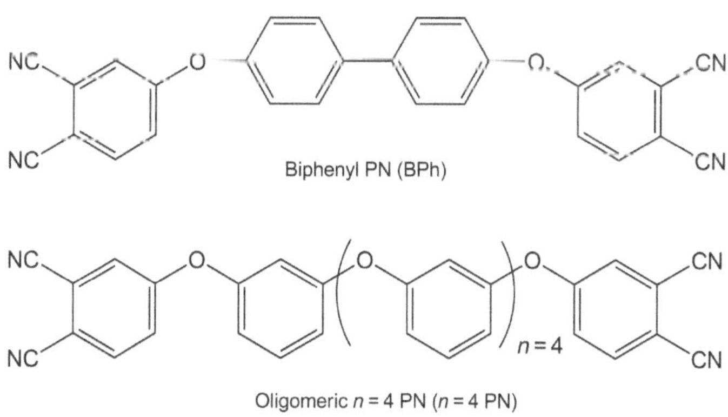

Figure 4.1 Chemical structures of the biphenyl PN (BPh) and oligomeric $n = 4$ PN. *PN*, phthalonitrile; *BPh*, biphenyl PN. Reprinted from Dominguez DD, Keller TM. Properties of phthalonitrile monomer blends and thermosetting phthalonitrile copolymers. Polymer 2007;48:91–7.

The curing behavior of the BPh/$n = 4$ PN blends in the presence of the curing agent was studied by dynamic scanning calorimetry (DSC). The data revealed that all the blends exhibited two melting points attributed to the melt of each phase. Interestingly, the melting point of the BPh monomers depended on their proportion in the blend and the $n = 4$ PN melting transition was not affected by the blend composition. In addition, all the blends showed one exothermic transition at around 262°C related to the initial reaction between the curing agent and the PN units. The exotherm maximum was independent from the blend composition because the polymerization of the studied PN monomers only depended on the amount and type of the curing agent. Another positive result is that the processing window of the blends was enlarged by more $n = 4$ PN contents and all the compositions showed larger processing windows than the BPh. For instance, the 25:75 BPh/$n = 4$ PN blend processing window reached 60°C which was 30°C higher than that of the neat BPh.

The processability of the neat PN monomers and their subsequent blends was analyzed by rheometric analysis. The complex viscosity changes as a function of temperature are shown in Fig. 4.2. As seen in this figure, the neat BPh monomers exhibited a sharp decrease in viscosity on melting. Conversely, the melt viscosity of the $n = 4$ PN monomers gradually decreased from 70 to about 175°C. For temperatures above 235°C, the viscosity of the neat PN monomers was minimal (0.01−1 Pa s). Moreover, it seemed that increasing the proportion of the $n = 4$ PN in the blends resulted in decreases in the temperatures corresponding to the blends initial melting. The 75:25 and 50:50 BPh/$n = 4$ PN blends appeared to melt in a one step process; however, the 25:75 blend presented a two-step melting process. In all instances, the minimum viscosity of 0.1 Pa s was noted for all the blends above 225°C. Further isothermal rheometric measurements at 200°C revealed that the slowest viscosity increases were depicted for the blend containing 75 wt.% of $n = 4$ PN, which was expected since the BPh monomers are well known for their high reactivity. In general, the rheometric measurements confirmed that the processability of the blends was improved in contrast with that of the neat BPh.

The thermomechanical properties, including the storage modulus and tan delta variations, were investigated by dynamic mechanical analysis (DMA) for the polymers and copolymers cured up to 375 and 425°C. The obtained storage modulus and tan delta curves are presented in Fig. 4.3. For the samples cured up 375°C, the storage modulus at 40°C ranged between 1.2 and 1.5 GPa. When moving from 40 to 450°C, the

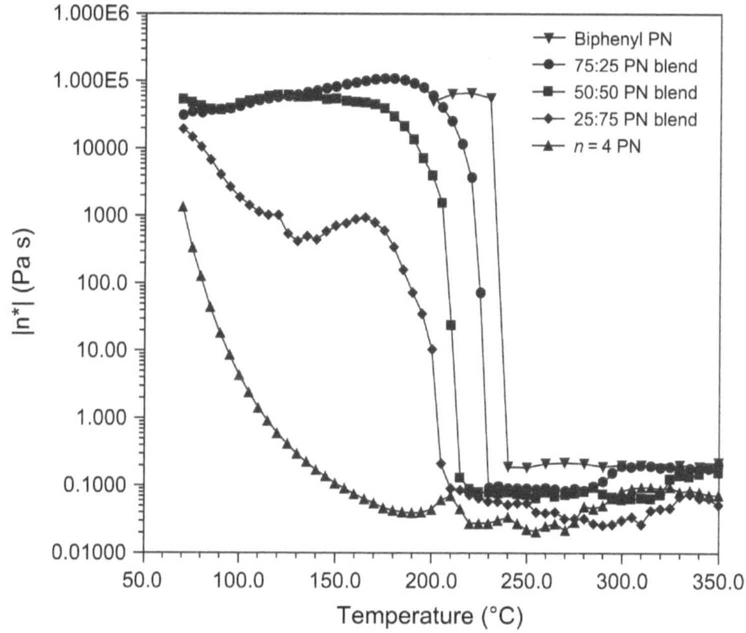

Figure 4.2 Complex viscosity variations over temperature of the neat the BPh monomers, $n = 4$ PN oligomers and their subsequent blends. *BPh*, biphenyl PN; *PN*, phthalonitrile. Reprinted from Dominguez DD, Keller TM. Properties of phthalonitrile monomer blends and thermosetting phthalonitrile copolymers. Polymer 2007;48:91–7.

BPh polymer only showed small changes in the storage modulus and tan delta. Conversely, the $n = 4$ PN polymer revealed consequent changes in the same interval of temperatures. Indeed, these observations confirmed that the $n = 4$ PN required higher temperatures to be fully cured. The BPh/$n = 4$ PN copolymers exhibited intermediate changes between those of their respective polymers. All the previous fluctuations were ascribed to the differences in the crosslinking degree and segmental motions of the polymeric chains. Independently from the proportion of each phase in the blend, all the copolymers showed one T_g between 225 and 230°C. Further postcuring the samples at 425°C resulted in the disappearance of the previously seen tan delta peaks which implied that a much stable network has been formed at higher temperatures with fewer segmental motions possibilities. The storage modulus values at 40°C were also slightly lowered and, upon heating to 450°C, only small decreases were noticed. These decreases were attributed to the relaxation process of the polymers/copolymers. Once more,

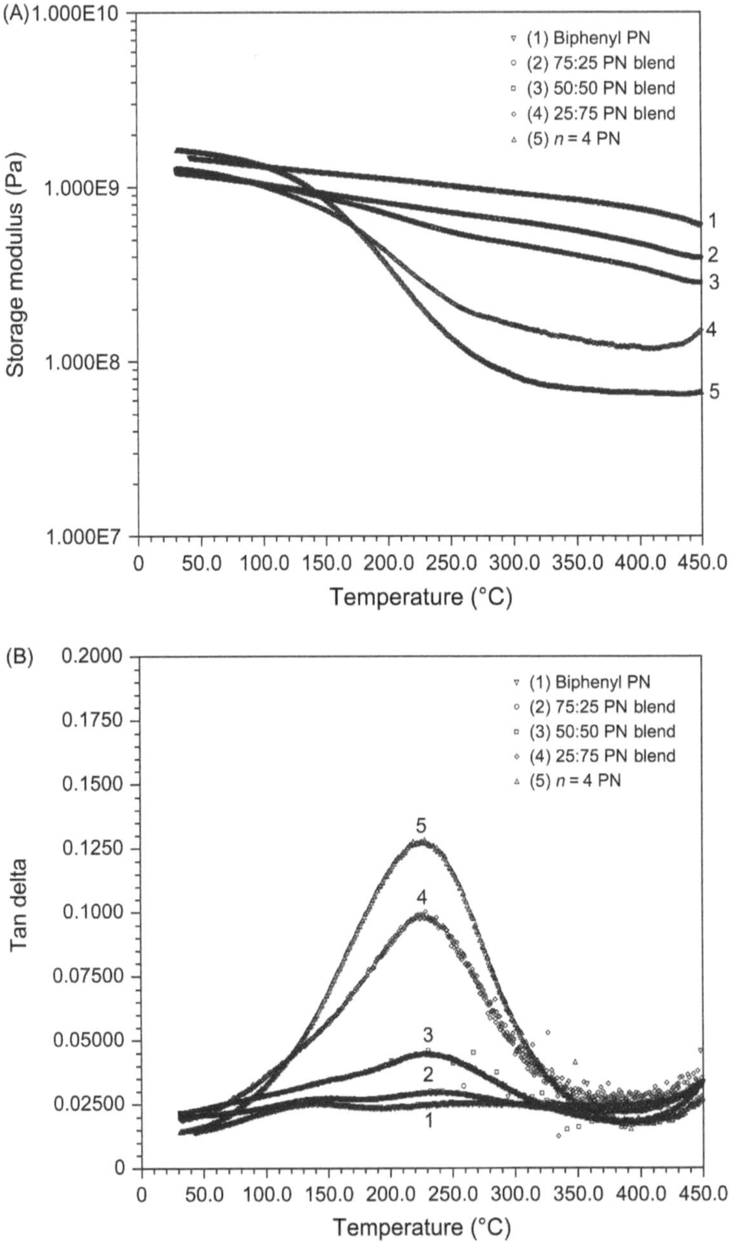

Figure 4.3 Storage modulus and loss factor (tan delta) variations for the BPh/ $n = 4$ PN blends, the neat BPh monomer and the $n = 4$ PN oligomer as a function of the maximum cure temperature: (A) Storage modulus after cure at 375°C, (B) tan delta after cure at 375°C, (C) storage modulus after cure at 425°C, and (D) tan delta after cure at 425°C. *BPh*, biphenyl PN; *PN*, phthalonitrile. Reprinted from Dominguez DD., Keller TM. Properties of phthalonitrile monomer blends and thermosetting phthalonitrile copolymers. Polymer 2007;48:91–7.

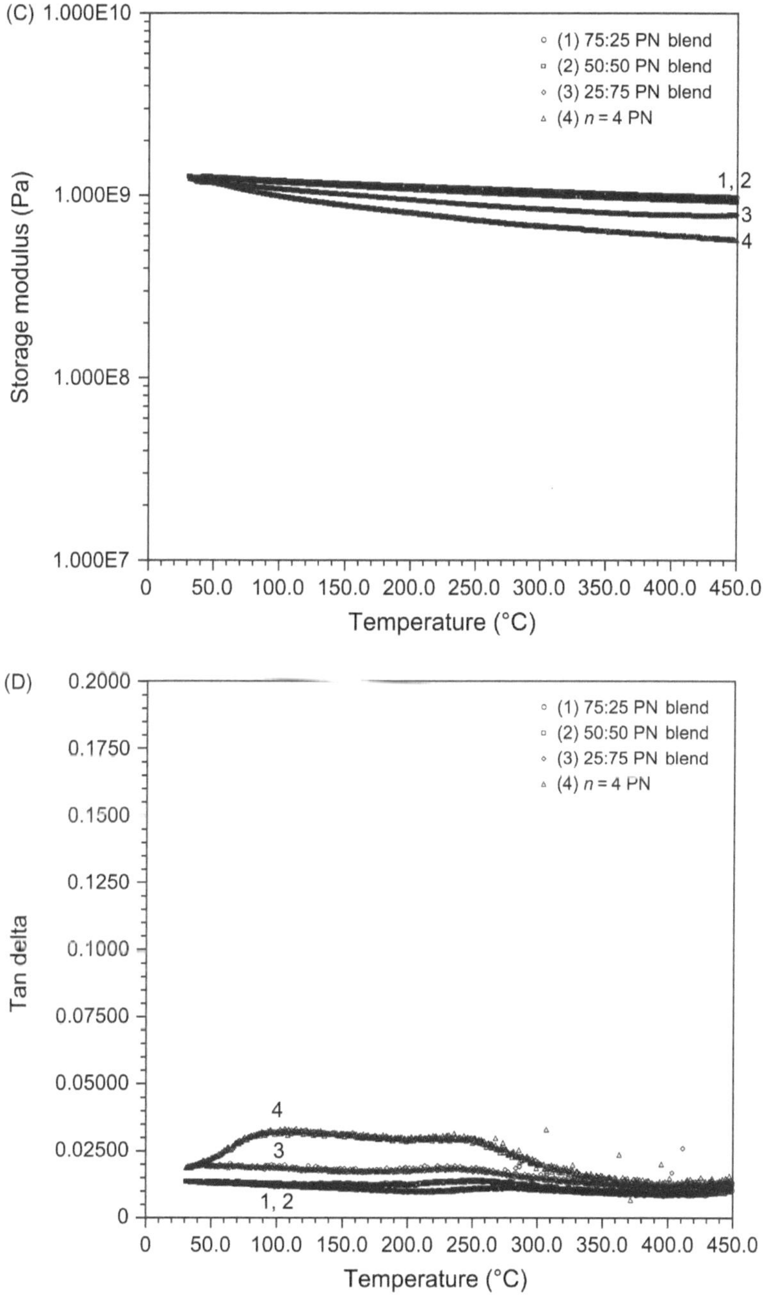

Figure 4.3 (Continued).

it was confirmed that the thermomechanical properties on the PN-based materials can be easily tailored by changes in the curing procedures.

The thermogravimetric analysis (TGA) revealed that the copolymers had an outstanding thermal stability in the range of the highly thermally stable PN resins. In fact, the 75:25, 50:50, and 25:75 BPh/$n=4$ PN copolymers, cured up to 375°C under nitrogen atmosphere, showed an excellent temperature of 5 wt.% weight loss ($T_{5\%}$) values of 550, 535, and 493°C, respectively. The char yield at 1000°C also reached up to 74% for the 75:25 BPh/$n=4$ PN copolymer. Further postcuring the copolymers up to 425°C ensued a superior thermal resistance. For instance, the $T_{5\%}$ reached the exceptional values of 580, 559, and 512°C for the copolymers of 75:25, 50:50, and 25:75 BPh/$n=4$ PN, respectively. Clearly, the thermal stability of these materials confirmed that the gain in the curing behavior and processability did not affect or slightly affect the thermal stability of these remarkable polymeric materials.

4.2.2 Phthalonitrile/Epoxy

Epoxy resins are a well-known class of high-performance thermosets widely used as the basis for adhesive compositions and particles/fibers reinforced composites. The typical epoxy monomers are cured in the presence of a curing agent to initiate the polymerization process. The resulted polymers present a highly crosslinked, amorphous network with many interesting properties such as good processability, high modulus, low creep, and globally reasonable thermal and mechanical properties between 100 and 150°C [30–32]. On the other hand, the epoxy resins have several shortcomings especially when being used as matrices for composites. For instance, many problems rise from the logistic of handling due to the low storage temperature requirements, weak toughness with a poor resistance to crack propagation, engineering limitations related to water diffusion between the epoxy matrices and their corresponding reinforcements, and most importantly their thermal stability is a way behind those of other high-performance thermosets [33]. Conversely, the PN prepegs can indefinitely be stored at room temperature without any inconvenient and their cured polymers exhibit outstanding thermal and thermo-oxidative stabilities. The PN resins are also better candidates for composites preparation because of their superior flame resistance and low water uptake. However, the PN resins require relatively higher curing temperatures and extended times even with the use of the curing additives. Moreover, the adhesive properties of the PN resins are below those of the epoxies. Therefore, it is advantageous to prepare polymeric systems benefiting from both interesting features of the PNs and epoxies. The resulting blends are expected to take advantage from the good processability, curing behavior, and adhesive properties of the epoxies and their resulting copolymers would exhibit ameliorated thermal resistance and lower water uptake.

Various PN monomers and oligomers were blended with epoxy resin (diglycidyl ether of bisphenol A) and interesting results were reported. According to the type of the PN precursors used to prepare the initial blends, the PN/epoxy blends and copolymers can be classified into three main categories. Thus, typical, self-catalyzed, and functional, PN monomers were effectively blended with epoxy and their resulting properties will be scrutinized for their curing behavior, processability, thermal resistance, mechanical properties, and eventually for their morphological behavior.

4.2.2.1 Typical Phthalonitrile/Epoxy Blends

Dominguez and Keller reported on the cure behavior and copolymers properties of two binary blends based on the BPh, $n = 4$ PN, and epoxy [34]. The binary blends were prepared by mixing the PN monomers/oligomers and the curing agent (*p*-BAPS or 1,4-bis(4-aminophenoxy) benzene (*p*-APB)) with the liquid epoxy resin under ambient conditions. The BPh and epoxy were blended under various molar ratios of BPh to epoxy (25:75, 50:50, and 75:25). On the other hand, the $n = 4$ PN were only blended with the epoxy with a 32:68 molar ratio of $n = 4$ PN to epoxy. Aiming to further study the impact of the curing additives, the same blends were prepared without the curing agents.

Investigations of the curing behavior for the neat BPh and epoxy as well as their subsequent blends were carried out by DSC. The neat BPh showed two endothermic transitions at 185 and 231°C related to the melting of the *p*-BAPS and the BPh monomers, respectively. An exothermic transition was also depicted at 260°C proving the ability of the curing agent to catalyze the cyano addition reaction of the BPh monomers. The neat epoxy cured with the amino-curing agent revealed the presence of two exothermic transitions located at 202 and 363°C. The peak at 202°C was attributed to the reaction of the epoxy monomers with the *p*-BAPS and the transition at 363°C was ascribed to the thermal decomposition of the epoxy, as cured polymer or as unreacted monomers/oligomers. The DSC curves of the blends only showed one endothermic transition between 207 and 223°C related to the melting of the BPh monomers and there was no evidence of the melting of the *p*-BAPS curing agent. Moreover, the exothermic transition depicted at 202°C was also not seen in the DSC curves of the blends which implied that the *p*-BAPS did not catalyze the cure reaction of the epoxy monomers in an appreciable manner. On the other hand, all the BPh/epoxy blends showed two exothermic transitions ranging between 332–339°C and 369–376°C, respectively. The first exotherm was attributed the reaction of the amino-curing agent with nitrile groups of the BPh monomers, which further triggered the epoxy monomers polymerization. This observation along with the disappearance of the curing peak, depicted at 202°C for the neat epoxy, confirmed the formation of an interpenetrating network between both phases. The second exothermic transition seen in the blends DSC curves was attributed the decomposition of the unreacted epoxy. In addition, by comparing the DSC curves of the blends in terms of their molar ratios, it was noticed that increasing the BPh amount resulted in both a decrease in the exothermic peak intensities and a separation between the two exothermic peaks. Further measuring the enthalpy of reaction reveled that more BPh contents reduced the heat of polymerization. It should be pointed out that a reduced curing enthalpy is preferred in the industry because it allows the fabrication of void-free thick polymeric sections. Increasing the proportion of the BPh in the blends also shifted the first exothermic transition to lower temperatures and the second exothermic transition moved to higher temperatures. These observations suggested that more BPh amount increased the reactivity and thermal stability of the blends.

The DSC measurements for the BPh/epoxy blends without the curing agent were also performed. All the studied blends showed a broad exothermic transition in the 341–356°C range attributed to the reaction of the BPh with the epoxy. Without the amino-curing agent, the cyano groups of the BPh monomers were only activated by the high temperature and the propagation of the reaction included the polymerization of both PN and epoxy forming an interpenetrating network. Moreover, the relatively high exothermic peaks seen in this case revealed that the uncatalyzed blends were less reactive than their catalyzed counterparts. The blends curves also showed a training shoulder after the exothermic transition attributed to either the continuation of the cure reaction or/ and the decomposition of the unreacted species.

Additionally, the curing behavior of the BPh/epoxy blends was also investigated by a rheological test. The complex viscosity evolutions over time for all the blends at 275°C with and without the addition of the curing agent were investigated. The general observation was that all the blends exhibited a gradual increase in the viscosity until gelation. Afterwards, a rapid increase in the viscosity was also noticed before the formation of a glassy material. These changes in the blends viscosities implied that the monomers were effectively reacting to form an interpenetrating network. The addition of more BPh contents resulted in an increase in the gelation time, which further confirmed the DSC findings regarding the decrease in the copolymerization rate by more BPh loading. In contrast with the uncatalyzed blends, the catalyzed one showed a reduced gelation time.

Following the same approach as the BPh/epoxy blends, the curing behavior of the $n = 4$ PN/epoxy blend was also studied with and without the presence of a curing agent. The chosen curing agent for this blend was the *p*-APB. The reason behind such choice was related to the reactivity of this catalyst. Indeed, the authors reported that the use of the *p*-BAPS did not achieve in a timely manner gelation, and as a result, a more reactive curing agent was used. The DSC curves of the $n = 4$ PN with *p*-APB showed a melting point at 42°C, also depicted in the DSC curve of the blend, and the reaction between the amino-curing agent and the $n = 4$ PN oligomers was confirmed by the exothermic transition at around 265°C. The epoxy monomers cured with the *p*-APB revealed the presence of two exothermic transitions at 181 and 341°C. The first exotherm was attributed to the polymerization of the epoxy and the second one was ascribed to the decomposition of the unreacted species. Interestingly, the blend curve also revealed the presence of two exothermic transitions. The first one was depicted at lower temperatures confirming the catalytic effect of the $n = 4$ PN oligomers on the epoxy and the second one shifted to higher temperatures suggesting more stability of the blend with respect to the neat epoxy. The DSC curve of the uncatalyzed blend only showed one exothermic transition at about 350°C which was ascribed to both the copolymerization and the decomposition reaction probably happening at the same time.

To further acknowledge about the curing behavior of the $n = 4$ PN/epoxy blends, a rheological study was performed on both the catalyzed and uncatalyzed blends to monitor the complex viscosity changes over time at various temperatures. The obtained data revealed that all the samples showed abrupt viscosity increases, implying that the copolymerization reaction was effectively progressing. It was also deduced that the appropriate curing temperatures for the blend with and without the curing additives were 100–200 and 250–300°C, respectively.

Considering the improved curing behavior and processability of the catalyzed blends in contrast with the uncatalyzed ones, the authors limited the thermal and thermomechanical properties investigations to the BPh/epoxy and $n = 4$ PN/epoxy cured with *p*-BAPS and *p*-APB, respectively. The thermal stability study of the neat polymers and their related copolymers was carried out by TGA analyses and the results are listed in Tables 4.1 and 4.2. By analyzing these data, it appeared that the neat epoxy cured with either *p*-BAPS or *p*-APB had the lowest thermal resistance. In fact, the $T_{5\%}$ was measured to be 378 and 356°C for the epoxy cured with *p*-BAPS and *p*-APB, respectively. Meanwhile, the $T_{5\%}$ of the BPh and $n = 4$ PN were 500 and 534°C, respectively. The blends thermal stabilities were superior to those of the epoxies and inferior to those of their respective PN resins. For instance, the $T_{5\%}$ of the copolymers was between 403 and 508°C. The blends char yield at 1000°C was also gradually enhanced by more PN contents. For

Table 4.1 Thermal Properties of the BPh/Epoxy Copolymers

Biphenyl PN/Epoxy (Mole Ratio)	*P*-BAPS (Mole %)	Maximum Cure (Temperature/Time)	$T_{5\%}$ (°C)	Char Yield at 1000°C (%)
0:100	25	380°C/1 h	378	19
25:75	2.5	325°C/1 h	403	61
50:50	2.5	325°C/1 h	419	70
75:25	2.5	325°C/1 h	479	74
100:0	2.5	375°C/8 h	500	71

Reprinted from Dominguez DD., Keller T.M. Phthalonitrile–epoxy blends: cure behavior and copolymer properties. J Appl Polym Sci 2008;110:2504–2515.

Table 4.2 Thermal Properties of the $n = 4$ PN/Epoxy Copolymers

$n = 4$ PN/Epoxy (Mole Ratio)	*p*-APB (Mole %)	Maximum Cure (Temperature/Time)	$T_{5\%}$ (°C)	Char Yield at 1000°C (%)
0:100	27	280°C/1 h	356	20
32:68	24	350°C/8 h	408	73
100:0	24	375°C/8 h	534	73

Reprinted from Dominguez DD, Keller TM. Phthalonitrile–epoxy blends: cure behavior and copolymer properties. J Appl Polym Sci 2008;110:2504–15.

example, the char yield of the copolymers reached 61%–74% which was markedly superior to that of the epoxies (19%–20%). The previous results confirmed the ability of the PN resins to highly increase the thermal stability of the epoxy through a simple blending technique.

The DMA was used to ascertain changes in the storage modulus and tan delta of the studied materials. The storage modulus gives important indications about the stiffness and the tan delta curves allow the determination of the T_g. The variations of these important properties for the BPh/epoxy copolymers are presented in Fig. 4.4. As clearly seen in this figure, the epoxy showed the lowest modulus and T_g values. Conversely, the BPh exhibited the highest modulus value with at least 400°C of T_g. The second peak seen in the tan delta of the epoxy was attributed to the decomposition of this resin. As for the thermal stability, the copolymers also presented thermomechanical properties intermediate between those of the epoxy and BPh. Moreover, all the cured blends showed only one tan delta peak indicating that the coreaction had effectively taken place resulting in an interpenetrating network with no phase separation. In addition, it was found that more BPh contents resulted in a decrease in the tan delta height, which was attributed to the reduction in the segmental motion due to the formation of oxazoline structure. In contrast with the neat epoxy, the BPh/epoxy copolymers showed a remarkable increase in the T_g up to 240°C for the molar ration of 75:25.

The variations of the storage modulus and tan delta for the $n = 4$ PN/epoxy copolymers are displayed in Fig. 4.5. Once more the neat epoxy presented the lowest thermomechanical properties with a room temperature storage modulus and T_g of 1260 MPa and 126°C, respectively. Also, the $n = 4$ PN resin showed the best combination of room temperature storage modulus and T_g with 1610 MPa and at least 400°C, respectively. As also expected, the thermomechanical properties of the cured blend of $n = 4$ PN/epoxy were between these two boundaries. The room temperature storage modulus was 1020 MPa and the T_g was found to be about 226°C.

Globally, the PN/epoxy blends discussed above showed ameliorated processability and curing behavior in contrast to the neat PN precursors, and their thermal and thermomechanical properties were way above those of the neat cured epoxies. Therefore, it was confirmed that both resins could be easily blended to improve some of each resin lowest performances.

Another interesting work performed by Zhao et al., investigated the effect of bisphenol A-based PN (BAPh) monomers of the flame

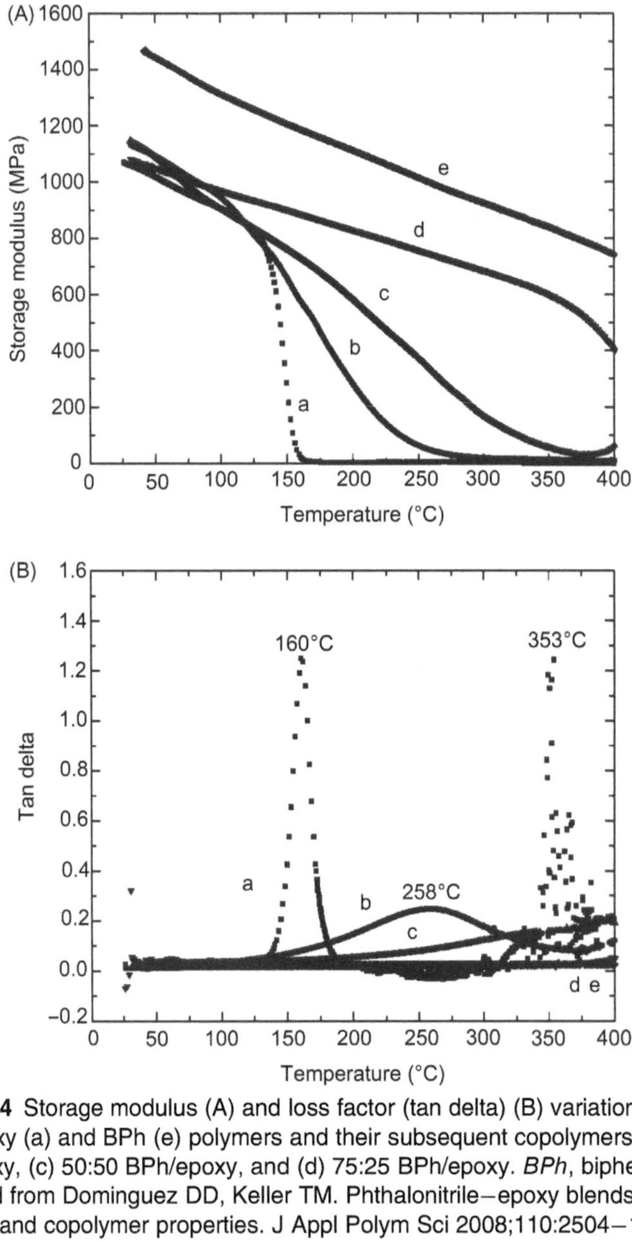

Figure 4.4 Storage modulus (A) and loss factor (tan delta) (B) variations for the neat epoxy (a) and BPh (e) polymers and their subsequent copolymers: (b) 25:75 BPh/epoxy, (c) 50:50 BPh/epoxy, and (d) 75:25 BPh/epoxy. *BPh*, biphenyl PN. Reprinted from Dominguez DD, Keller TM. Phthalonitrile–epoxy blends: cure behavior and copolymer properties. J Appl Polym Sci 2008;110:2504–15.

retardancy, thermal, and mechanical properties of the epoxy resin [35]. The blends were prepared by adding a mixture of epoxy and 27 wt.% of 4,4′-diaminodiphenyl sulfone (DDS) to different amounts of melted BAPh monomers. The relative weight ratio of BAPh to epoxy ranged

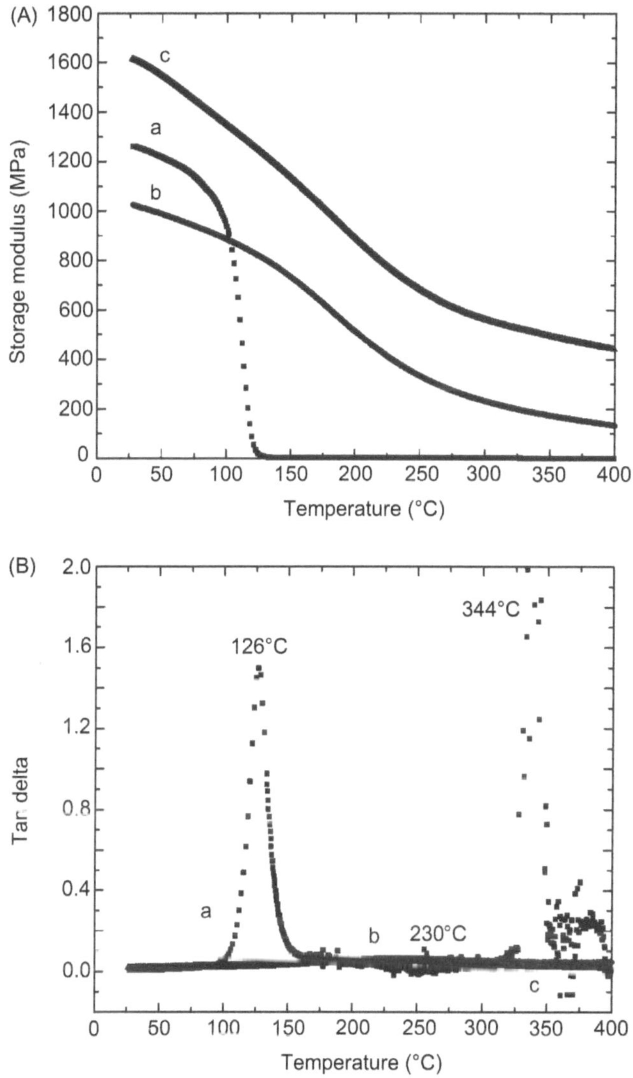

Figure 4.5 Storage modulus (A) and loss factor (tan delta) (B) variations for the neat epoxy (a) and $n = 4$ PN (c) polymers and their subsequent copolymer of 32:68 $n = 4$ PN/epoxy (b). *PN*, phthalonitrile. Reprinted from Dominguez DD, Keller TM. Phthalonitrile–epoxy blends: cure behavior and copolymer properties. J Appl Polym Sci 2008;110:2504–15.

between 10 and 50 wt.% with increments of 10 wt.%. The blends were placed in adequate molds and cured following the cure schedule of 190°C for 4 h, 210°C for 2 h, 230°C for 2 h, 250°C for 2 h, and 270°C for 2 h.

The neat cured epoxy and its related BAPh-based copolymers were analyzed by Fourier transform infrared spectroscopy (FTIR). The bands depicted at 1180 and 1290 cm^{-1} were the characteristic absorptions of the epoxy. The introduction of the BAPh resulted in the appearance of new bands at 1010 and 1360 cm^{-1}. Increasing the proportion of the BAPh in the blends led to the increase in intensity of the BAPh related absorptions and simultaneously the epoxy absorptions decreased in intensity. Although these observations are correct, the authors did not report about the bands related to the reaction between the epoxy and BAPh, especially those of oxazoline structure which is expected to increase in intensity with more BAPh contents.

The impact of the BAPh amount on the curing behavior and processability of the epoxy was studied by DSC and rheological measurements. The DSC results confirmed that the neat epoxy had one exothermic peak related to the reaction between the epoxy monomers and the DDS curing agent. Similarly, all the blends showed an exothermic peak which position shifted to higher temperatures with more BAPh amounts. The presence of one exothermic peak confirmed the reaction between the blends constituents. The T_g variations also revealed that more BAPh contents ensued higher T_g values. For instance, the T_g of the 50:50 BAPh/epoxy reached 167°C which is 28.7°C higher than that of the neat epoxy. In addition, the presence of only one T_g implied that there was no phase separation and both BAPh and epoxy coreacted to form one network. Moreover, the increase in the T_g values can be attributed to the formation of the oxazoline structure and to the reduced segmental motions due to more BAPh contents.

The rheological measurements were only carried out for the 20:80 BAPh/epoxy blend. The time-sweep viscosity curves under different temperatures and prepolymerization conditions were reported. It was observed that before the initiation of the cure reaction, the viscosity was very low. However, a sharp increase in viscosity was noticed upon moving forward with the cure reaction until gelation occurred. For a similar prepolymerization time, the viscosity of the blend was sharply increased and the gelation time was markedly reduced with higher prepolymerization temperatures. These observations were in good agreements with the curing behavior and processability of both BAPh and epoxy. Indeed, both resins showed accelerated crosslinking reaction either alone or as a blend.

To overcome the weak flame retardancy of the epoxy resins, two approaches were reported: either by the physical incorporation of flame retardant or by the addition of chemically bonded flame retardant. The first approach is now obsolete since it gives rise to toxic gases and smokes. The preparation of halogen-free flame retardant has recently spurred much of the attention considering its advantage over the first method. The PN resins are well-known for their superior flame resistance and can be good candidates to be used as chemically bonded additives for epoxy flame retardancy improvements. The flame retardancy properties of the neat epoxy and its subsequent BAPh/epoxy copolymers were investigated by the limiting oxygen index (LOI) testing and the vertical burning test (UL-94) and the obtained resulted are listed in Table 4.3. It should be pointed that the LOI value is defined as the concentration in which no self-extinction happens within half of a minute after the removal of the flame. In other words, higher LOI values indicate better flame resistance. The UL-94 classifies the materials by ranks according to the time frame needed before self-extinction occurs. The best NR rank means no rank and V-0 is the most desired. By analyzing the data from both tests, it was first confirmed that the neat epoxy presented the lowest grades. Meanwhile, adding more BAPh contents gradually improved the LOI values and UL-94 rank. For instance, the 50:50 PN/epoxy had an LOI value of 34.5% and was ranked V-0. These improvements in the epoxy flame resistance were attributed to the formation of a highly crosslinked and thermal resistant polymeric network afforded by the addition of the BAPh.

Some insights about the mechanical properties of the BAPh/epoxy blends were also discussed in terms of the effect of the PN contents on the epoxy resin flexural properties. The flexural strength variations in function of the BAPh loadings are reproduced in Fig. 4.6 and the stress/strain curves of the BAPh/epoxy copolymers are shown in Fig. 4.7. The

Table 4.3 Flame Retardancy Properties of the BAPh/Epoxy Copolymers

BAPh/Epoxy (Weight Ratio)	LOI (%)	UL-94
0:100	29.5	NR
10:90	30.5	V-1
20:80	31.5	V-0
30:70	32.5	V-0
40:60	33.5	V-0
50:50	34.5	V-0

Reprinted from Zhao X, Lei Y, Zhao R, Zhong J, Liu X. Preparation and properties of halogen-free flame-retarded phthalonitrile–epoxy blends. J Appl Polym Sci 2012;123:3580–86.

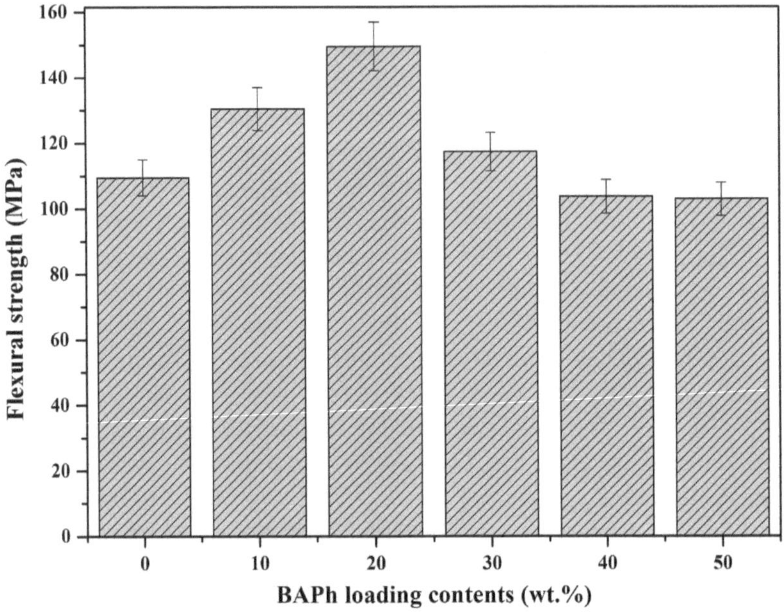

Figure 4.6 Flexural strength variations of the epoxy as a function of the BAPh loading contents. *BAPh*, bisphenol A-based phthalonitrile. Reprinted from Zhao X, Lei Y, Zhao R, Zhong J, Liu X. Preparation and properties of halogen-free flame-retarded phthalonitrile–epoxy blends. J Appl Polym Sci 2012;123:3580–86.

neat epoxy resin exhibited a flexural strength of about 109.5 MPa. Increasing the amount of the BAPh in the blends resulted in an increase in the bending strength to a certain extent. In fact, the 20:80 PN/epoxy copolymer presented the highest strength value of 149.5 MPa and further increasing the amount of the BAPh resulted in a negative impact on the strength. Moreover, high BAPh contents also reduced the ductility of the pure epoxy resin. Obviously, the high rigidity and brittleness of the BAPh resins was behind the previous changes in the epoxy bending properties.

4.2.2.2 Self-Catalyzed Phthalonitrile/Epoxy Blends

In the previously discussed blends, the use of a curing agent was required to perform the curing in a cost-effective and a timely manner. However, the use of curing agents always has many side-effects some of them related to the volatility of the unreacted molecules which may affect the thermal resistance of the desired materials. Indeed, since the

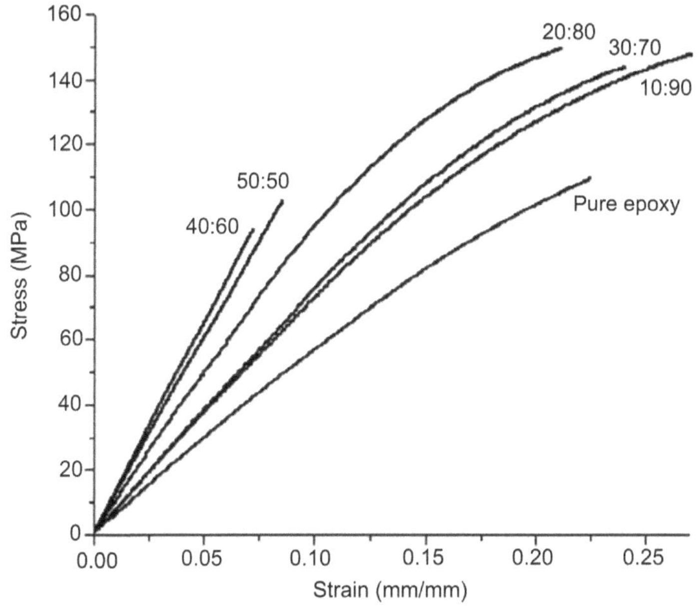

Figure 4.7 Stress/strain curves of the neat epoxy resin and the BAPh/epoxy copolymers. *BAPh*, bisphenol A-based phthalonitrile. Reprinted from Zhao X, Lei Y, Zhao R, Zhong J, Liu X. Preparation and properties of halogen-free flame-retarded phthalonitrile–epoxy blends. J Appl Polym Sci 2012;123:3580–6.

emergence of the PN resins, the preparation of self-cured, PN monomers/oligomers has been one of the most explored areas of research. Thus, many self-catalyzed, PN monomers were prepared by introducing, within the monomers backbones, organic functions able to catalyze the cyano addition reaction [36–39]. Another interesting fact is that the self-catalyzed, PN monomers have an excess amount of activating functions which is ideal for chemically bonding with other monomers/oligomers.

Following this point of view, Guo et al. prepared series of thermosetting blends from the self-catalyzed, 4-aminophenoxy PN (4-APN) and epoxy and studied their curing behavior, processability as well as the thermal stability of their resulting copolymers [40]. The PN/epoxy blends were prepared in three weight ratios of 4-APN to epoxy, namely 25:75, 50:50, and 75:25. Basically, for each ratio, the 4-APN and epoxy were mixed and mechanically stirred at 50–70°C for 30–45 min. The mixtures were melted, placed in preheated molds and then thermally cured following the curing procedure of 200°C for 1 h, 240°C for 3 h, 260°C for 2 h, and 280°C for 1 h.

DSC investigations were carried out to study the self-curing behavior of the blends. The 4-APN monomers showed one clear endothermic peak at 134°C related to the melting and one exothermic transition at around 293°C, confirming the self-curing behavior of these monomers. Each of the three blends also showed one endothermic transition which position varied according to the 4-APN amount. For instance, with more 4-APN contents, the melting related peak shifted to higher temperatures. The presence of only one endothermic transition implied that the blends already started to form prepolymers at the early stage of the preparation process. All the studied blends exhibited two exothermic transitions. The first exotherm was ascribed to the epoxy ring opening thanks to the amino groups of the 4-APN monomers and further increasing the amount of the latter resulted in shifting the first exotherm to lower temperatures due to the presence of more active hydrogens. The second exotherm was ascribed to the polymerization reaction of the 4-APN monomers and the formation of an interpenetrating network with the epoxy. The position of the second exotherm was affected by the amount of the 4-APN in a way similar to that of the first exotherm. Hence, the 4-APN/epoxy blends were confirmed to be able to copolymerize without the addition of any external catalyst.

The processability of the blends was also studied by monitoring the changes in the complex viscosity over various temperatures and times. All the blends showed a large decrease in the viscosity at 120°C. Afterwards, all the compositions revealed a two-step increase in the viscosity related to ring opening of the epoxy and to the polymerization reaction of the PN monomers. Moreover, further investigating the viscosity evolution over time at various temperatures revealed that higher curing temperatures ensued faster rate of viscosity increase. It was also confirmed that both polymerization reactions can be accelerated by higher curing temperatures.

The thermal stability measurements of the 4-APN/epoxy copolymers were performed by TGA and the starting decomposition temperatures of 5 and 30 wt.% ($T_{5\%}$, $T_{30\%}$) as well as the char yield (Y_c) at 800°C were used to compare the thermal resistances of the copolymers. The TGA curves of the cured blends under nitrogen atmosphere highlighted the fact that the thermal stability of the copolymers was closely related to the amount of the 4-APN. Indeed, the 4-APN/epoxy ratio of 75:25 presented the highest thermal stability with a $T_{5\%}$, $T_{30\%}$, and Y_c of 397°C, 698°C, and 68%, respectively. Once more, the highly interconnected and crosslinked network afforded by the 4-APN was behind the improvements of the epoxy thermal stability. Thus, it is clear that the self-catalyzed, PN/epoxy blends and copolymers are promising materials with a self-curing behavior and intermediate properties between the PN and epoxy.

4.2.2.3 Functional-Phthalonitrile/Epoxy Blends

Kaliavaradhan et al. synthesized a new functional, PN monomers based from an incompletely condensed trisilanol phenylpolyhedral oligomeric silsesquioxane (POSS) [41]. The synthetic route was a typical nucleophilic displacement of a labile nitro group from 4-nitrophthalonitrile. The afforded new PN monomers, referred to as TPnPh-POSS, were used to enhance the thermal, thermomechanical, and dielectric properties of the neat epoxy. The blends were prepared by mixing the epoxy and 1, 3, and 5 wt.% of PN additives in the presence of DDS as curing agent. Then, the mixtures were poured into preheated molds and cured following the cure schedule of 150°C for 3 h, 200°C for 2 h, 230°C for 2 h, 250°C for 2 h, and 270°C for 2 h.

The DSC curves of the neat epoxy and its subsequent blends in the range of 200–300°C were presented. All the studied samples exhibited only one large exothermic transition which position depended on the proportion of the PN in the blend. These exothermic transitions were attributed to the DDS catalyzed reaction between both PN and epoxy monomers. Moreover, no other transitions were depicted up to 300°C which implied that the studied materials were thermally resistant. In contrast with the self-catalyzed PN/epoxy in which the DSC of the blends revealed the presence of two exothermic peaks, the main difference is in the amount of the activators in the reaction. Indeed, for the self-catalyzed, PN/epoxy blends, the ratio of the PN was at least 25 wt. % and, therefore the number of the amine functions was enough to quickly activate the ring opening of the epoxy and the cyano addition reaction of the PN at distinct temperatures.

The impact of the newly developed, PN monomers on the stiffness and T_g of the neat epoxy was studied by DMA and the obtained results are plotted in Fig. 4.8. It can be clearly seen in Fig. 4.8A that increasing the amount of the PN was found to impart higher storage modulus for the neat epoxy. For example, the epoxy reinforced with 5 wt.% PN reached up to 3.2 GPa of modulus at 50°C. On the other hand, the tan delta variations, presented in Fig. 4.8B, only showed one peak for each PN/epoxy copolymer which indicated that both systems perfectly reacted to form one crosslinked network. The T_g also gradually increased with more PN amounts to reach 136°C at 5 wt.% PN. The

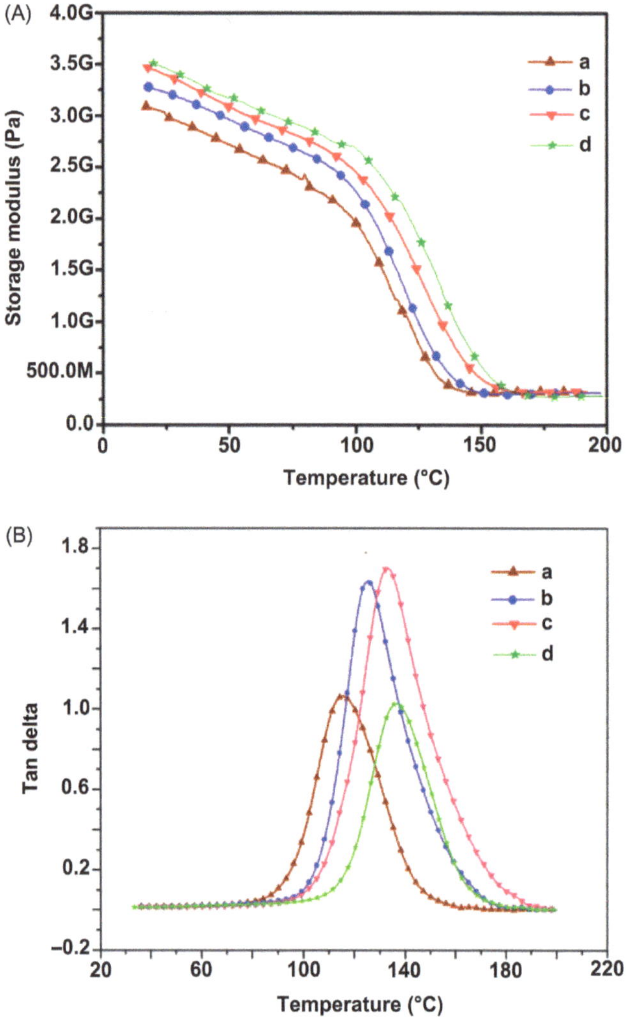

Figure 4.8 Storage modulus (A) and loss factor (tan delta) (B) variations for the epoxy/POSS-functional, PN copolymers: (a) 0 wt.% of PN, (b) 1 wt.% of PN, (c) 3 wt.% of PN, and (d) 5 wt.% of PN. *POSS*, polyhedral oligomeric silsesquioxane; *PN*, phthalonitrile. Reprinted from Kaliavaradhan K, Muthusamy S. Studies on novel tri-phthalonitrile phenyl polyhedral oligomeric silsesquioxane and the phthalonitrile–epoxy blends. High Perform Polym 2016;28:296–308.

improvements obtained for both the stiffness and T_g were logically attributed to the reduction in segmental motions afforded by the rigid POSS-based PN monomers.

The dielectric properties of the PN/epoxy copolymers were discussed in terms of the influence of the frequency on the dielectric constant and the dielectric loss. Globally, both dielectric properties were reduced by increasing the frequency for all the studied materials. At 1 MHz, the neat epoxy had a dielectric constant and dielectric loss of 2.54 and 0.86, respectively. At the same frequency, the PN/epoxy copolymers exhibited reduced dielectric properties as far as the proportion of the PN in the blends increased. For instance, the dielectric constant of the copolymers containing 1, 3, and 5 wt.% PN was 2.47, 2.37, and 2.27, respectively. The dielectric loss reached its lowest value of 0.66 at 5 wt. % PN. The decrease in the dielectric constant upon adding the PN was attributed to the increase in the free volume because of the high steric volume of the POSS functionality in the PN structure.

The thermal stability of the neat epoxy and its PN-based copolymers was studied by TGA under nitrogen atmosphere. The neat epoxy resin showed the decomposition temperatures of $T_{5\%}$ and $T_{10\%}$ of 365 and 380°C, respectively, with Y_c at 800°C of 21.3%. It should be understood that the neat epoxy or PN resins properties may vary according to different parameters, such as the amount and type of the used curing agent, curing schedule, and postcure conditions. All the PN/epoxy copolymers presented ameliorated thermal stability parameters compared to the neat epoxy. For example, at the maximum PN amount of 5 wt.%, the copolymer had a $T_{5\%}$ and $T_{10\%}$ of 389 and 410°C, respectively, with a Y_c at 800°C of 27.1%.

Similarly to the typical and self-catalyzed, PN monomers, the functional, PN ones also allowed consequent improvements in the thermal stability and thermochemical properties of the neat epoxy. Through the previously discussed blends and copolymers, it clearly appears that the combination of the PN and epoxy resulted in advanced materials with many interesting properties. Although the PN/epoxy blends were mainly prepared from the diglycidyl ether of bisphenol A epoxy and considering the huge potential of these materials, the next few years may show an increasing number of studies on multiple combinations between the PN resins and epoxies.

4.2.3 Phthalonitrile/Benzoxazine

Polybenzoxazines, as recently developed phenolic resins, are synthesized by the cationic ring-opening polymerization of their corresponding benzoxazine monomers without the addition of any extra catalysts and without generating by-products or volatiles. These new materials seek to combine the thermal properties of phenolic resins with the mechanical performance and design flexibility of advanced epoxy resins. Physical and mechanical characterization has revealed that this family of resins possess many interesting features, such as near-zero volumetric shrinkage or expansion upon curing, good processability (due to low melting viscosity before polymerization), low-water absorption, outstanding T_g, high char yield, and low coefficient of thermal expansion. In addition, low dielectric constant and dissipation loss, high mechanical performance, and great molecular design flexibility make polybenzoxazine attractive for many industrial applications [42–45].

Although the polybenzoxazines have a good thermal stability, further improvements are needed to fulfill the requirements of advanced applications. On the other hand, the PN resins have a far more better thermal stability and their thermomechanical properties are also superior. Thus, the preparation of blends and copolymers from benzoxazine and PN can improve the thermal and thermomechanical properties of the benzoxazine. Meanwhile, the PN can take advantage from the oxazine ring opening of the benzoxazine which produces the Mannich bridge ($-CH_2-NR-CH_2-$) and structures with phenolic hydroxyl functional groups. The presence of these active hydroxyls can catalyze the cyano addition reaction of the PN monomers resulting in an interpenetrating network. As a result, the PN/Benzoxazine blends and copolymers can be cured without the addition of any catalyst and they are expected to exhibit an improved curing behavior and interesting properties.

It is also important to point out that the typical polybenzoxazines suffer from the instability of their Mannich bridge, which is in fact the weakest point of their crosslinked network [46]. Therefore, end-capping or functionalizing the benzoxazine with pendent polymerizable groups is a current practice to improve the degree of crosslinking of the polybenzoxazines. In fact, the only studied PN/benzoxazine blends were prepared from functional benzoxazine monomers because of their superior curing behavior and properties over the nonmodified/typical ones. In the following parts, the majority of the published works on the PN/benzoxazine blends will be reviewed and discussed in terms of their curing behavior and the properties of their related copolymers.

4.2.3.1 Phthalonitrile-Functional Benzoxazine/Phenylnitrile-Functional Benzoxazine

The first report on the combination of PN and benzoxazine was reported by Brunovska and Ishida in 1999 [47]. The authors synthetized various PN-functional benzoxazine (BZPN) and phenylnitrile-functional benzoxazine and studied the curing behavior and the properties of their resulting blends and copolymers. The prepared monomers are presented in Fig. 4.9. It should be understood that, in this case, the BZPN

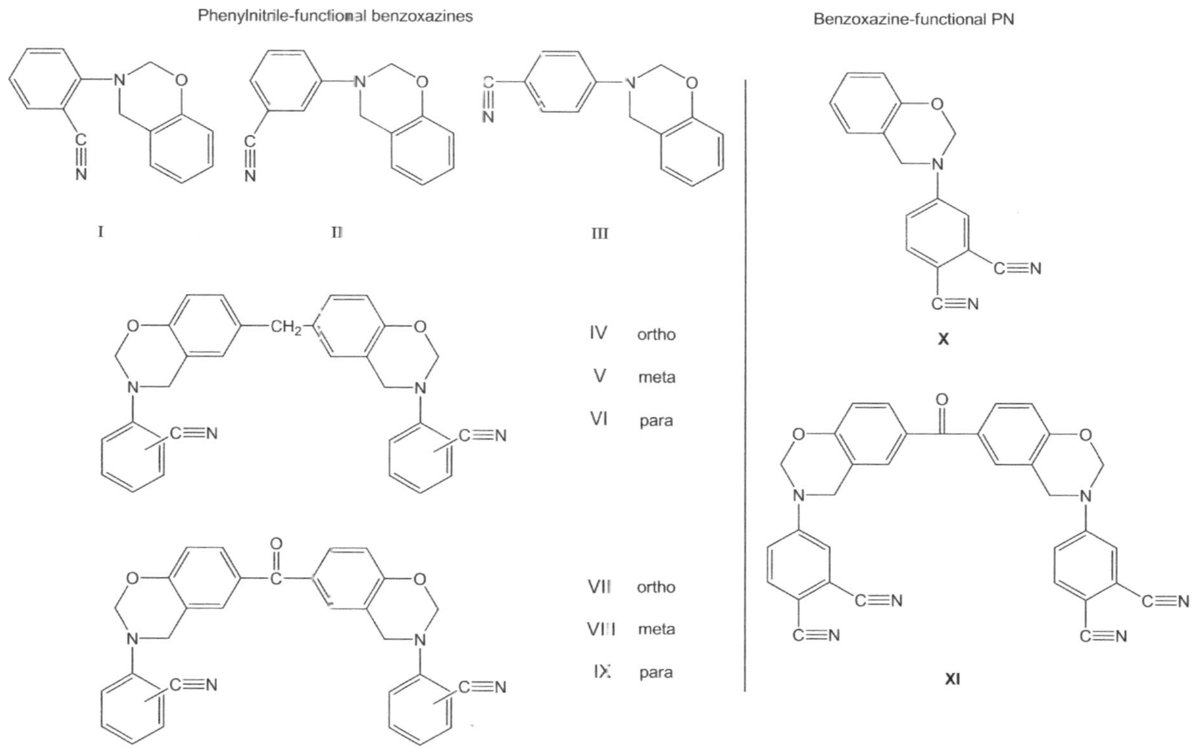

Figure 4.9 Chemical structures of the phenylnitrile-functional benzoxazine and benzoxazine-functional, PN monomers. *PN*, phthalonitrile. Reprinted from Brunovska Z, Ishida H. Thermal study on the copolymers of phthalonitrile and phenylnitrile-functional benzoxazines. J Appl Polym Sci 1999;73:2937–49.

monomers were seen as PN monomers, hence their presentation in this part. The blends were formulated by direct mixing the PN and benzoxazine monomers in acetone at various molar ratios of PN to benzoxazine. The resulting mixtures were heated at 40°C in a vacuum oven to evaporate the solvent. The curing was performed by heating the mixtures at 220/250°C in an air circulating oven.

The DSC investigations were carried out to acknowledge about the curing mechanism of the blends. The DSC curves of the neat PN and benzoxazine monomers as well as the curves of the 30:70 PN/benzoxazine blends are displayed in Fig. 4.10. The DSC curves of the blends showed two exothermic transitions at 250 and 315°C for the *ortho*-nitrile-functional benzoxazine-based blend and at 225 and 292°C for the blend prepared from the *meta*-compound. On the other hand, the neat PN and the benzoxazine monomers clearly exhibited different curing profiles. To further understand the nature of these exothermic transitions, the authors analyzed the polymerization products of each blend by FTIR. From the obtained results, it was deducted that the first exotherm was related to both the oxazine ring opening from both monomers and to the intramolecular reaction between the nitrile groups and the oxazine rings. Furthermore, evidences of the presence of the triazine

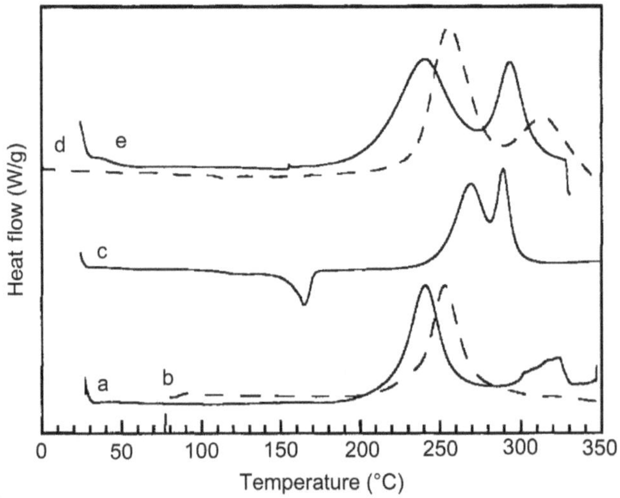

Figure 4.10 DSC thermograms of phenylnitrile-functional benzoxazine I (a), phenylnitrile-functional benzoxazine II (b), benzoxazine-functional, PN X (c), 30:70 mixture of I and X (d), and mixture of II and X (e). *DSC*, dynamic scanning calorimetry; *PN*, phthalonitrile. Reprinted from Brunovska Z, Ishida H. Thermal study on the copolymers of phthalonitrile and phenylnitrile-functional benzoxazines. J Appl Polym Sci 1999;73:2937–49.

structure were also found in the FTIR spectra. The triazine formation may be ascribed to the addition reaction of the cyano groups obviously activated by the hydroxyl groups resulting from the oxazine ring opening. The second exotherm was attributed to the degradation of the methylene units in the Mannich bridge. Subsequently, evidences of the continuation of the cyano addition reaction were also confirmed at higher temperatures (above 400°C).

The TGA deduced thermal properties of the PN- and phenylnitrile-functional polybenzoxazines and their copolymers are listed in Table 4.4. In all instances, the polymers and their related copolymers showed good thermal resistances. It can be also seen that the monofunctional polymers showed a reduced thermal stability compared to their bifunctional counterparts. Moreover, all the copolymers of the *ortho*-phenylnitrile functional benzoxazine and BZPN exhibited improved starting decomposition temperatures and char yields compared to the *ortho*-phenylnitrile homopolymers. The best thermal resistance was achieved with the copolymer of 30:70 PN/*ortho*-phenylnitrile-bifunctional benzoxazines with $T_{5\%}$ and $T_{10\%}$ under nitrogen atmosphere of 496 and 520°C, respectively, and a char yield at 800°C of 80%. As we stated before, the Mannich bridge constitutes the weakest point of the crosslinked network. Thus, the authors attributed the improvements in the thermal stability to the *ortho*-phenylnitrile compound, which underwent a special reaction resulting in a restricted Mannich bridge not in the char formation but during the polymerization. As a result, the starting decomposition temperatures ($T_{5\%}$ and $T_{10\%}$), also influenced by the polymeric structure involving the Mannich bridge, were also improved. Moreover, through the obtained results, it clearly appears that the "end-capping" of the benzoxazine with PN units was advantageous in enhancing the overall properties of the neat benzoxazines and their related copolymers.

4.2.3.2 Typical Phthalonitrile/Phthalonitrile-Functional Benzoxazine

Following the same path, Xu et al., studied the self-curing behavior of the typical BAPh oligomers blended with the well-known, BZPN [48]. The aim was to take advantage from the ring-opening polymerization of the benzoxazine to catalyze the cyano addition reaction in both systems. The chemical structure of the studied monomers/oligomers and the possible processes of polymerization are resumed in Scheme 4.1. The oligomeric BAPh were obtained by melting the BAPh monomers at

Table 4.4 **Thermal Properties of the PN and Phenylnitrile-Functional Polybenzoxazines and Their Copolymers**

Monomer Unit	$T_{5\%}$ (°C) in N_2	$T_{10\%}$ (°C) in N_2	Char Yield (wt.%) at 800°C in N_2	$T_{5\%}$ (°C) in Air	$T_{10\%}$ (°C) in Air	Char Yield (wt.%) at 700°C in Air
I	332	371	60	365	407	41
II	317	355	56	357	398	18
III	331	369	55	343	374	30
50:50 of I and X	430	515	77	–	–	–
IV	356	410	60	366	421	48
50:50 of IV and X	438	508	76	–	–	–
VIII	410	537	74	448	528	46
IX	384	436	68	418	475	55
50:50 of VII and X	410	510	76	445	526	21
30:70 of VII and XI	496	520	80	–	–	–

Reprinted from Brunovska Z, Ishida H. Thermal study on the copolymers of phthalonitrile and phenylnitrile-functional benzoxazines. J Appl Polym Sci 1999;73:2937–49.

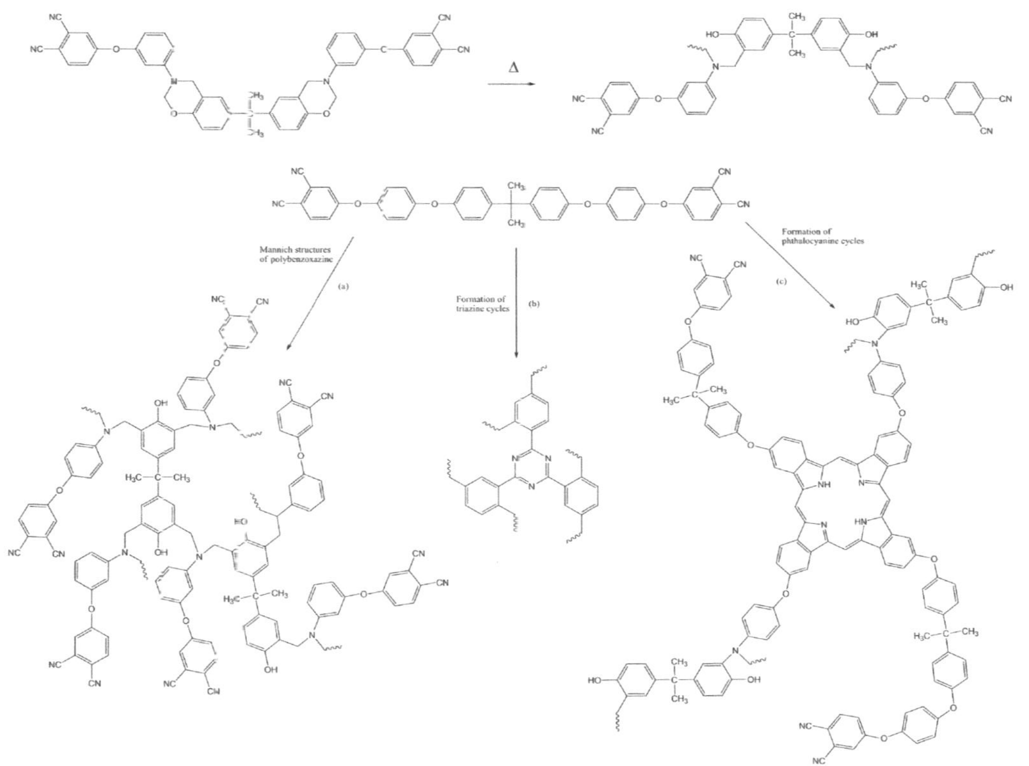

Scheme 4.1 Possible reaction pathways between the BAPh and BZPN monomers. *BAPh*, bisphenol A-based PN; *BZPN*, PN-functional benzoxazine. Reprinted from Xu M, Liu M, Dong S, Liu X. Design of low temperature self-cured phthalonitrile-based polymers for advanced glass fiber composite laminates. J Mater Sci 2013;48:8108–116.

200°C for 30 min. Then, the oligomeric BAPh were transformed to powder before starting the blending process. The blends were prepared by direct mixing the BAPh and BZPN powders at room temperature. Different weight ratios of BAPh to BZPN were chosen for the study. The prepolymers were then obtained by heating the blends at 200°C for 15 and 60 min, respectively.

The DSC investigations were performed for the neat BAPh oligomers and their related BAPh/BZPN blends. The neat oligomeric BAPh had an endothermic transition ranging between 170 and 190°C corresponding to the melting points of these oligomers. The relatively reduced melting points of these oligomers in contrast with their initial monomers is due to the low degree of prepolymerization. Thereafter, no exothermic transition was depicted, even at high temperatures, indicating that the polymerization of these oligomers is quite difficult to achieve without the use of curing additives. All the BAPh/BZPN blends showed endothermic transitions lower than that of the neat PN oligomers and it appeared that increasing the proportion of the benzoxazine in the blends further shifted the melting points toward lower temperatures. At temperatures ranging between 250 and 280°C evident exothermic transitions, ascribed to the cyclotrimerization of the nitrile groups triggered by the active hydrogen issued from the ring opening of the benzoxazine units, were depicted for all the blends. Interestingly, it was observed that with more benzoxazine contents the exothermic transitions shifted to lower temperatures and a new exotherm appeared and gradually increased at temperatures ranging between 200 and 240°C. The reason behind that is related to the coexistence of both the self-polymerization of the benzoxazine monomers and the cyano addition reaction of the PN oligomers. In fact, increasing the proportion of the benzoxazine in the blends resulted in less active hydrogen and therefore the cyano addition reaction was limited to a certain extent.

Isothermal DSC scans were also carried out at 200°C on the neat BAPh oligomers and their related BAPh/BZPN blends. The isothermal DSC investigations were performed at 200°C because all the tested samples started polymerizing at this temperature. The BAPh oligomers exhibited large and weak exotherm starting at about 30 min confirming the challenging curing character of these oligomers. All the studied blends showed a double curing behavior with two distinct exotherms occurring at 0.5−15 min and 30−50 min. Moreover, the intensity of the fist exotherm was closely related to the amount of the benzoxazine in the blend. Thus, with more benzoxazine contents, the intensity of this exotherm markedly increased. Such a high intensity was indicative of a higher activity mainly attributed to the Mannich bridge formation from the ring-opening polymerization of the benzoxazine monomers. The second exotherm was attributed to the cyano addition reaction which existence and intensity depended on the number of the active hydrogens.

In order to further confirm the self-curing behavior of the BAPh/BZPN blends, FTIR analysis was performed on the copolymers cured at 200°C for 15 and 60 min. In Fig. 4.11 is presented the FTIR spectra of the BAPh/BZPN copolymers cured for 15 min while Fig. 4.12 shows the FTIR spectra of those cured for 60 min. By analyzing the spectra from Fig. 4.11, Xu et al., attributed the band at 1478 cm^{-1} to the in plane carbon–carbon stretching of the tetra-substituted benzene, revealing the formation of methylene bridges in the free *ortho* positions of the phenolic units. Also, the band at 946 cm^{-1} was ascribed to the tri-substituted benzenes attached with the oxazine rings. The intense band depicted at 2230 cm^{-1} indicated the presence of large number of nitrile groups, and the low absorption at 1015 cm^{-1} resulting from the phthalocyanine structure confirmed that, for the chosen curing time and temperature, the main reaction was the benzoxazine, ring-opening

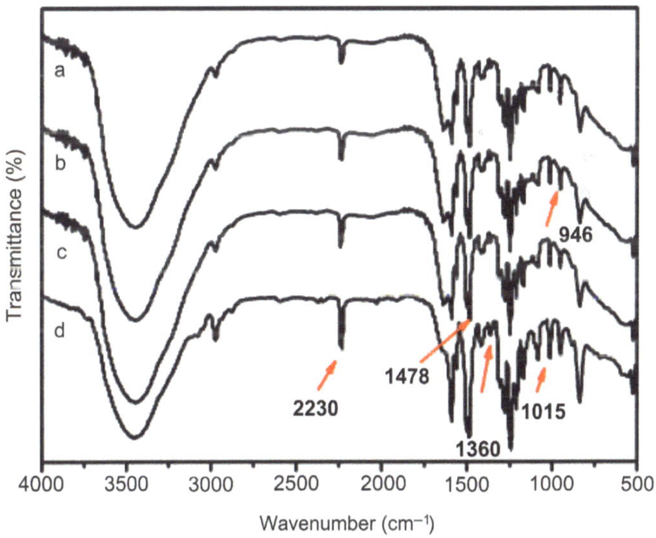

Figure 4.11 FTIR of BAPh/BZPN blends acquired after heating at 200°C for 15 min: (a) 30:70, (b) 50:50, (c) 70:30, and (d) pure BAPh. *FTIR*, Fourier transform infrared; *BAPh*, bisphenol A-based phthalonitrile; *BZPN*, phthalonitrile-functional benzoxazine. Reprinted from Xu M, Liu M, Dong S, Liu X. Design of low temperature self-cured phthalonitrile-based polymers for advanced glass fiber composite laminates. J Mater Sci 2013;48:8108–16.

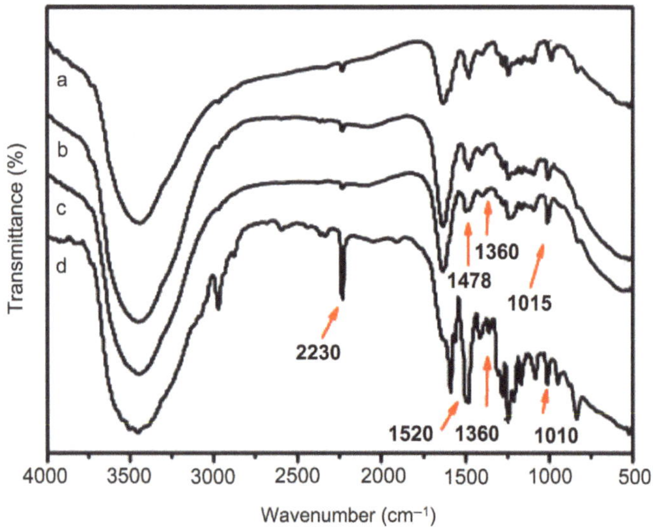

Figure 4.12 FTIR of BAPh/BZPN blends acquired after heating at 200°C for 60 min: (a) 30:70, (b) 50:50, (c) 70:30, and (d) pure BAPh. *FTIR*, Fourier transform infrared; *BAPh*, bisphenol A-based phthalonitrile; *BZPN*, phthalonitrile-functional benzoxazine. Reprinted from Xu M, Liu M, Dong S, Liu X. Design of low temperature self-cured phthalonitrile-based polymers for advanced glass fiber composite laminates. J Mater Sci 2013;48:8108–16.

polymerization. For a longer curing time, the FTIR spectra underwent many changes. For instance, the intensity of the nitriles absorption was largely weakened while the bands attributed to the phthalocyanine and triazine increased in intensity. Furthermore, the band at 1478 cm^{-1} was still present; however, the one at 946 cm^{-1} completely disappeared implying that the benzoxazine rings had to a large extent been involved in the reaction. Globally, the studied BAPh/BZPN blends showed the desired self-curing behavior and the polymerization degree as well as the copolymers structures can be easily tuned by changing the relative amount of BAPh and BZPN. The authors also reported about the use of these blends as matrices for composites and the obtained results will be discussed in Chapter 5, Fiber-Reinforced Phthalonitrile Composites concerned with the PN-based fiber-reinforced composites.

4.2.3.3 Allyl-Functional Phthalonitrile/Phthalonitrile-Functional Benzoxazine

The allyl-functional, PN monomer was first synthetized by Zou et al., and the authors reported that in contrast with the typical PN monomers, this new monomer had better processability with a lower melting point and a wider processing window [49]. The newly developed monomers also showed a two-stage curing behavior involving the allyl moieties polymerization and the cyano addition reaction of the nitrile groups. Aiming to further improve the curing behavior and the processability at low temperatures, these new PN monomers were blended with the BZPN monomers [50]. The chosen benzoxazine monomer is well-known for its good processability and low melting viscosity. Moreover, its cured polymers possess an enhanced thermal stability afforded by the polymerization of the PN ends. Thus, the introduction of the benzoxazine was expected to improve the polymerization behavior of the allyl-functional PN. The chemical structures of these monomers have been previously presented elsewhere. The blends were prepared by mixing a specified amount of allyl-functional, PN monomers with various weight ratios of BZPN ranging from 10 to 50 wt.% with increments of 10 wt.%. The prepolymers were obtained after heating the mixtures at 140°C for 1 h and the allyl-functional, PN/BZPN copolymers were afforded after curing the prepolymers in an air circulating oven following the curing schedule of 200°C for 2 h, 240°C for 4 h, 280°C for 4 h, and 320°C for 4 h.

The DSC investigations for the neat monomers revealed that both allyl-functional, PN and BZPN had two distinct exothermic transitions. For the BZPN, the first exotherm, located at around 228°C, was related to the ring-opening of the oxazine rings and the second one, depicted at about 265°C, was due to the cyano addition reaction. For the allyl-functional, PN monomers, the exotherms at 280 and 340°C were ascribed to the polymerization of the allyl moieties and to the cyano addition reaction, respectively. The DSC curves of the allyl-functional, PN/BZPN blends are shown in Fig. 4.13. It was evident that when the amount of the benzoxazine in the blends exceeded 30 wt.%, two exothermic transitions were observed at around 237 and 271°C. These exotherms were logically ascribed to the oxazine rings opening and to the simultaneous polymerization of the allyl moieties and cyano groups, respectively. On the other hand, decreasing the amount of the benzoxazine reduced the intensity of the previously discussed exotherms and, with less than 20 wt.% of benzoxazine in the blend, another exothermic transition appeared at 329°C, implying that the cyano addition reaction was hindered by the proprietary polymerization of the benzoxazine. Therefore, these results clearly confirmed that the reactivity of the blend is closely related to the amount of benzoxazine. Furthermore, the reactivity of the allyl-functional, PN monomers was also positively affected by the introduction of the benzoxazine.

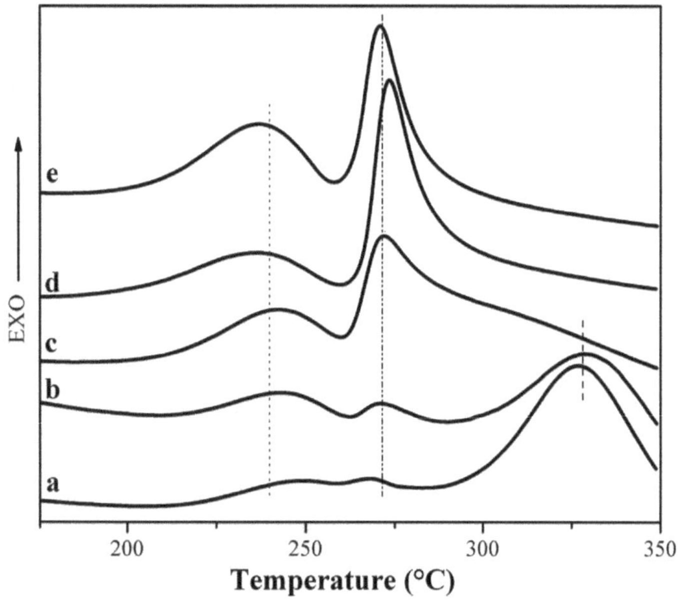

Figure 4.13 DSC curves of the allyl-functional, PN/BZPN blends respectively containing (a) 10, (b) 20, (c) 30, (d) 40, and (e) 50 wt.% of BZPN content. *DSC*, dynamic scanning calorimetry; *PN*, phthalonitrile; *BZPN*, PN-functional benzoxazine. Reprinted from Zou X, Yang X, Xu M, Jia K, Liu X. Curing behaviors and properties of allyl-and benzoxazine-functional phthalonitrile with improved processability. J Polym Res 2016;23:2.

The DSC results were also corroborated with those from the rheological tests. As seen from Fig. 4.14, where are displayed the complex viscosity changes of the allyl-functional, PN/BZPN blends as a function of temperature, the increase in viscosity indicated the starting of the polymerization and the viscosity further increased from 180 to 235°C. It was also confirmed that the polymerization temperatures of the blends decreased with more benzoxazine contents. Thus, improvements in the copolymerization reactivity can be achieved by increasing the benzoxazine weight ratio.

The thermal stability of the cured allyl-functional, PN/BZPN copolymers was studied by TGA and the obtained results are summarized in Table 4.5. The allyl functional, PN thermal stability parameters, under nitrogen atmosphere, have been previously determined to be about 467°C for $T_{5\%}$ and 73.4% for the char yield at 800°C. Interestingly, the neat BZPN also showed quite similar thermal stability parameters with 438°C for $T_{5\%}$ and 73.7% for the char yield at 800°C. Similarly, all the tested copolymers also showed closer thermal stability performances. For example, the $T_{5\%}$ ranged from 437 to 451°C, $T_{10\%}$ from 478 to

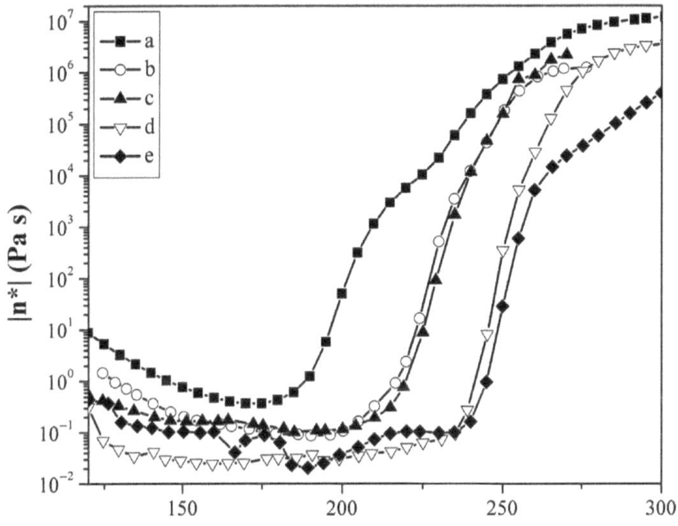

Figure 4.14 Complex viscosity variations as a function of temperature for the allyl-functional, PN/BZPN blends respectively containing (a) 10, (b) 20, (c) 30, (d) 40, and (e) 50 wt.% of BZPN content. *PN*, phthalonitrile; *BZPN*, PN-functional benzoxazine. Reprinted from Zou X, Yang X, Xu M, Jia K, Liu X. Curing behaviors and properties of allyl-and benzoxazine-functional phthalonitrile with improved processability. J Polym Res 2016;23:2.

Table 4.5 Thermal Properties of the Allyl-Functional, PN/BZPN Copolymers Under Nitrogen Atmosphere

PN/Benzoxazine (Weight Ratio)	$T_{5\%}$ (°C)	$T_{10\%}$ (°C)	Char Yield at 800°C (%)
10:90	451	480	69.6
20:80	450	485	70.9
30:70	446	483	71.5
40:60	441	479	71.2
50:50	437	478	71

Reprinted from Zou X, Yang X, Xu M, Jia K, Liu X. Curing behaviors and properties of allyl-and benzoxazine-functional phthalonitrile with improved processability. J Polym Res 2016;23:2.

483°C, and the char yield at 800°C was between 69.6% and 71.5%. Globally, the allyl-functional, PN/BZPN blends were proven to possess ameliorated curing behavior and processability, and their cured copolymers preserved the excellent thermal stability of their respective precursors.

4.2.4 Phthalonitrile/Novolac

Novolac resins are a class of phenolic polymers characterized by their good flame resistance and low cost. As a result, the novolac-based materials have found many applications in civil structures and transport vehicles. The good flame resistance of the novolac polymers is attributed to their highly aromatic structures and possibly to the inhibiting effect of their phenolic compounds. However, similarly to any other polymer, the novolac resins have many shortcomings such as limited thermal and thermo-oxidative stabilities and void problems caused by the evacuation of volatiles during the curing reaction [26]. These two shortcomings were found to be easily overcome by the preparation of physical blends with PN resins. Indeed, the PN resins have superior flame resistance, outstanding thermal stability, and the cure reaction of the PN monomers can be initiated by the phenolic hydroxyl groups of the novolac. Thus, aiming to improve the properties of the novolac resins, many PN/novolac blends and copolymers were studied and hereafter are discussed some of the most important reported works.

Sumner et al. reported about the mechanical, thermal, and fire characteristics of several copolymers prepared from the typical BPh and the commercial phenol formaldehyde novolac resin with an average phenol functionality of 7.1 [51]. The blends were prepared by first melting the novolac at 200°C. Then, the BPh monomers, in form of powder, were added in different weight proportions to the melt. The mixtures were thoroughly mixed, degassed, and then poured into a preheated mold for curing under air atmosphere. The chosen curing schedule was 200°C for 1 h and 220°C for 3 h. The curing reaction was assumed and further confirmed to start by the nucleophilic phenols attacking the nitrile groups of the BPh monomers. As a result, a phenoxy substituted imine was formed and the reaction propagated by attacking the other nitrile groups. The initialization of the polymerization reaction is believed to be similar to any other PN monomers/oligomers catalyzed by hydroxyl groups. More information about the hydroxyl catalyzed cyano addition reaction can be found in Chapter 1, Introduction to the Phthalonitrile Resins and Their Curing Behavior.

The cured PN/novolac copolymers were analyzed by SEM to see whether the addition of the BPh effectively resulted in void-free materials. The copolymers of 15:85 and 20:80 BPh to novolac weight ratios were analyzed and the results confirmed the formation of a void-free network. Thus it appeared that the most important shortcoming of the novolac resins was easily overcome by a simple physical blend with a typical PN resin. Since the voids are the main reason behind the propagation of the cracks, their absences indicated improvements in the mechanical properties and precisely the toughness was intended to be consequently ameliorated. Therefore in order to judge about this important mechanical property, the critical stress intensity factors (K_{IC}) were measured and the obtained results are presented in Fig. 4.15. By analyzing the data, it was found that the toughness of the copolymers ranged from 0.8 to >1 MPa $m^{1/2}$ and these values were in the range of those of epoxy polymers. Interestingly, the improvement in the crosslinking degree for the BPh/novolac copolymers did not affect their toughness.

The thermal stability of the copolymers was investigated by TGA under both nitrogen and air atmospheres. Globally, all the studied copolymers exhibited an improved thermal stability compared to the neat novolac resin. For instance, the copolymers char yields at 700°C in air and nitrogen ranged from 60% to 80%. Moreover, increasing the amount of the BPh in the blends resulted in higher char yields and starting decomposition temperatures. The $T_{5\%}$ of the copolymers were only

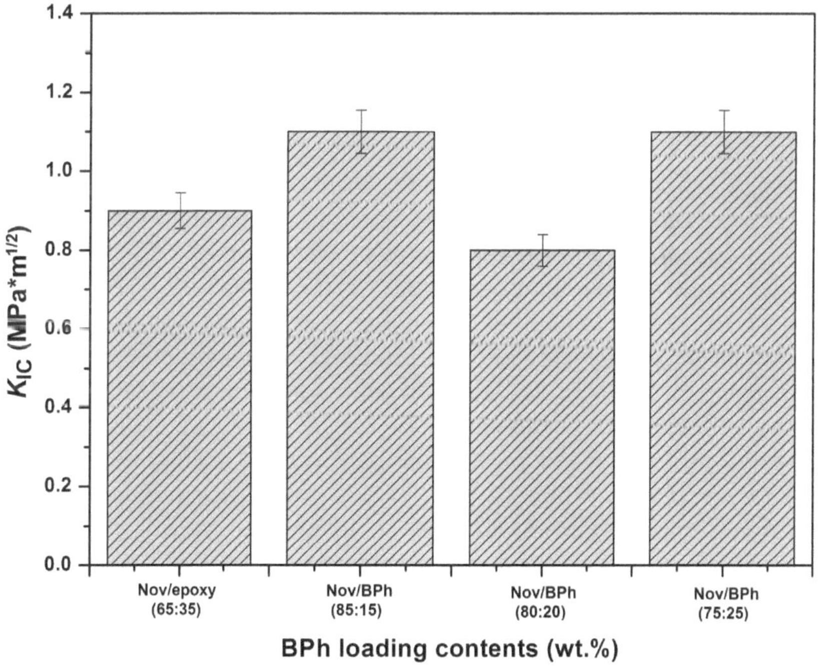

Figure 4.15 Critical stress intensity factors (K_{IC}) of the novolac/BPh and novolac/epoxy copolymers. *BPh*, biphenyl phthalonitrile. Reprinted from Sumner M, Sankarapandian M, McGrath J, Riffle J, Sorathia U. Flame retardant novolac–bisphthalonitrile structural thermosets. Polymer 2002;43:5069–76.

depicted until 480−500°C which is comparable to the thermal stability of some advanced PN resins. In all instances, the starting decomposition temperatures and char yields were similar in nitrogen and air atmospheres, implying that the oxygen was not playing any major role in the thermal degradation of these copolymers.

The flammability measurements were carried out using a cone calorimeter. Basically, radiant heat fluxes were applied on the samples surface. The surfaces were continually sparked until ignition occurred, then the samples were allowed to cool down. Meanwhile, the cone calorimeter was recording the flammability properties. The cone calorimeter obtained results and those of other polymeric systems are listed in Table 4.6. By analyzing the data, it appeared that the BPh/novolac copolymers exhibited an outstanding flame resistance which was comparable to that of other advanced flame retardant systems. For instance, for an incident heat flux of 50 kW/m^2, the peak heat release of the 20:80 BPh/novolac was measured to be 137 kW/m^2 which was in the range of resole control material with a value of 116 kW/m^2. Moreover, increasing the BPh contents from 10 to 20 wt.% reduced the ignition time by almost 40%. Thus it is important to restate that PN resins are one the few polymers that satisfy the highest standard of safety in board of airplanes, submarines, and other military vessels. The outstanding flame resistance of the PN resins, with peak heat release of less than 150 at 100 kW/m^2 of heat irradiance along with an ignition time superior to 60 s, allows the typical PN resins to satisfy the US-Navy extremely stringent MIL-STD-2031 norm [5,6]. Through the presented results, it clearly appears that the novolac was positively affected by the incorporation of the PN units and most of the limitations of the novolac resins were proved to be easily overcome with small amounts of PN.

Guo et al., also studied another kind of PN/novolac blends prepared from the typical BAPh and novolac [52]. The authors investigated the curing behavior and processability of the blends and prepolymers, and the resulting thermal and mechanical properties of their resulting copolymers were also explored. The blends were prepared by powder-mixing various weight ratios of BAPh to novolac ranging from 20 to 80 wt.%. On the other hand, the prepolymers were obtained after first melting the novolac monomers at 120−170°C, then the desired BAPh amount was added. The mixtures were vigorously stirred for 10 min before being cooled down to room temperature. Finally, the cured copolymers were obtained by curing the prepolymers following the curing schedule of 200°C for 1 h, 230°C for 3 h, 260°C for 2 h, and 290°C for 1 h.

Table 4.6 Cone Calorimeter Results Measured at an Incident Heat Flux of 50 kW/m^2 for the 10:90 and 20:80 BPh/Novolac Copolymers Compared with Other Results from Various Polymeric Systems

Samples	Time to Ignition	Total Heat Release (MJ/m^2)	Peak Heat Release Rate (kW/m^2)	Co/Co_2 (kg/kg)	Char Yield (%)
Novolac/epoxy (65:35)	74	156	263	0.02	26
Commercial resol network	–	450	116	0.011	65
Polyetherimide	70	125	128	0.08	79
Polysulfone	67	297	431	0.06	56
Fluorine and phosphorus polyether	91	87	120	5.0	19
BPh/novolac (10:90)	73	85	212	0.015	53
BPh/novolac (20:80)	102	49	137	0.019	54

Reprinted from Sumner M, Sankarapandian M, McGrath J, Riffle J, Sorathia U. Flame retardant novolac–bisphthalonitrile structural thermosets. Polymer 2002;43:5069–76.

The curing behavior of these BAPh/novolac blends was investigated by DSC in the range of 50–350°C. All the blends revealed the presence of one endothermic transition between 156 and 192°C which was ascribed to the melting of the blends. In each case, the position of the melting peak depended on the amount of the BAPh monomers, and increasing this amount resulted in higher melting point. The presence of only one endothermic transition was ascribed to the characteristic of the novolac which is a mixture of linear oligomers with no melting points but do display a softening temperature in the range of 110–180°C. Importantly, all the blends showed large processing windows up to 118°C. Through these observations, it can be confirmed that the BAPh/novolac blends are particularly appealing for industrial applications where large processing windows allow the preparation of advanced composites using the resin transfer molding (RTM) or vacuum assisted RTM.

Further investigations of the PN/novolac blends were carried out by rheological analyses. The variations of the complex viscosity were measured as function of temperature. In all instances, the viscosity was found to decrease between 175 and 190°C. All the blends also showed sharp increases in the viscosity at higher temperatures which indicated that the curing reaction was happening. Moreover, the good processability of the blends was also confirmed by the low viscosity values at moderate temperature. Subsequently, the viscosities of the BAPh/novolac prepolymers were also evaluated and their gelation times were also measured. The gelation times of the blends were found to be closely related to the amount of BAPh. In fact, increasing the BAPh contents up to 50 wt.% reduced the gelation point of the prepolymer at 200°C to 8 min. Thus, the overall processing properties of the blends and prepolymers can be easily tailored by changing the relative proportion of both phases.

The thermal stability measurements for the BAPh/novolac copolymers were performed on TGA under nitrogen and air atmospheres. The resulting thermal stability parameters, such as the starting decomposition temperatures and the char yields at 800°C are listed in Table 4.7. The copolymer of 20:80 PN/novolac showed a $T_{5\%}$ and $T_{30\%}$ of 415 and 647°C, respectively, with a char yield at 800°C of 65%. Increasing the proportion of the BAPh resulted in consequent improvements in all the thermal stability parameters. For instance, the copolymer of 80:20 BAPh/novolac showed a $T_{5\%}$ and $T_{30\%}$ of 487 and 789°C, respectively, with a char yield at 800°C of 71%. The higher rigidity and crosslinking density of the PN along with the formation of an interpenetrating

Table 4.7 Thermal Properties of the BAPh/Novolac Copolymers Cured up to 290°C

	Nitrogen			Air		
PN/Novolac (Weight Ratio)	$T_{5\%}$ (°C)	$T_{30\%}$ (°C)	Char Yield at 800°C (%)	$T_{5\%}$ (°C)	$T_{30\%}$ (°C)	Char Yield at 800°C (%)
20:80	415	647	65	477	630	1
40:60	467	789	70	480	666	23
50:50	470	790	70	484	668	27
60:40	484	791	71	497	670	29
80:20	487	789	71	497	674	35

Reprinted from Guo H, Lei Y, Zhao X, Yang X, Zhao R, Liu X. Curing behaviors and properties of novolac/bisphthalonitrile blends. J Appl Polym Sci 2012;125:649–56.

network with the novolac units are the main reasons behind such amelioration in the global thermal stability of the copolymers. Besides the proportion of each phase in the blend, other parameters such as the chosen curing schedule can influence the thermal stability of the copolymers. To clarify that, the authors added 330°C for 2 h to the previous curing program and investigated the resulting thermal stability. The obtained data revealed that all the thermal stability parameters were improved confirming that the thermal resistance of the BAPh/novolac copolymers can be tailored by several means.

The effects of the BAPh contents and the maximum curing temperature on the flexural properties of the novolac resin were also discussed. The variations of the flexural strength and modulus are shown in Fig. 4.16A and B, respectively. The BAPh/novolac copolymers exhibited flexural strengths ranging between 49 and 91 MPa and flexural modulus in the 4–6 GPa range. Although the flexural properties were dependent of the BAPh amount, the best match was depicted for the 50:50 BAPh/novolac copolymer with a flexural strength and modulus of 91.4 MPa and 5.8 GPa, respectively. Meanwhile, the impact of varying the maximum curing temperature did not show any consequent change in the flexural properties. What is interesting is that the 50:50 PN/novolac had better bending properties than neat BAPh resin. Therefore, in this case, the novolac improved some of the BAPh resin mechanical properties.

Generally, the PN/novolac blends and copolymers prepared from different precursors were proven to be high-performance materials and

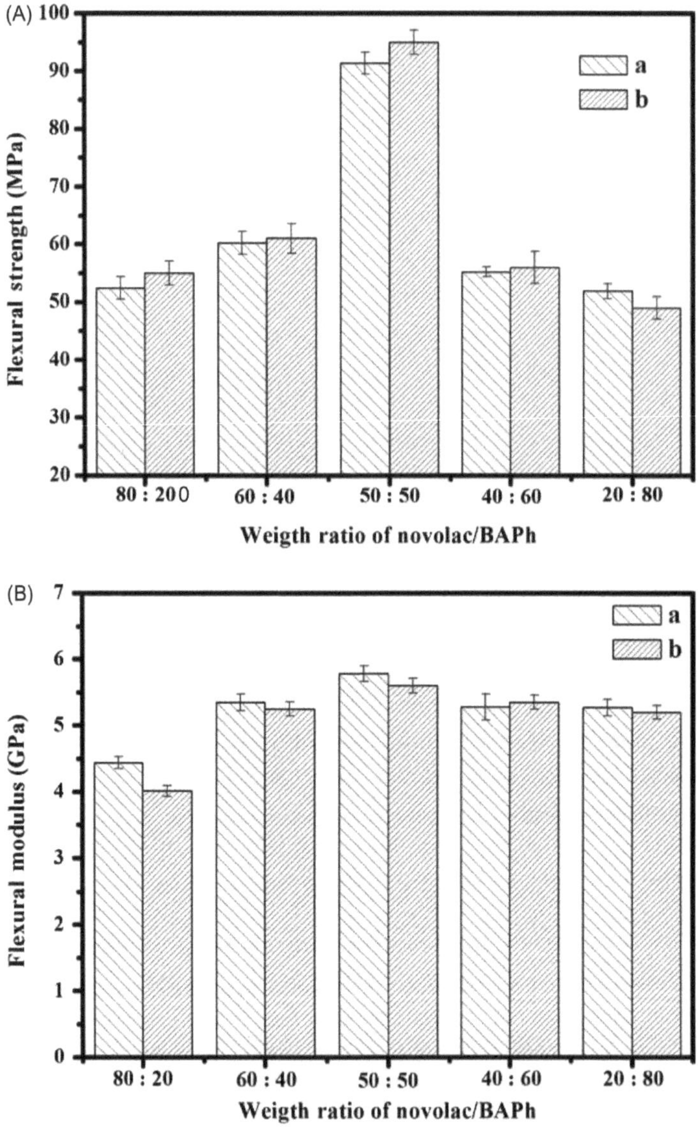

Figure 4.16 Flexural strength (A) and modulus (B) of the novolac/BAPh copolymers. *BAPh*, bisphenol A-based phthalonitrile. Reprinted from Guo H, Lei Y, Zhao X, Yang X, Zhao R, Liu X. Curing behaviors and properties of novolac/bisphthalonitrile blends. J Appl Polym Sci 2012;125:649–56.

they are expected to be used as advanced polymeric matrices. Furthermore, it should be pointed out that till the date this book was written no report was found on the use of any PN/novolac blend to prepare high-performance composites. Therefore, considering the numerous benefits of these materials, we expect a rapid increase in the number of studies in this area for the next few years.

4.2.5 Phthalonitrile/Bismaleimide

Before the emergence of the PN resins, the bismaleimide polymers dominated the field of high-performance thermosets mainly owing to their advanced properties including excellent thermomechanical properties, good fatigue resistance, high-performance to cost ratio, low moisture absorption, and superior flame resistance [53]. The bismaleimide monomers are characterized by their carbon–carbon double bonds which allow higher activity because of the electron withdrawing ability of these carbonyl groups. Hence, many possible addition reactions can be envisaged with compounds containing active hydrogens. Besides being both advanced thermosets, the typical PN and bismaleimide polymers have also many similar shortcomings related to their brittleness and curing difficulties. In fact, both types of monomers need to be activated in order to achieve the polymerization. Therefore, direct blending of the typical PN and bismaleimide monomers will not resolve any of the previous issues. As a result, the physical blending of these monomers is limited to the use of functional monomers able to provide the active source of hydrogen necessary to activate the curing process. In the open literature, the only reports concerned with the PN and bismaleimide are those prepared from benzoxazine- and allyl-functional PN with the typical *N,N′*-(4,4′-methylenediphenyl) bismaleimide. In the next two sections, these PN/bismaleimide blends and copolymers are discussed in details.

4.2.5.1 Benzoxazine-Functional Phthalonitrile/Bismaleimide

Jiang et al. also choose the BZPN monomers as a new kind of additive for the typical *N,N′*-(4,4′-methylenediphenyl) bismaleimide [54]. This choice was supported by the advanced processability and properties of these functional, PN monomers. In fact, the BZPN monomers showed a double-stage curing behavior starting with the oxazine ring-opening and the formation of active phenolic hydroxyls, which later triggered the cyano addition reaction. Thus, the authors were interested in obtaining an interpenetration network by activating the addition reactions between these hydroxyls with the nitrile and carbonyl groups belonging to the PN and bismaleimide, respectively. Thus, the blends were formulated in 40:60, 50:50, and 60:40 molar ratios of BZPN to bismaleimide. The cured copolymers were obtained after heating the blends following the curing schedule of 160°C for 2 h, 180°C for 2 h, 200°C for 2 h, 240°C for 2 h, and 280°C for 2 h.

The DSC investigations were performed on the neat BZPN and bismaleimide monomers as well as their related blends. The BZPN PN had two distinct exotherms at 223 and 260°C confirming the double-stage polymerization of these monomers. The first exotherm was related to the ring-opening of the oxazine rings and the second exotherm was ascribed to the nitrile groups activated polymerization. In a different manner, the neat bismaleimide showed a sharp endotherm at 157°C followed by an exotherm at 251°C corresponding to the thermal curing of the bismaleimide units. On the other hand, all the blends had a double-stage curing behavior which was explained as follows: first the ring-opening of the oxazine rings resulted in the formation of the phenolic hydroxyls which initiated the addition reaction of the bismaleimide monomers and the cyano addition reaction of the nitrile groups. Interestingly, it was also noticed that increasing the amount of the bismaleimide resulted in lower initial curing temperature and the peak corresponding to the nitriles polymerization shifted to higher temperatures. Therefore, to further understand the curing process, a model blend formed from the typical BAPh monomers and the previous bismaleimide was studied by DSC. By analyzing the results, it appeared that the 50:50 model blend only showed one endothermic transition and no exothermic peak was depicted even at high temperatures. Thus, it was concluded that the bismaleimide double bond cannot activate the PN cyano addition reaction. As a result, the plausible explanation for the BZPN/bismaleimide curing behavior at high amount of bismaleimide was due to the decrease in the active hydroxyls due to their consumption by the bismaleimide monomers resulting in difficulties in initiating the curing of the nitrile groups.

By taking into account the DSC results, the authors followed the structural changes accompanying the curing by FTIR. The spectra of the 50:50 BZPN/bismaleimide blend heated at 200°C at various curing times revealed that the apparition of the first hydroxyl groups was relatively quick. Indeed, after only 1 min of curing at 200°C the band at 1250 cm^{-1} related to the phenolic hydroxyls was depicted along with the characteristic bands of bismaleimide, benzoxazine, and PN. Moving forward in time until 20 min resulted in a decrease in all these characteristic absorptions, which was essentially due to the reaction between the phenolic hydroxyls and bismaleimide. After 100 min of curing, the phthalocyanine characteristic absorptions were depicted confirming that the cyano addition reaction was the last reaction in this curing process. In summary, the FTIR results corroborated those of DSC concerning the polymerization reaction steps.

The thermal and thermo-oxidative stabilities were analyzed by TGA. Under nitrogen atmosphere, the copolymers thermal stabilities were found to be improved with more bismaleimide contents, which were ascribed to the improved crosslinking degree from the reaction between the bismaleimide and the BZPN. By analyzing the degradation process, it appeared that in the range of 50–500°C the bridging methylene and methyl groups were the first to decompose followed by the deterioration of the maleamic moieties. For higher temperatures (>600°C), the main degradation mechanism was the chain scission. Meanwhile, the lower thermal stability parameters obtained in air atmosphere were attributed to the thermal oxidative scission of the Mannich bridges and the phenolic degradations in the polymer backbone.

4.2.5.2 Allyl-Functional Phthalonitrile/Bismaleimide

Similarly to the BZPN, the allyl-functional, PN monomers also exhibit a self-curing behavior with a two-stage reaction mechanism. Therefore, Zou et al., aimed to take advantage from the active hydrogens resulting from the polymerization of the allyl moieties to catalyze both the bismaleimide and PN polymerizations [55]. The allyl-functional, PN/bismaleimide prepolymer was obtained by first dissolving an equivalent molar ratio of both phases in a specific solvent. Then, the resulting mixture was heated until its boiling point and remained at the same temperature for 5 h. Afterwards, the mixture was precipitated in distilled water and dried at 100°C overnight. The proposed reaction mechanism was the early ene reaction between the allyl and imide groups as shown in Scheme 4.2. The allyl-functional, PN/bismaleimide copolymer was finally obtained by curing the prepolymer following the curing procedure of 180°C for 2 h, 220°C for 3 h, 260°C for 3 h, 300°C for 2 h, and 320°C for 3 h.

DSC analyzes were used to study the curing behavior of the neat allyl-functional, PN and bismaleimide monomers and their related prepolymer. The neat allyl-functional, PN monomers showed one endothermic transition at about 122°C ascribed to the melting transition. The neat allyl-functional, PN monomers also presented two exothermic transition attributed to the allyl polymerization and to the cyano addition reaction. The neat bismaleimide monomers had one melting peak and one exotherm corresponding to the polymerization of the imides. The

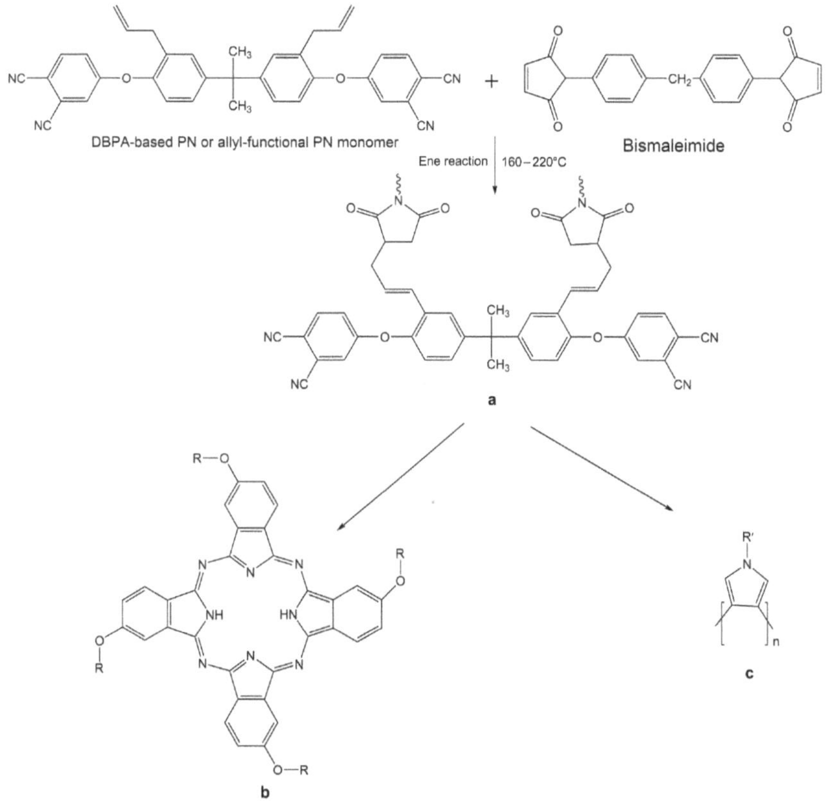

Scheme 4.2 Possible reaction pathways between the allyl-functional, PN and bismaleimide monomers. *PN*, phthalonitrile. Reprinted from Zou X, Xu M, Jia K, Liu X. Copolymerizing behavior and processability of allyl-functional bisphthalonitrile/bismaleimide system. Polym Compos 2015.

DSC scan of the allyl-functional, PN/bismaleimide prepolymer revealed the presence of two clear exothermic transitions at 263 and 332°C. The first exotherm was attributed to the ene reaction between the allyl of the PN with the imide groups of the bismaleimide as well as the self-polymerization of the allyl moieties and imide units. The second exotherm corresponded to the cyano addition reaction of the nitrile groups catalyzed by the active hydroxyls resulted from the addition reaction of the allyl groups at lower temperatures.

The FTIR spectra of the allyl-functional, PN/bismaleimide copolymers at various curing temperatures are displayed in Fig. 4.17. The absorption bands at 1710 and 680 cm^{-1} attributed to the imide groups were depicted along with the absorption band at 1175 cm^{-1} corresponding to the succinimide. Moreover, the band at 1010 cm^{-1} evidenced the

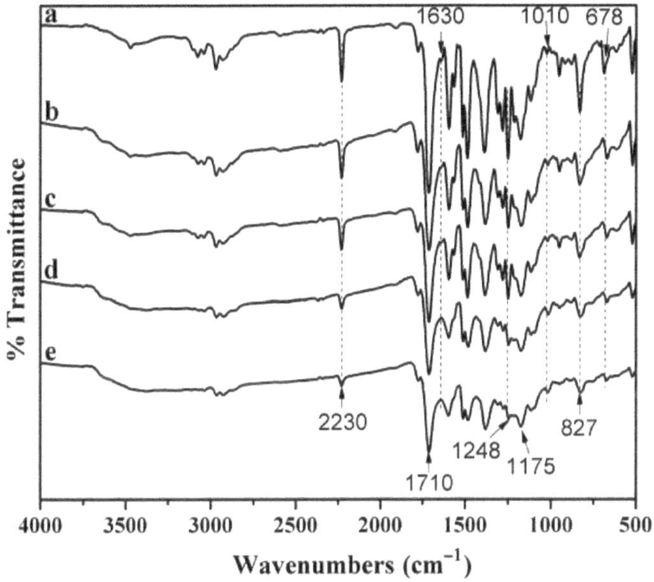

Figure 4.17 FTIR spectra of allyl-functional, PN/bismaleimide copolymers cured at: (a) 180°C, (b) 220°C, (c) 260°C, (d) 300°C, and (e) 320°C. *FTIR*, Fourier transform infrared; *PN*, phthalonitrile. Reprinted from Zou X, Xu M, Jia K, Liu X. Copolymerizing behavior and processability of allyl-functional bisphthalonitrile/bismaleimide system. Polym Compos 2015.

presence of the phthalocyanine structures. It was also found that increasing the curing temperatures resulted in the decrease in the intensity of the bands at 2230, 1630, and 827 cm^{-1} ascribed to the nitrile groups, allyl groups, and the maleimide double bond, respectively. Thus, it was once more advanced that the curing process of the copolymers involved the ene reaction between the allyl groups of the functional PN and the bismaleimide, the homopolymerization of the bismaleimide, the addition reaction of the allyl moieties, and finally the activated cyano addition reaction of the PN units.

The thermomechanical properties of the studied allyl-functional, PN/bismaleimide copolymer were investigated by DMA up to 400°C in nitrogen atmosphere. The storage modulus and tan delta variations as function of temperature are shown in Fig. 4.18. The storage modulus at 30°C was measured to be 3.8 GPa and this value decreased to 1.4 GPa at 400°C. The tan delta curve showed one large relaxation peak ranging from 80 to 170°C and another peak starting at 350°C. The first peak was attributed to the movement of oligomers or packing defects in the glassy state and the second peak confirmed that the T_g of the copolymer was above 350°C. In contrast with the neat allyl-functional PN, the

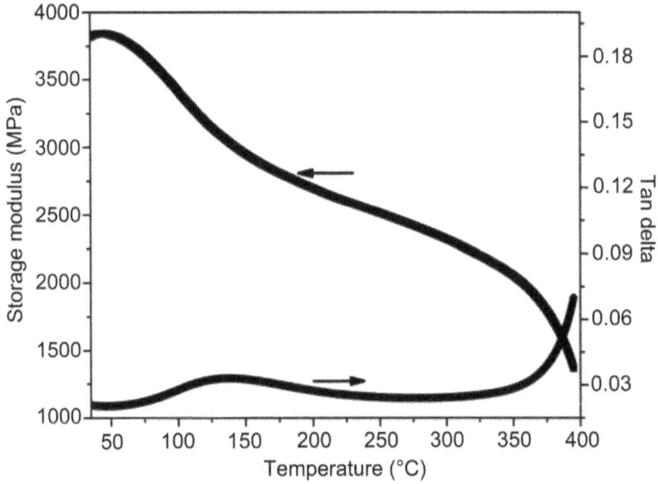

Figure 4.18 Storage modulus and loss factor (tan delta) variations for the allyl-functional, PN/bismaleimide copolymers. *PN*, phthalonitrile. Reprinted from Zou X, Xu M, Jia K, Liu X. Copolymerizing behavior and processability of allyl-functional bisphthalonitrile/bismaleimide system. Polym Compos 2015.

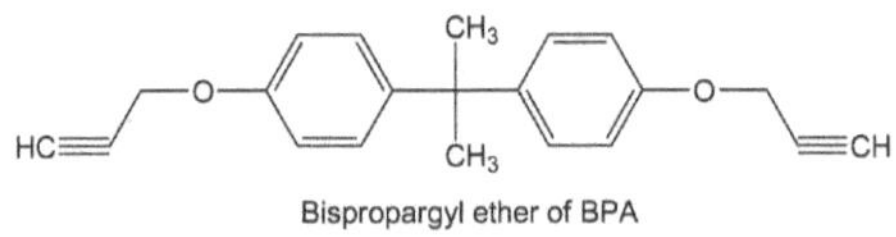

Figure 4.19 Chemical structure of the bispropargyl ether of bisphenol A.

studied copolymer thermomechanical properties was slightly lower. It should be understood that the thermomechanical properties can be tailored to a certain extent by modifying the curing procedure or postcuring the samples at high temperatures.

In all instances, the PN/benzoxazine blends and copolymers prepared from various precursors were proven to exhibit outstanding mechanical and thermal properties. Meanwhile, their improved curing behavior and processability as well as their ease of preparation make these materials particularly appealing in some particularly exigent applications.

4.2.6 Phthalonitrile/Propargyl

Augustine et al., studied the influence of the propargyl groups, from the bispropargyl ether of BPA, on the curing behavior, processability, and thermal properties of the typical BAPh resin [56]. The chemical structure of the bispropargyl ether of BPA is shown in Fig. 4.19. The propargyl monomers were chosen because of their interesting curing pattern and the intermediates involved in their curing process (Scheme 4.3). Indeed, the sigmatropic/Claisen rearrangement of the propargyl ethers results in the formation of active hydroxyl groups followed by the production of chromene structures, which can be both used to catalyze the addition reaction of the PN units. The BAPh/propargyl blends were formulated by direct mixing both monomers at various molar ratios of BAPh to propargyl (1:1, 1.5:1, 3:1, and 5.7:1) in the presence of small quantity of solvent. The solvent was then removed and the blends were cured in an inert atmosphere following the curing schedule of 200°C for 2 h, 250°C for 3 h, 280°C for 4 h, and 330°C for 5 h.

Investigations on the curing behavior and possibilities of coreaction between the BAPh and propargyl were carried out by DSC and the obtained thermograms for the neat monomers and their blends are displayed in Fig. 4.20A and B, respectively. From Fig. 4.20A, it was clear that the neat BAPh monomers only exhibited one endothermic transition at 195°C related to their melting points and no exothermic transition was seen. Such behavior is commonly observed for the typical PN monomers and confirms their difficult polymerization process. On the other hand, the propargyl monomers showed two transitions; the first one was ascribed to the melting at 80°C, and the second one peaking at 280°C was attributed to the homopolymerization of the propargyl to afford polychromene structures. The DSC curves of the blends of 1.5:1 and 3:1 BAPh/propargyl, shown in Fig. 4.20B, revealed to the presence of two distinct endothermic transitions related to the melting of each phase, and two overlapped exothermic transitions peaking at about 280 and 310°C. The fist exotherm, which position was similar to that of the neat propargyl monomers, was once more ascribed to the Claisen rearrangement or self-polymerization of the propargyl units. Subsequently, the second exotherm corresponded to the cyano addition reaction of the PN activated by the chromene units. Interestingly, a discontinuity was also depicted at around 290°C and was assigned to the cyano addition reaction triggered, not by the chromene units, but by the active hydroxyl groups preceding the formation of the chromene. The limited effect of these hydroxyls on the nitriles polymerization is mainly due to their prior participation in the formation of the chromene units. Hence, the previous results clearly confirmed the catalytic nature of the cure reaction prevailing in the BAPh/propargyl blends.

The DSC results were also validated by rheometric analyzes of the blends. Globally, the copolymerization of the blends is more difficult to achieve than the homopolymerization of the propargyl monomers

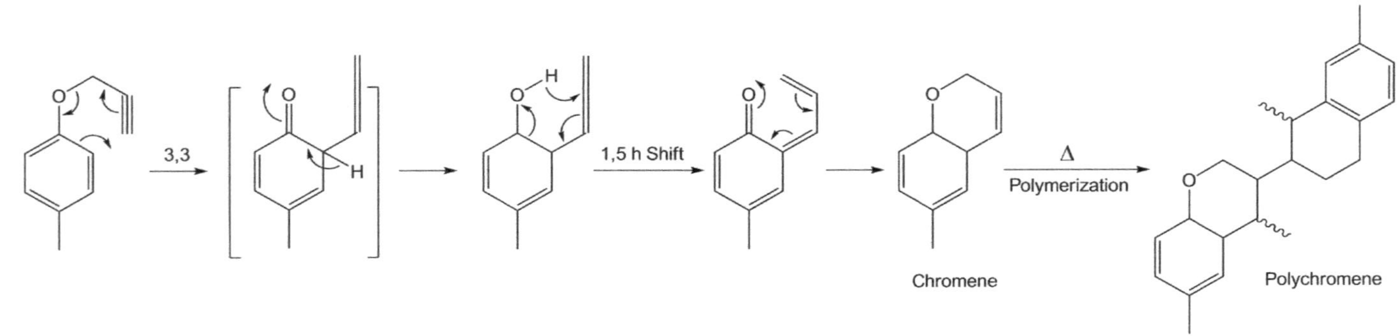

Scheme 4.3 Polymerization mechanism of propargyl ether functions. Reprinted from Augustine D, Mathew D, Nair CR. Mechanistic and kinetic aspects of the curing of phthalonitrile monomers in the presence of propargyl groups. Polymer 2015;60:308–17.

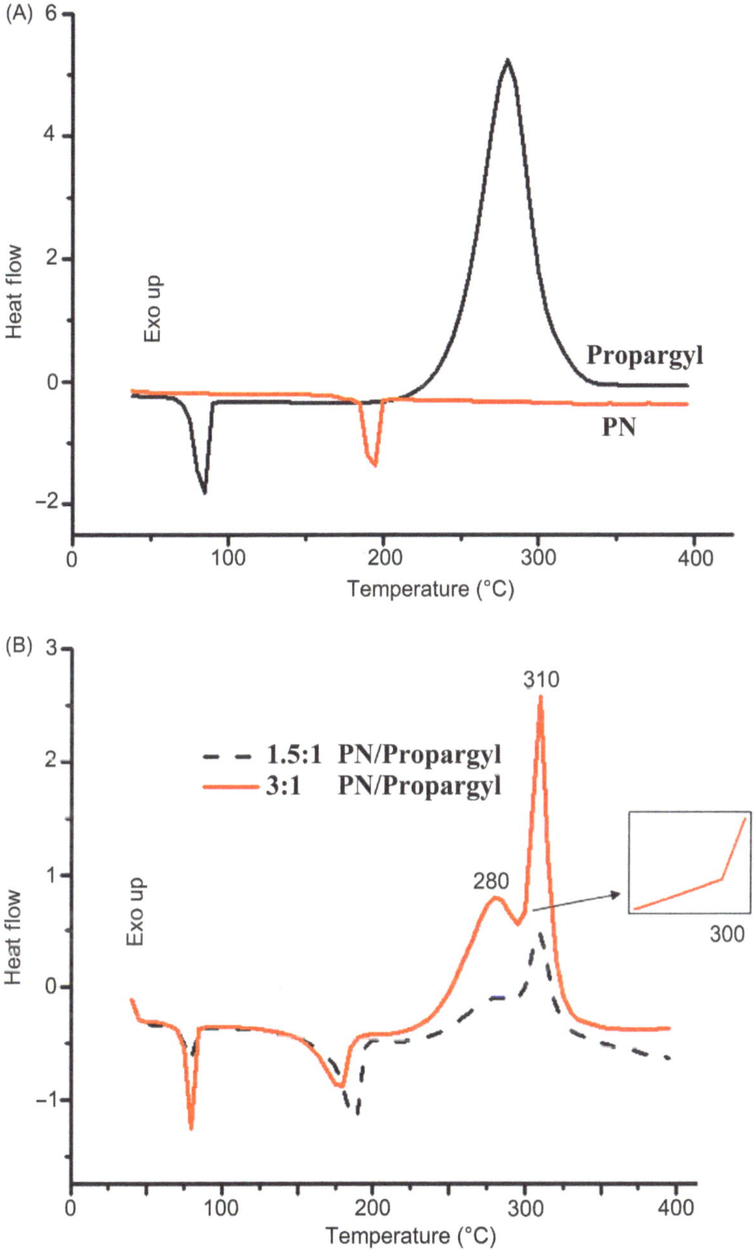

Figure 4.20 DSC curves of (A) the neat BAPh and bispropargyl monomers and (B) their respective blends of various proportions. *DSC*, dynamic scanning calorimetry; *BAPh*, bisphenol A-based phthalonitrile. Reprinted from Augustine D, Mathew D, Nair CR. Mechanistic and kinetic aspects of the curing of phthalonitrile monomers in the presence of propargyl groups. Polymer 2015;60:308–17.

especially at high PN contents. In fact, the FTIR results revealed that after curing, the majority of the nitriles remained unreacted. It should be understood that the cyano addition reaction of the PN monomers depends on the hydroxyls and chromene units provided during the homopolymerization of the propargyl monomers. Thus, fewer amount of the propargyl in the blends resulted in difficulties to achieve the initiation of the nitriles polymerization. Meanwhile, increasing the amount of the propargyl in the blends ensued lower ultimate curing temperature. Finally, taking into account all the reaction possibilities, the BAPh cure reaction in the presence of propargyl units can be resumed as seen in Scheme 4.4.

The thermal resistances of the neat BAPh and propargyl polymers as well as their subsequent copolymers were investigated by TGA and the obtained curves are shown in Fig. 4.21. All the BAPh/propargyl copolymers were thermally stable up to 400°C and the peak decomposition temperatures were depicted around 450°C. Although the cured networks were composed of homopolymers and copolymers, their thermal stabilities were still in the range of highly thermally stable polymers. The neat propargyl polymer exhibited the highest char yield at 900°C of 58% and the copolymers presented values slightly lower. Conversely to what we have seen in all the blends prepared from PN materials, increasing the proportion of the PN in the blends beyond 35% negatively affected the char yield. The reason behind this observation is once more related to the activators of the cyano addition reaction: with less propargyl contents, the hydroxyl and chromene groups are not enough to initiate the cure of the PN and, as a result, the cured network contains partially cured PN units with lower thermal stability.

The use of propargyl as a new kind of curing activator for the PN monomers presented some new insights. However, the priority of the propargyl polymerization limited, to a certain extent, the cyano addition reaction and the resulting thermal properties seemed to be negatively affected. Hence, further research needs, to be performed on the mechanical and thermomechanical properties of these copolymers to effectively judge about their performances.

4.2.7 Phthalonitrile/Poly(Arylene Ether Nitrile)

In the previous sections, we discussed the blends and copolymers prepared from PN and other thermosets. Generally, these materials were shown to exhibit ameliorated curing behavior, processability, and thermomechanical properties; however, the brittleness remained one of the

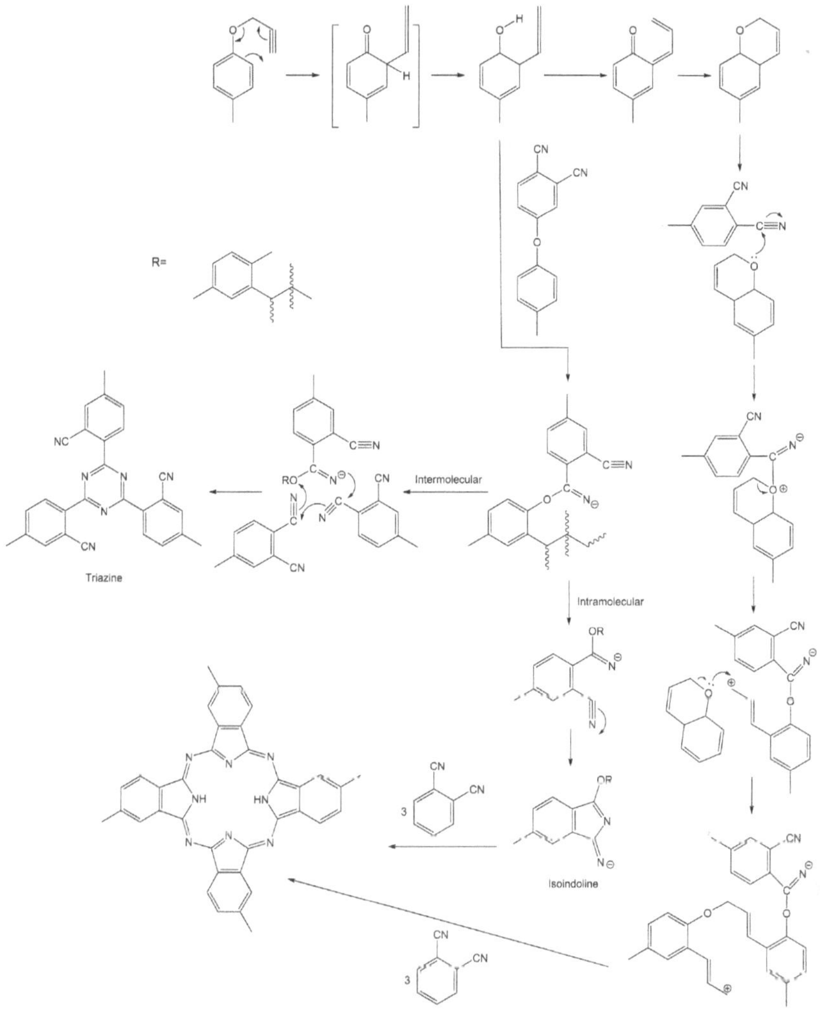

Scheme 4.4 PN cure reaction pathways in the presence of propargyl functions. *PN*, phthalonitrile. Reprinted from Augustine D, Mathew D, Nair CR. Mechanistic and kinetic aspects of the curing of phthalonitrile monomers in the presence of propargyl groups. Polymer 2015;60:308–17.

biggest shortcomings of these PN-based materials. To overcome this limitation, some researchers focused their attention on preparing interpenetrating polymer networks from PN and thermoplastics, such as the poly(arylene ether nitrile) (PEN) polymers. PEN polymers are a kind of high-performance, engineering plastics with a large panel of interesting features, such as excellent mechanical strength and toughness, good flame and radiation resistance, and high thermo-oxidative stability [57].

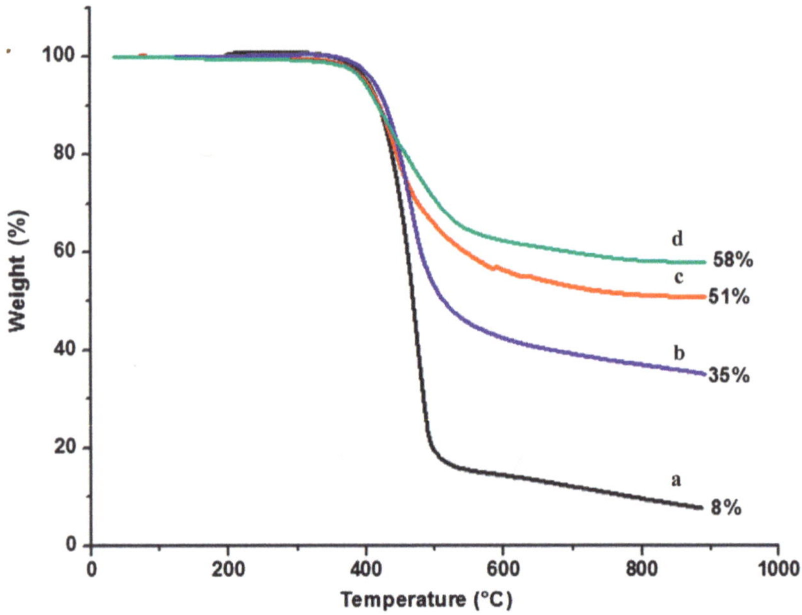

Figure 4.21 TGA curves under nitrogen atmosphere of the (a) neat BAPh polymer, (b) copolymer of 1:5.7 molar ratio of bispropargyl and BAPh, (c) copolymer of 1:1 molar ratio of bispropargyl and BAPh, and (d) neat bispropargyl polymer. *TGA*, thermogravimetric analysis; *BAPh*, bisphenol A-based phthalonitrile. Reprinted from Augustine D, Mathew D, Nair CR. Mechanistic and kinetic aspects of the curing of phthalonitrile monomers in the presence of propargyl groups. Polymer 2015;60:308–17.

Thus, it was believed that in case of successful preparation on an interpenetrating network between the PN and PEN, the resulted materials would show improved mechanical properties, including a highly ameliorated toughness than that of the neat PN resins.

Lei et al., reported about a thermoset/thermoplastic blend prepared from the well-known BPh and phenolphthalein-based PEN [57]. The resulting materials were mechanically and thermally characterized in terms of their relative amounts of PEN which ranged between 5 and 15 wt.%. The BPh/PEN blends were prepared by melting a specific amount of BPh monomers at 250°C followed by the addition of the desired PEN amount. The resulting mixtures were stirred for 10 min prior to the addition of the 2 wt.% of DDS curing agent. The BPh/PEN copolymers were afforded by curing the mixtures at 250°C for 4 h, 280°C for 4 h, 300°C for 4 h, 350°C for 6 h, and 375°C for 2 h.

The FTIR spectra of the copolymers revealed that the nitrile absorption, depicted at 2230 cm^{-1}, was markedly reduced in intensity after curing. At the same time, new absorption bands emerged at 1360, 1520, and 1010 cm^{-1} confirming the formation of phthalocyanine and triazine structures. The FTIR results also indicated that the PN polymerization effectively happened and the PEN degree of crosslinking was increased. Although the authors did not fully study the curing mechanism and processability of the blends, they expected that the cyano groups present on the PEN participated in the formation of the desired interpenetrating network.

The effect of the PEN on the thermal stability of the BPh polymer was studied by TGA and the resulting thermal stability parameters are summarized in Table 4.8. According to the chosen curing schedule, the BPh polymer exhibited the highest thermal resistance with $T_{5\%}$ and $T_{10\%}$ of 543 and 595°C, respectively, and a char yield at 800°C of 79%. Such an outstanding thermal performance is what gives the typical PN resins their superiority on the other kinds of high-performance thermosets. As expected, the introduction of the PEN thermoplastic resulted in a gradual decrease in the thermal stability parameters. For instance, at 15 wt.% of PEN, the resulting copolymer had a $T_{5\%}$ and $T_{10\%}$ of 476 and 518°C, respectively, and a char yield at 800°C of 63%. The reason behind these decreases is obvious since the soft polymeric chains of the PEN reduce the thermal stability of the PN polymer.

The flexural properties of the neat resins and their related copolymers were measured and the obtained results are shown in Fig. 4.22. The cured PEN showed the best match in term of toughness with a flexural strain of 9.1%. On the other hand, the brittle BPh polymer exhibited a flexural strength and strain of only 83 MPa and 2.1%, respectively.

Table 4.8 Thermal Properties of the BPh/PEN Copolymers Under Nitrogen Atmosphere

PN/PEN (Weight Ratio)	$T_{5\%}$ (°C)	$T_{10\%}$ (°C)	Char Yield at 800°C (%)
100:0	543	595	79
95:05	539	586	78
90:10	514	557	67
85:15	476	518	63
0:100	452	483	57

Reprinted from Lei Y, Zhao R, Yang X, Liu X. Mechanical and thermal properties of graphite nanoplatelets reinforced polyarylene ether nitriles/bisphthalonitrile IPN system. J Appl Polym Sci 2013;127:3595–600.

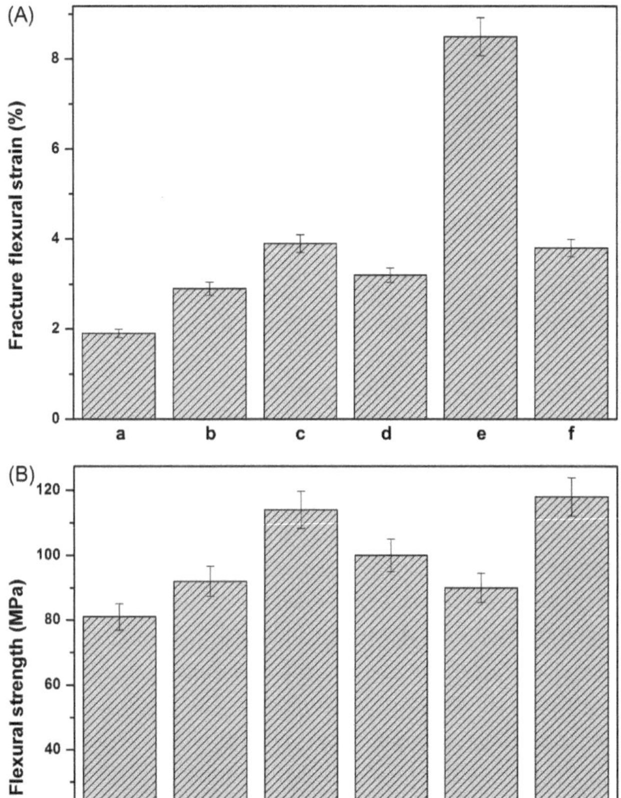

Figure 4.22 Flexural strain (A), strength (B), and modulus (C) of the (a) pure BPh, (b) BPh with 5 wt.% of PEN, (c) BPh with 10 wt.% of PEN, (d) BPh with 15 wt.% of PEN, (e) pure PEN, and (f) BPh with 10 wt.% of graphite nanoplatelets. *BPh*, biphenyl phthalonitrile; *PEN*, poly(arylene ether nitrile). Reprinted from Lei Y, Zhao R, Yang X, Liu X. Mechanical and thermal properties of graphite nanoplatelets reinforced polyarylene ether nitriles/bisphthalonitrile IPN system. J Appl Polym Sci 2013;127:3595–600.

Moreover, by increasing the PEN contents, both flexural strength and strain were improved to a certain extent. In fact, at 10 wt.% of PEN contents, these two properties reached their maximum values of 108 MPa and 4.1%, respectively. The positive impact of the PEN on the BPh polymer was attributed to the formation of a highly interpenetrating polymer network. Thus, it was assumed that the rigid frame of the PN network was surrounded by soft and flexible PEN chains which acted as a buffer zone for ambient pressure. Conversely, the flexural modulus of the neat BPh resin was reduced by adding more PEN contents. For example, the neat BPh polymer had a flexural modulus of 3.6 GPa and this value was reduced up to 3.2 GPa for the copolymer of 10 wt.% PEN. It was also found that when the PEN amount reached 15 wt.%, the phase segregation problems and low interpenetrating degree between the two phases negatively affected all the flexural properties.

Overall, blending PN monomers with thermoplastics, like the PEN, seems to be a cost-effective way to improve the mechanical properties of the typical PN resins with small sacrifices on the thermal resistance. Beside the work of Lei et al., Yang et al., prepared functional, PN oligomers containing biphenyl ether nitrile moieties and studied their blends with the BPh monomers [58]. Briefly, it was found that these prepared PEN-functional, PN monomers could improve the processability of the BPh monomers without scarifying the outstanding thermal stability of the polymer. Zhao et al., also studied the influence of the PEN on the curing behavior and properties of the PN/epoxy blends [59]. The curing behavior study of the ternary blends highlighted the fact that more PEN contents resulted in lower curing temperatures. The prepared ternary blends were also used as matrices for glass fiber-reinforced composites. A detailed discussion about the obtained results is provided in Chapter 5, Fiber-Reinforced Phthalonitrile Composites concerned with the fiber-reinforced PN composites.

4.2.8 Phthalonitrile/Poly(Ether Ether Ketone)

As another high-performance, engineering thermoplastic, the poly (ether ether ketone) (PEEK) can be used to improve the mechanical properties of the PN resins through the formation of an effective interpenetrating network. The challenge for such combinations is always about compromises. For instance, the improvements in the toughness are traded for less thermal stability. Thus, the choice of the blends precursors is always done by keeping in objective the targeted applications with a minimum decrease in the exceptional thermal properties of the PN resins.

Following this point of view, Augustine et al. studied the effect of two types of PEEK on the curing behavior and thermal properties of a novolac-functional PN [60]. The chosen PEEKs were the hydroxyl-terminated PEEK (PEEKOH) and the PN-terminated PEEK (PEEKPN). The cured novolac-functional, PN polymers possess an excellent thermal stability along with good mechanical properties and high flame resistance. However, the curing of the novolac-functional, PN monomers is similar to other PN monomers/oligomers, sluggish and requires high temperatures and extended time, while the toughness of its cured polymers is also limited. Thus, the PEEKOH were expected to improve the curing behavior of the monomers and the PEEKPN were supposed to improve the integration of the soft PEEK moieties in the PN network for toughness improvements. The PN blends with PEEKOH or PEEKPN were prepared by mixing the two phases in various weight ratios in tetrahydrofuran solvent. To the mixtures was also added 1 wt.% of DDS curing agent. After the removal of the solvent, the blends were cured according to the following schedule: 120°C for 30 min, 150 and 200°C for 1 h each, 250°C for 2 h, 300°C for 3 h, and 330°C for 4 h.

The curing reaction between the PN/PEEK blends in the presence of DDS was studied by DSC. Prior to the DSC investigations, the FTIR analyses revealed the presence of multiple reaction pathways resulting in isoindoline, phthalocyanine, and triazine structures. The presence of all these structures was due to the amine and phenol-mediated cyano addition reaction of the PN units. As seen in Chapter 1, Introduction to the Phthalonitrile Resins and Their Curing Behavior, the type of the curing agent impacts on the reaction pathway and final polymeric structures. The DSC curves of the PN/PEEKOH blends were composed of one prominent exotherm peaking at 250°C followed by a shallow peak at about 350°C. The first exotherm was ascribed to the amino and phenol-activated cyano addition reaction, while the second exotherm was attributed to the self-polymerization of the PN units at high temperatures. Moreover, it appeared that increasing the amount of the PEEKOH in the blends up to 30 wt.% resulted in 51% decrease in the heat of polymerization. Indeed, with more PEEKOH contents, the steric hindrance rendered the access to the crosslinking sites difficult to achieve, resulting in difficulties in initiating the polymerization reaction. For the PN/PEEKPN blend, the reaction was only triggered by the DDS, as a result, the first exothermic peak was smaller and the second one was more pronounced. Overall, the PN/PEEKOH blends showed a better curing behavior.

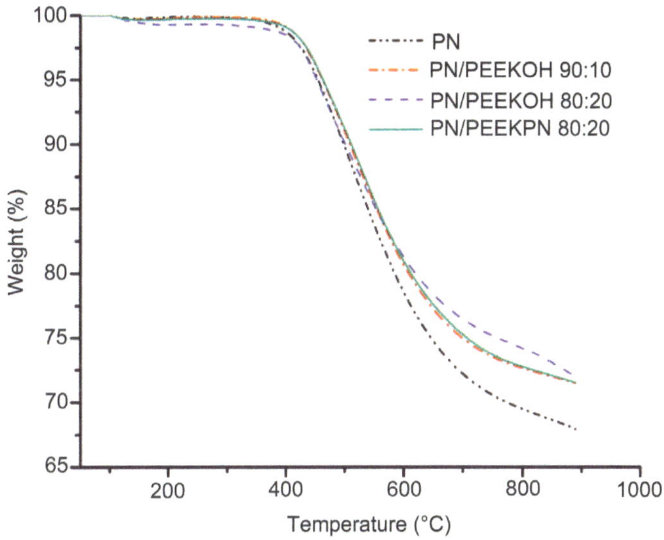

Figure 4.23 TGA curves under nitrogen atmosphere for the neat novolac-functional, PN polymer and its related copolymers with PEEKOH and PEEKPN. *TGA*, thermogravimetric analysis; PN, phthalonitrile; *PEEKOH*, hydroxyl-terminated poly(ether ether ketone); *PEEKPN*, PN-terminated poly(ether ether ketone). Reprinted from Augustine D, Mathew D, Nair CR. End-functionalized thermoplastic-toughned phthalonitrile composites: influence on cure reaction and mechanical and thermal properties. Polym Int 2015;64:146–53.

The thermal stability of the copolymers was excellent as seen from the TGA curves presented in Fig. 4.23. All the tested samples were thermally stable up to 400°C. The thermal stability parameters including the $T_{5\%}$ and char yield at 900°C were slightly improved with more PEEKOH contents. Thus, the first goal of this combination was clearly achieved since the introduction of the thermoplastic ameliorated the curing behavior and did not decrease the original thermal stability of the neat PN resin. To further judge about the toughening effect, the authors prepared carbon fabric reinforced composites and studied their mechanical properties. A proper discussion about those findings can be found in Chapter 5, Fiber-Reinforced Phthalonitrile Composites related to the PN-based fiber-reinforced composites.

4.2.9 Phthalonitrile/Carboxyl-Terminated Butadiene Acrylonitrile Rubber

Similarly to the thermoplastics, rubbers can also be used as toughening agents for the PN resins. The objective is also to prepare an interpenetrating network with a high resistance to the crack growth and

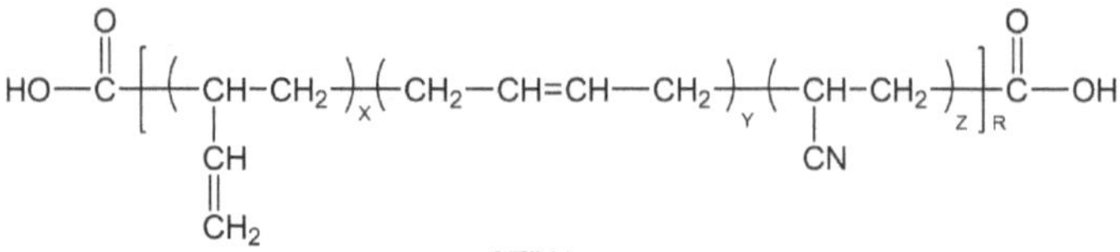

CTBN

Figure 4.24 Chemical structure of the CTBN rubber. *CTBN*, carboxyl-terminated butadiene acrylonitrile.

propagation. The only report about the use of rubber as curing catalyzed as well as toughening agent was reported by Pan et al. The authors studied the effect of various amounts of carboxyl-terminated butadiene acrylonitrile (CTBN) rubber (Fig. 4.24) on the curing behavior and processability of a BAPh resin [61]. The investigations also included the use of the BAPh/CTBN blends as advanced matrices for glass fiber-reinforced composites. The blends were formulated as follows: to a melt of BAPh monomers was consecutively added 2 wt.% DDS curing agent and 10, 20, 30, and 40 wt.% weight ratios of CTBN. The mixtures were then mechanically stirred for 30 min at 200°C to finally afford the BAPh/CTBN prepolymers.

The ability of the active hydroxyl groups of CTBN to catalyze the cure reaction of the PN units was evidenced by DSC. In fact, all the blends showed a prominent curing exotherm peaking at around 360°C. Moreover, increasing the CTBN amount resulted in increases in the heat of polymerizations. Meanwhile, the curing exotherms slightly shifted to lower temperatures. Thus it was proven that the CTBN could effectively catalyze the cyano addition reaction of PN monomers and since the hydroxyl groups were also attached to the CTBN units, the resulting copolymers were mainly composed of interpenetrating polymers chains.

The complex viscosity changes of the BAPh/CTBN prepolymer with 20 wt.% of CTBN as function of time revealed that the increase in the viscosity was slower at 180°C and extremely faster at 220°C. As a result, the initial curing temperature was set to 200°C as the most appropriate temperature in terms of viscosity rate increases. In close range with the DSC results, the BAPh/CTBN prepolymers were cured according the curing schedule of 200°C for 2 h, 250°C for 2 h, and 300°C for 3 h. Further studying the complex viscosity changes as a function of temperature showed that the melt viscosity increased with more CTBN contents. In fact, the lowest melt viscosity of about 6 Pa s was depicted for the prepolymer of 10 wt.% CTBN. Moreover, in the temperature range of 160–195°C, all the prepolymers exhibited lower melt viscosity. Overall, it was confirmed that the curing behavior and processability of the PN can be easily tailored by varying the amount of CTBN and curing temperatures.

4.3 Conclusions

The outstanding thermal and thermomechanical performances of the PN resins were proven to be easily shared with many other polymers. Meanwhile, the curing behavior and processability of the PN monomers/oligomers can be improved by a judicious choice of the blends precursors. Moreover, some PN resins mechanical properties, including the toughness, can also be simply enhanced by blending with thermoplastics or rubbers. Considering the numerous possible blend compositions, the resulting materials properties can be tailored to meet the requirement of many exigent industries. This chapter also reveals that many other PN blends and copolymers are still yet to be investigated.

References

[1] Ting RY, Keller TM, Price TR, Poranski JCF. Characterization of the cure of diether-linked phthalonitrile resins. ACS Symp Ser 1982;195:337–50.
[2] Achar BN, Fohlen GM, Parker JA. Synthesis and characterization of the novel type of polymerizable bisphthalonitrile monomers. J Polym Sci A: Polym Chem 1986;24:1997–2010.
[3] Keller TM. Phthalonitrile-based high temperature resin. J Polym Sci A: Polym Chem 1988;26:3199–212.
[4] Warzel M, Keller TM. Tensile and fracture properties of a phthalonitrile polymer. Polymer 1993;34:663–6.
[5] Sastri SB, Armistead JP, Keller TM. Phthalonitrile–carbon fiber composites. Polym Compos 1996;17:816–22.
[6] Sastri SB, Armistead JP, Keller TM. Phthalonitrile–glass fabric composites. Polym Compos 1997;18:48–54.
[7] Sastri SB, Keller TM. Phthalonitrile polymers: cure behavior and properties. J Polym Sci A: Polym Chem 1998;37:2105–11.
[8] Keller TM, Dominguez DD. High temperature resorcinol-based phthalonitrile polymer. Polymer 2005;46:4614–18.
[9] Dominguez DD, Jones HN, Keller TM. The effect of curing additive on the mechanical properties of phthalonitrile–carbon fiber composites. Polym Compos 2004;25:554–61.

[10] Derradji M, Ramdani N, Gong LD, Wang J, Xu XD, Lin ZW, et al. Mechanical, thermal, and UV-shielding behavior of silane surface modified ZnO-reinforced phthalonitrile nanocomposites. Polym Adv Technol 2016;27:882–6.

[11] Derradji M, Ramdani N, Zhang T, Wang J, Gong LD, Xu XD, et al. Effect of silane surface modified titania nanoparticles on the thermal, mechanical, and corrosion protective properties of a bisphenol-A based phthalonitrile resin. Prog Org Coat 2016;90:34–43.

[12] Keller T.M., Griffith J.R. Polyphthalocyanine resins. United States Patent US 4234712. 1980 Nov 18.

[13] Keller T.M., Griffith J.R. Polyphthalocyanine resins. United States Patent US 4304896. 1981 Dec 8.

[14] Keller T.M., Griffith J.R. Aliphatic phenoxy polyphthalocyanine. United States Patent US 4223123. 1980 Sep 16.

[15] Keller T.M. Phenolic-cured phthalonitrile resins. United States Patent US 4410676. 1983 Oct 18.

[16] Chen Z, Guo H, Tang H, Yang X, Xu M, Liu X. Preparation and properties of bisphenol A-based bis-phthalonitrile composite laminates. J Appl Polym Sci 2013;129:2621–8.

[17] Xu M, Jia K, Liu X. Effect of bisphenol-A on the structures and properties of phthalonitrile-based resin containing benzoxazine. Exp Polym Lett 2015;9:567–81.

[18] Keller T.M. Phthalonitrile resin from diphthalonitrile monomer and amine. United States Patent US 4408035. 1983 Oct 4.

[19] Keller T.M. Synthesis and polymerization of phthalonitrile monomers containing multiple phenoxy and sulfone linkages. United States Patent US 4409382. 1983 Oct 11.

[20] Keller TM. Phthalonitrile – based conductive polymer. J Polym Sci A: Polym Chem 1987;25:2569–76.

[21] Burchilo P.J., Keller T.M. Curing phthalonitrile resins with acid and amine. United States Patent US 5237045. 1993 Aug 17.

[22] Keller T.M. Curing phthalonitriles with acid. United States Patent US 5247060. 1993 Sep 21.

[23] Keller TM. Imide-containing phthalonitrile resin. Polymer 1993;34:952–5.

[24] Zhang J, Liu X, Wen H, Xie M, Cai X. Investigation of the properties of phthalocyanine resin containing bismaleimide groups. Polym Int 1997;42:363–6.

[25] Zuo F, Liu X. Synthesis and curing behavior of a novel benzoxazine-based bisphthalonitrile monomer. J Appl Polym Sci 2010;117:1469–75.

[26] Yang Y, Min Z, Yi L. A novel addition curable novolac bearing phthalonitrile groups: Synthesis, characterization and thermal properties. Polym Bull 2007;59:185–94.

[27] Sumner MJ, Weyers RY, Rosario AC, Riffle JS, Sorathia U. Synthesis and characterization of vinyl ester networks containing phthalonitrile moieties. Polymer 2004;45:5199–206.

[28] Wu D, Zhao Y, Zeng K, Yang G. A novel benzimidazole-containing phthalonitrile monomer with unique polymerization behavior. J Polym Sci A: Polym Chem 2012;50:4977–82.
[29] Dominguez DD, Keller TM. Properties of phthalonitrile monomer blends and thermosetting phthalonitrile copolymers. Polymer 2007;48:91–7.
[30] Keller T.M. Epoxy phthalonitrile polymers. United States Patent US 4619986. 1986 Oct 28.
[31] Lee H., Neville K. Handbook of epoxy resins; 1967, New York: McGraw-Hill.
[32] May C. Epoxy resins: chemistry and technology. New York: CRC Press; 1987.
[33] Keller T.M. High temperature epoxy-phthalonitrile blends. United States Patent US 5939508. 1999 Aug 17.
[34] Dominguez DD, Keller TM. Phthalonitrile–epoxy blends: cure behavior and copolymer properties. J Appl Polym Sci 2008;110:2504–15.
[35] Zhao X, Lei Y, Zhao R, Zhong J, Liu X. Preparation and properties of halogen-free flame-retarded phthalonitrile–epoxy blends. J Appl Polym Sci 2012;123:3580–6.
[36] Zeng K, Zhou K, Zhou S, Hong H, Zhou H, Wang Y, et al. Studies on self-promoted cure behaviors of hydroxy-containing phthalonitrile model compounds. Eur Polym J 2009;45:1328–35.
[37] Amir B, Zhou H, Liu F, Aurangzeb H. Synthesis and characterization of self-catalyzed imide-containing pthalonitrile resins. J Polym Sci A: Polym Chem 2010;48:5916–20.
[38] Guo H, Chen Z, Zhang J, Yang X, Zhao R, Liu X. Self-promoted curing phthalonitrile with high glass transition temperature for advanced composites. J Polym Res 2012;19;9918–8.
[39] Zhang T, Wang J, Derradji M, Ramdani N, Wang H, Lin ZW, et al. Synthesis, curing kinetics and thermal properties of a novel self-promoted fluorene-based bisphthalonitrile monomer. Thermochim Acta 2015;602:22–9.
[40] Guo H, Zou Y, Chen Z, Zhang J, Zhan Y, Yang J, et al. Effects of self-promoted curing behaviors on properties of phthalonitrile/epoxy copolymer. High Perform Polym 2012;24:571–9.
[41] Kaliavaradhan K, Muthusamy S. Studies on novel tri-phthalonitrile phenyl polyhedral oligomeric silsesquioxane and the phthalonitrile–epoxy blends. High Perform Polym 2016;28:296–308.
[42] Ishida H, Rimdusit S. Very high thermal conductivity obtained by boron nitride-filled polybenzoxazine. Thermochim Acta 1998;320:177–86.
[43] Ishida H, Agag T. Handbook of benzoxazine resins. Amsterdam: Elsevier; 2011.
[44] Zhang T, Wang J, Feng T, Wang H, Ramdani N, Derradji M, et al. A novel high performance oxazine derivative: design of tetrafunctional

monomer, step-wise ring-opening polymerization, improved thermal property and broadened processing window. RSC Adv 2015;5:33623–31.
[45] Ramdani N, Derradji M, Feng T-t, Tong Z, Wang J, Mokhnache E-O, et al. Preparation and characterization of thermally-conductive silane-treated silicon nitride filled polybenzoxazine nanocomposites. Mater Lett 2015;155:34–7.
[46] Rimdusit S, Jubsilp C, Tiptipakorn S. Alloys and composites of polybenzoxazines. Singapore: Springer; 2013.
[47] Brunovska Z, Ishida H. Thermal study on the copolymers of phthalonitrile and phenylnitrile-functional benzoxazines. J Appl Polym Sci 1999;73:2937–49.
[48] Xu M, Liu M, Dong S, Liu X. Design of low temperature self-cured phthalonitrile-based polymers for advanced glass fiber composite laminates. J Mater Sci 2013;48:8108–16.
[49] Zou X, Xu M, Jia K, Liu X. Synthesis, polymerization, and properties of the allyl-functional phthalonitrile. J Appl Polym Sci 2014;131:2937–49.
[50] Zou X, Yang X, Xu M, Jia K, Liu X. Curing behaviors and properties of allyl-and benzoxazine-functional phthalonitrile with improved processability. J Polym Res 2016;23:2.
[51] Sumner M, Sankarapandian M, McGrath J, Riffle J, Sorathia U. Flame retardant novolac–bisphthalonitrile structural thermosets. Polymer 2002;43:5069–76.
[52] Guo H, Lei Y, Zhao X, Yang X, Zhao R, Liu X. Curing behaviors and properties of novolac/bisphthalonitrile blends. J Appl Polym Sci 2012;125:649–56.
[53] Kumar KS, Nair CR, Sadhana R, Ninan K. Benzoxazine–bismaleimide blends: curing and thermal properties. Eur Polym J 2007;43:5084–96.
[54] Jiang M, Xu M, Jia K, Liu X. Copolymerization of self-catalyzed phthalonitrile with bismaleimide toward high-temperature-resistant polymers with improved processability. High Perform Polym 2016;28:895–907.
[55] Zou X, Xu M, Jia K, Liu X. Copolymerizing behavior and processability of allyl-functional bisphthalonitrile/bismaleimide system. Polym Compos 2017;38:1591–9.
[56] Augustine D, Mathew D, Nair CR. Mechanistic and kinetic aspects of the curing of phthalonitrile monomers in the presence of propargyl groups. Polymer 2015;60:308–17.
[57] Lei Y, Zhao R, Yang X, Liu X. Mechanical and thermal properties of graphite nanoplatelets reinforced polyarylene ether nitriles/bisphthalonitrile IPN system. J Appl Polym Sci 2013;127:3595–600.
[58] Yang X, Lei Y, Zhong J, Zhao R, Liu X. Preparation and thermal properties of novel phthalonitrile oligomer containing biphenyl ethernitrile/bisphthalonitrile blends. J Appl Polym Sci 2011;119:882–7.
[59] Zhao X, Guo H, Lei Y, Zhao R, Zhong J, Liu X. Effect of polyarylene ether nitriles on processing and mechanical behaviors of phthalonitrile-

epoxy copolymers and glass fiber laminated composites. J Appl Polym Sci 2013;127:4873–8.

[60] Augustine D, Mathew D, Nair CR. End-functionalized thermoplastic-toughened phthalonitrile composites: influence on cure reaction and mechanical and thermal properties. Polym Int 2015;64:146–53.

[61] Pan H, Xu M, Lei X, Jia K, Liu X. Curing behaviors and performance of a carboxyl-terminated butadiene acrylonitrile rubber/bisphthalonitrile resin system. High Perform Polym 2016;28:581–90.

5 Fiber-Reinforced Phthalonitrile Composites

5.1 Introduction

Fiber-reinforced composites (FRC) are hybrid materials consisting of polymeric, metallic, or ceramic matrices in which fibers of high strength (HT) and modulus are embedded in or bonded to. By combining the performances of each phase, these materials intend to achieve a combination of properties that cannot be obtained with the constituent acting alone. The main fibers in use as reinforcements are the various types of glass and carbon fibers (CF), as well as the Kevlar 49. Other fibers based on boron, silicon carbide, aluminum oxide, and zirconium oxide are also used in limited applications. The FRC are generally fabricated by stacking a number of layered fibers in a chosen direction/sequence and a matrix prior to the consolidation step. The as such manufactured FRC are known as laminates.

Among all the FRC, the fiber-reinforced polymers (FRP) offer the best combination of cost and performances, and often exhibit comparable or even better properties than the traditional metallic materials. Indeed, the low densities of the FRP along with their HT-to-weight and modulus-to-weight ratios make them suitable to be used as replacement to some metallic materials in exigent applications requiring high-performance (HP), lightweight structures. In contrast with the traditional metallic structures which are isotropic, the properties of the FRP largely depend of the direction of measurement. For instance, the tensile properties of the FRP are maximal in the longitudinal direction, while the lowest performances are obtained in the transversal direction. The nonisotropicity of the FRP allows more flexibility in tailoring the properties of the composites according to the design requirements. Thus any direction of major stress in a specific structure can be overcome by locally improving the stiffness through a proper orientation of the FRP. Moreover, the FRP also allow the design of complex geometrical structures with negligible coefficients of thermal expansion. In addition, the failure mechanisms of metals and FRP are quite different. Typically, a metallic structure will fail more likely in a catastrophic manner while the FRP generally show a gradual deterioration process. The FRP are also known for their high internal

Phthalonitrile Resins and Composites.
DOI: https://doi.org/10.1016/B978-0-12-812966-1.00005-6

damping which allows a reduced transmission of sound and vibrations to the neighboring structures. Nowadays, the FRP are gradually replacing metallic materials in many applications, such as aerospace, marine, automotive, sporting goods, and infrastructure. The FRP are also used in electronics, medical industry, building construction, and so on.

Considering the large number of available polymeric matrices and fibers, the materials selection requires a full knowledge of the performances of each constituent. If the materials selection is not effectively done, the resulting FRP will certainly fail in exhibiting the required performances. Generally, the fibers are the main constituent of the FRP and the performances of the resulting composites depend to a large extent on the fibers properties. The fibers are chosen in accordance with their densities, mechanical properties, electrical and thermal conductivity, and, most importantly, their costs. The fibers also play the major role in the tensile load carrying capacity of the composite structure. On the other hand, the role of the polymeric matrix includes keeping the fibers together, transferring the stress, and protecting the fibers from the harsh environmental conditions and from the mechanical degradations. In terms of mechanical properties of the composite, the polymeric matrix has the major influence over the compressive, flexural, and shears properties. Additionally, the quality of the FRP depends on the processing characteristics of the polymeric matrix which include melt viscosity, curing schedule, and amount of defects in the cured form.

Both thermosets and thermoplastics matrices were extensively used in the fabrication of the FRP. The thermosets are mainly used as matrices for long or continuous FRC and the thermoplastics are generally reinforced with short fibers which can be processed by injection molding. For HP applications. such as aerospace and aircraft industries, it has been established that an ideal polymeric system for metal replacement should satisfy to following conditions: easy tapes and prepregs preparation, good shelf-life, void-free cured structure, good mechanical performances, and excellent thermal stability at high temperatures. Nowadays, the currently in-use, HP, polymeric matrices, such epoxies, polyimides, polybenzimidazoles, and cyanate ester resins can only afford a compromise of these properties. For instance, the epoxy-based composites have a poor thermal resistance, weak toughness, and readily absorb water. Although the cyanate ester resins are less hydrophilic than epoxies, problems related to hydrolytic stability can greatly weaken their strength and glass transition temperatures. Materials from polyimides, such as polymerization of monomeric reactants (PMRs), are high temperature resistant, but they absorb water and have some engineering limitations. Out of the above, the phthalonitrile (PN) resins present the best match in terms of combination of properties. PN resins are recognized for their outstanding thermal resistance with starting decomposition temperatures over 450°C for the typical PN polymers. The good thermal stability of the PN resins is accompanied by an ameliorated processability, superior flame resistance, absence of a glass transition temperature (T_g) before their thermal decomposition, low water absorption, advanced UV shielding, and excellent corrosion protective properties [1–10]. Such a combination of properties promotes the PN resins over the traditional, HP thermosets and makes them suitable to be used as advanced polymeric matrices for FRC.

Although the PN resins are a relatively new kind of HP thermosets, they have been extensively studied in the last few decades as advanced polymeric matrices for carbon and glass FRC. The superior properties of the PN precursors make them suitable to be processed by many cost-effective techniques, such the conventional prepregs consolidation [11], resin transfer molding [12–14], resin infusion molding (RIM) [15], filament winding [15], and vacuum infusion process (VIP) [16]. Some of the early reported, HP applications of the PN matrices included PN as matrices constituent for fiber-reinforced, ablative nozzle components [17] and high-temperature, sizing materials for fibers [18]. Nowadays, the PN resins have gained a lot of importance in the industry and they are currently in use or under consideration in many civilian and military applications. Thus, the objective of this chapter is to review all the available literature about the fiber-reinforced, PN composites aiming to reveal the exceptional properties of these materials.

5.2 Carbon Fibers/Phthalonitrile Composites

5.2.1 Properties of Carbon Fibers

As indicated by their name, CF refer to fibers containing at least 92% of carbon in their composition. These fibers are well-known for their superior, tensile, strength-to-weight and modulus-to-weight ratios, excellent structural stability with minor coefficient of linear thermal expansion, high fatigue resistance, and thermal conductivity performances even higher than those of some metals. As a result, the CF allow the preparation of composites with metal/ceramic equivalent or better properties. Considering their appealing features, the CF have been extensively used in extremely exigent applications, such as aerospace, automobiles, sporting goods, and any other applications requiring high modulus (HM) and strength along with a low density.

The fabrication of the CF is a costly process involving several steps and curing at high temperatures. Generally, the CF are manufactured from either polyacrylonitrile (PAN) or pitch precursors [19–21]. The synthetic routes from both starting materials are displayed in Scheme 5.1. The first process starts by wet spinning filaments from a PAN solution. The as-such, obtained filaments are then stretched at high temperatures to align the polymer chains in the filament direction. Afterwards, the stretched filaments are heated at 200–300°C in air atmosphere for a certain period of time to achieve the combination of the nitrile groups into a more stable and rigid ladder structure. Follows the carbonization of the filaments at controlled rate of heating from 1000 to 2000°C in an inert atmosphere. It should be pointed out that the filaments should remain stretched during the whole process to avoid their shrinking and to improve the molecular orientation. At this point, the oxygen and nitrogen atoms should already be eliminated and the filaments are mainly composed of carbon atoms. Nonetheless, the carbon atoms in the neighboring planes are not yet perfectly ordered which means that if the process is stopped at this step, the resulting fibers will have a low tensile modulus. The next step is the graphitization of the carbonized filaments at temperatures above 2000°C. The as-such, obtained fibers should exhibit high, tensile modulus while their

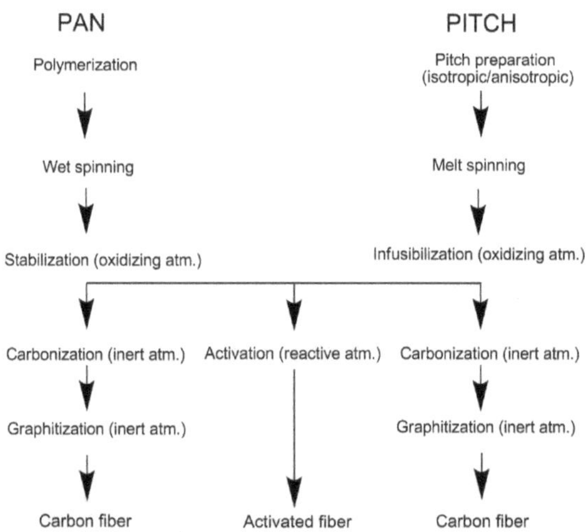

Scheme 5.1 Processes of making CF from pitch and PAN precursors. *CF*, carbon fibers; *PAN*, polyacrylonitrile. Reprinted from Chung D. Carbon fiber composites. Butterworth-Heinemann; 2012.

tensile strength can be further improved by hot stretching them above 2000°C as well. Subsequently, other properties including the coefficient of thermal expansion and oxidation resistance can be improved by a proper control of the degree of crystallinity and amount of structural defects.

The second process is based on the pitch which is a byproduct of petroleum refining or coal coking. Therefore, the CF prepared following via this procedure are relatively less expensive than their PAN-based counterparts. Considering that the carbon atoms in the pith are arranged in low molecular weight, aromatic rings, heating to temperatures above 300°C is necessary to increase the length of the molecular chains thorough a polymerization process. From the obtained, highly viscous, state of pitch, commonly known as mesophage, filaments are melt spanned by passing through a spinner and then cooled down to freeze the molecular orientation. Immediately after, the filaments are heated between 200 and 300°C in air atmosphere to stabilize and render them infusible. Next, the filaments are carbonized at about 2000°C and the following steps are akin to those of the PAN process.

According to their overall performances, the CF can be classified as general purpose (GP) or HP. Table 5.1 lists the tensile and diameter properties of various CF from different manufacturers [19]. By comparing these properties, it appears that the HP-CF based on pitch exhibit higher tensile, modulus values than those prepared from PAN. For instance, the E-130 CF from DuPont have a tensile modulus of about 894 GPa. The reason behind the superior tensile modulus of the pitch-based CF is that the pitch is more graphitizable than the PAN. Meanwhile, higher modulus values are generally accompanied with lower elongations at break as seen from the tensile properties of the pitch-based HP-CF regardless of their manufacturer. Nonetheless, for quite similar modulus values (for the E-130 and P-120S), the DuPont manufactured CF showed improved tensile strength and elongation at break. On the other hand, the PAN-based CF can reach higher tensile strengths and larger elongations at break because the shear is easier between the carbon layers in a graphitized fiber. For example, the T-1000 CF made by Toray have a tensile strength up to 7060 MPa with an elongation at break of 2.4%. Moreover, it seems that the PAN-based CF can be prepared with smaller diameters that those achieved with pitch. Regardless of their precursors, the GP-CF refers to fibers with lower tensile modulus and strength values but with larger elongations at break.

Another class of CF known as "short carbon fibers" recently emerged and is now under rapid development. These fibers can be as small as

Table 5.1 Tensile and Modulus Properties of Various Commercial CF

CF Type	Designation	Tensile Strength (MPa)	Tensile Modulus (GPa)	Elongation at Break (%)	Diameter (μm)	Manufacturer
GP	T-101S	720	32	2.2	14.5	Kureha Chem.
	T-201S	690	30	2.1	14.5	Kureha Chem.
	S-210	784	39	2	13	Donac
	P-400	690	48	1.4	10	Ashland Petroleum
HP (PAN)	T-300	3530	230	1.5	7	Toray
	T-400H	4410	250	1.8	7	Toray
	T-800H	5590	294	1.9	5.2	Toray
	T-1000	7060	294	2.4	5.3	Toray
	MR 50	5490	294	1.9	5	Mitsubishi Rayon
	MRE 50	5490	323	1.7	6	Mitsubishi Rayon
	HMS-40	3430	392	0.87	6.2	Toho Rayon
	HMS-40X	4700	392	1.2	4.7	Toho Rayon
	HMS-60X	3820	588	0.65	4	Toho Rayon
	AS-1	3105	228	1.32	8	Hercules
	AS-2	2760	228	1.2	8	Hercules
	AS-4	3795	235	1.53	8	Hercules
	AS-6	4140	242	1.65	5	Hercules
	IM-6	4382	276	1.5	5	Hercules
	HMS4	2484	338	0.7	8	Hercules
	HMU	2760	380	0.7	8	Hercules
HP (Pitch)	P-25	1400	160	0.9	11	Amoco
	P-75S	2100	520	0.4	10	Amoco
	P-120S	2200	827	0.27	10	Amoco
	E-35	2800	241	1.03	9.6	DuPont
	E-75	3100	516	0.56	9.4	DuPont
	E-130	3900	894	0.55	9.2	DuPont
	F-140	1800	140	1.3	10	Donac
	F-600	3000	600	0.52	9	Donac

CF, carbon fibers; *PAN*, polyacrylonitrile; *GP*, general purpose; *HP*, high-performance.
Reprinted from Chung D. Carbon fiber composites. Butterworth-Heinemann; 2012.

0.1 μm and are obtained via the process of catalytic growth using solid catalyst particles such as Fe. Subsequently, the filaments are grown by chemical vapor deposition from a carbonaceous gas. However, from an economic point of view, the pitch-based CF remain less expensive, although for some application, such as aerospace the HP of the PAN-based CF are preferred regardless of their relatively higher cost.

It is also relevant to point out that another classification of the CF was adopted by the IUPAC organization. Basically, the CF are categorized according to their tensile strength and modulus properties. Following this classification, CF with modulus in excess of 500 GPa are known as ultrahigh modulus, those with modulus greater than 300 GPa and strength-to-modulus ratio less than 1% are referred as HM, IM (intermediate modulus) corresponds to CF with modulus up to 300 GPa and strength-to-modulus above 1×10^{-2}, HT describes CF with strength superior to 3 GPa and strength-to-modulus ratio between 1.5 and 2×10^{-2}, and the last category is known as the low-modulus type which includes CF with modulus lower than 100 GPa and low strength.

5.2.2 Carbon Fibers Reinforced Phthalonitrile Composites

According to the type of the PN resin, three (03) categories can be distinguished. The first category englobes the PN matrices prepared from either typical or common, PN monomers/oligomers and the second category includes the PN matrices prepared from X-functional, PN monomers/oligomers. The last category regroups the PN matrices made from physical blends of typical/functional PN alloyed with other polymeric systems including thermosets and thermoplastics.

5.2.2.1 Typical Phthalonitrile/Carbon Fibers Composites

The first report about the PN/CF composites was performed by Ting and Nash in the earliest eighties [22]. The aliphatic diamide-linked, PN monomers, presented in Chapter 1, Introduction to the Phthalonitrile Resins and their Curing Behavior, were chosen as the matrices to prepare the PN/CF composites. These monomers, previously reported by Griffith et al., [23], were synthetized by the reaction of 4-aminophthalonitrile with an aliphatic diacide chloride. According to the number (n) of carbons in the diacide chloride, the resulting PN monomers were labeled as Cn-PN. The Thornel 300 (T-300) CF, manufactured by Union Carbide Company, were chosen as the reinforcing phase. Prior to the prepregs preparation, the authors studied the influence of the length of the aliphatic spacer on the T_g and the room temperature shear modulus of the cured Cn-PN polymers. The tests, performed by torsional pendulum analyzes, revealed that as the length of the aliphatic spacer increased, the T_g and shear modulus decreased. For instance, the T_g ranged from 90°C for the C36-PN to 350°C for the C6-PN. Considering that the T_g of a material determines its application and since the PN/CF composites were designated for exigent applications, the authors chose the C10-PN as the suitable PN matrix for the preparation of their composites.

The authors also performed several other tests to find the proper curing cycle for these monomers. It should be pointed out that at that time, the curing behavior of the PN resins was not fully acknowledged. Indeed, one of the reasons behind the preparation of PN monomers with long aliphatic linkage was to achieve a high molecular weight, since the polymerization of PN monomers containing highly rigid, aromatic structures was not resolved at that time. Regardless of the historical facts, the curing temperatures of the C10-PN monomers were determined to be somehow between 230 and 260°C. The processability of these monomers alone and in the presence of the CF was studied by dynamic dielectric analysis (DDA) by continuously measuring the capacitance and dielectric dissipation factor as a function of the cure time and temperature. The DDA technique is similar to the dynamic mechanical analysis (DMA) used nowadays. The DDA results revealed that the activation energy of the studied prepregs followed an Arrhenius relationship with an activation energy slightly higher than that of the epoxy resin, which means that these PN monomers required higher temperatures and time to achieve the polymerization. These facts along with other observations and investigations allowed other researchers to develop curing agents and self-cured, PN monomers, as it is for the epoxies.

The C10-PN/T-300 prepregs in which the proportion of the resin was 32 wt.% were prepared by the conventional hot melt method. For the mechanical measurements including the tensile and bending tests, 16 quasi-isotropic and angle plies were manufactured. The quasi-isotropic laminates had the configuration of (90, ±45, 0°) while the angles plies consisted of two orientations namely the ±15 and ±45°, respectively. The obtained results along with those of epoxy (Namco 5208) and polyimide (Hexcel F-178)/T-300 composites are presented in Table 5.2. By analyzing the data, it appeared that the C10-PN/T-300 composites had quite similar tensile and flexural modulus to those of the F-178/T-300 composites. Conversely, the epoxy-based composites showed a superior strength compared to the other systems and especially for the $2\theta = 30°$, the gap was the largest. In all instances, further increasing the angle resulted in a reduction in both tensile and flexural properties. Indeed, with larger angles, the contribution of the fillers became less significant, while that of the polymeric matrices increased in importance. For the quasi-isotropic orientation the overall performance of the C10-PN/T-300

Table 5.2 Room Temperature Mechanical Properties of the C10-PN, Epoxy, and Polyimide Reinforced with the T-300 CF

Orientation	Properties (GPa)		C10-PN/ T-300	Epoxy (Namco 5208)/T-300	Polyimide (Hexcel F-178)/T-300
Quasi-isotropic	Strength:	Tensile	0.47	0.48	0.39
		Flexural	0.46	–	0.52
	Modulus:	Tensile	75.2	48.3	62.8
		Flexural	33.8	–	37.3
$2\theta = 30°$	Strength:	Tensile	0.56	0.75	0.61
		Flexural	0.66	1.03	0.81
	Modulus:	Tensile	93.2	104.9	97.3
		Flexural	75.2	107.0	91.1
$2\theta = 90°$	Strength:	Tensile	0.11	0.16	0.13
		Flexural	0.18	–	0.28
	Modulus:	Tensile	37.3	26.9	40.7
		Flexural	9.7	–	17.3
$2\theta = 150°$	Strength:	Tensile	0.031	0.041	0.035
	Modulus:	Tensile	22.8	10.4	26.2

PN, phthalonitrile.
Reprinted from Ting R, Nash H. The development of C-10 phthalocyanine for composite matrix applications. Polym Eng Sci 1981;21:441–8.

surpassed those of the epoxy- and polyimide-based composites. Further investigations regarding the fracture behavior of these composites revealed that the PN matrix can provide performances as good as those obtained with other HP thermosets.

In summary, the work reported by Ting and Nash set the basis for further investigations. Through their results, it was confirmed that the PN resins even with flexile aliphatic linkages can result in advanced composites with performances akin or superior to those from epoxy and polyimide systems.

Few years later, the advances in the curing techniques and monomers design allowed the preparation of highly, thermally stable, PN resins which can be easily cured by the addition of small amounts of amine-curing agents [2,24–34]. These developments also afforded the preparation of CF-reinforced PN composites with even better performances. In 1996, Sastri et al., reported about CF-reinforced PN composites prepared from the typical and HP, biphenyl PN (BPh) and Hercules unsized IM-7 12 K CF [11]. The chosen PN monomers were cured in the presence of small amounts (2.1 wt.%) of 1,3-bis(3-aminophenox)-benzene (*m*-APB) to afford the polymerization of the PN units in a timely manner. It should be reminded that the effects of the *m*-APB curing agent on the curing behavior and processability of the BPh resin were previously discussed in Chapter 1, Introduction to the Phthalonitrile Resins and their Curing Behavior, and herein priority is given to the composites properties. Briefly, the curing behavior investigations revealed that the addition of the *m*-APB curing agent allowed the initiation of the cyano addition reaction of the PN units at a rate dependent on the amount of the catalyst. It was also found that at low amine amounts (1.5−1.7 wt.%), the viscosity gradually increased from 0.05 to 0.25 Pa s over a period of 2 h at 250°C. When a high amount of *m*-APB was added (up to 2.3 wt.%), the viscosity increased at a higher rate. For instance, after heating the prepolymers containing 2.1 and 2.3 wt.% *m*-APB for 1 h at 250°C an increase in viscosity up to six- and fortyfold, respectively, was noticed. Further addition of the curing agent resulted in a glassy thermosets after 5 min of reaction time at 250°C. Therefore, the curing conditions of the BPh can be easily controlled by a proper choice of the amount of the *m*-APB curing agent.

The BPh/IM-7 prepregs were obtained by a solution impregnation process using a 90/10 mixture of dimethyl acetamide and *N*-methylpyrrolidinone (NMP) solvents and about 36/64 weight ratio of BPh resin to CF. Unidirectional composites panels were then afforded by stacking 8 or 16 plies of prepregs between 3-mil-thick DuPont Kapton release films. Prior to the consolidation of the prepregs by autoclave cure, the authors investigated the cure conditions of the prepregs by dynamic scanning calorimetry (DSC) and rheology measurements to check whether the viscosity of the resin was affected during the prepregs preparations. The obtained prepregs viscosity variations as a function of the dwell time at 250°C revealed a rapid increase in viscosity after 40 min. Therefore, the authors chose to apply a 200-psi pressure just after the resin melted at 230°C. In the meantime, based on the overall curing behavior and processability of the neat BPh resin and the prepregs, a curing cycle including dwells for 1 h at 250°C and 3 h at 325°C was adopted. The as-such, obtained, BPh/IM-7 composites were confirmed by ultrasound examinations to be void and defeat free panels.

In addition to the previous curing schedule, some panels were postcured at the elevated temperature of 350 and 375°C at various times to assess the postcure effect on the thermomechanical properties of the BPh/IM-7 composites. The storage modulus changes as a function of the postcuring conditions were investigated by DMA. In all instances, the storage modulus values at room temperature were quite similar. The non-post-cured samples had a T_g in excess of 350°C and retained 75% of their initial modulus at 400°C. Interestingly, postcuring these panels for only 2 h at 350°C shifted the T_g to more than 400°C. Further postcuring the panels for 2, 4, and 8 h at 375°C resulted in T_g values in excess of 450°C and the storage modulus maintained up to 83% of its initial value. More importantly, with extended postcuring conditions (8 h at 350 and 375°C) only 10% of the room temperature storage modulus was lost at 450°C. Clearly, these results highlighted the positive effect of the post-curing cycles on the storage modulus of the BPh/IM-7 composites.

The tensile and flexural properties of these PN-based, CF composites were compared with those of PMR-15 composites reinforced with the same type of fillers. The obtained transversal and longitudinal tensile and flexural properties are listed in Table 5.3. It should be pointed out

Table 5.3 Room Temperature Mechanical Properties of the BPh/IM-7 and PMR-15/IM-7 Composites

Property	BPh/IM-7	PMR-15/IM-7
0° Tension		
Strength (MPa)	2000	2500
Modulus (GPa)	183	146
Strain (%)	1.0	1.6
90° Tension		
Strength (MPa)	41	29
Modulus (GPa)	10	9
Strain (%)	0.4	0.5
0° Flexure		
Strength (MPa)	2350	1530
Modulus (GPa)	174	122
Strain (%)	1.3	1.3
90° Flexure		
Strength (MPa)	80	–
Modulus (GPa)	11	–
Strain (%)	0.6	–
Short beam shear (MPa)	85	105

IM, intermediate modulus.
Reprinted from Sastri SB, Armistead JP, Keller TM. Phthalonitrile-carbon fiber composites. Polym Compos 1996;17:816–22.

that except for the longitudinal, tensile measurements which were performed for the 8 plies panels, all the other mechanical properties were investigated for the 16 plies panels. Globally, the PN composites showed performances equivalent or even better than those of the PMR-15 composites. By analyzing the data, it appeared that, as far as the modulus was concerned, the PN composites exhibited the best match either for the tensile or bending regardless of the orientation. Moreover, for the longitudinal orientation, the tensile strength of the PN composites was slightly lower than that of the PMR-15 composites, but largely surpassed the latter in the transversal direction. On the other hand, the PN composites also exhibited a 54% higher flexural strength than their PMR-15 counterparts. Nonetheless, the short beam shear strength of the PN composites was measured to be 85 MPa while that of the PMR-15 composites was 20 MPa higher.

Similarly to the storage modulus, the postcure effect on the PN composites' mechanical properties was also investigated and the obtained results are summarized in Table 5.4. Obviously, short, postcure conditions (8 h at 325°C) had a quite similar effect to the longer ones (8 h at 350 and 375°C). For instance, the tensile, longitudinal strength and modulus were measured to be 2.1 and 172 GPa, respectively, for the short, postcuring cycle while the values of 2 and 183 GPa were noted for the extended postcuring cycle, respectively. Generally, the postcure plays a crucial role in improving the thermal and thermos-mechanical properties of the PN resins. Although only slight improvements in the mechanical properties can be expected, performing an appropriate postcure remains necessary to achieve HP, PN materials.

To be able to meet the requirements of marine applications, the polymer materials need to conform to low water uptake over extended period of immersion. Therefore, to acknowledge about the water resistance of the BPh/IM-7 composites, the weight gain percentage was measured for up to 16 months. The obtained results clearly confirmed that the PN composites were able to satisfy the water resistance requirements with less than 1% water uptake after 16 months of immersion.

The flame resistance is another stringent requirement that often cannot be overcome by many polymeric materials. For instance, to meet the US Navy flammability requirements MIL-STD-2031 for applications in hull, machinery, and structural applications, polymeric materials should satisfy a time to ignition (T_{ig}) greater than 60 s and the peak heat release (PHR) should be less than 150 at 100 kW/m^2 during a 300-s heat treatment. Cone calorimeter investigations in according with the US Navy standards revealed that the CF-reinforced, PN composites were the

Table 5.4 Influence of the Postcure Conditions of the Mechanical Properties of the BPh/IM-7 Composites

Property	BPh/IM-7 Postcure Cycle: 350°C 8 h, 375°C 8 h	PMR-15/IM-7 Postcure Cycle: 325°C 8 h
0° Tension		
Strength (MPa)	2000	2100
Modulus (GPa)	183	172
Strain (%)	1.0	1.14
90° Tension		
Strength (MPa)	41	33
Modulus (GPa)	10	10
Strain (%)	0.4	0.33
0° Flexure		
Strength (MPa)	2350	2380
Modulus (GPa)	174	183
Strain (%)	1.3	1.3
90° Flexure		
Strength (MPa)	80	79
Modulus (GPa)	11	10
Strain (%)	0.6	0.5
Short beam shear (MPa)	85	80

IM, intermediate modulus.

Reprinted from Sastri SB, Armistead JP, Keller TM. Phthalonitrile-carbon fiber composites. Polym Compos 1996;17:816–22.

only candidate to fully satisfy the navy requirements. As seen in Table 5.5, among all the presented systems the PN composites showed the largest T_{ig} of 75 s and a PHR of 118 kW/m^2. On the other hand, other systems such as the PMR-15/CF composites only satisfied the PHR condition while their T_{ig} was only 55 s.

In the last discussed research, *m*-APB was added to the melt of BPh monomers mainly because of the volatility of this curing agent. Indeed, if added to the powdered form, the volatilization could occur for temperatures over 200°C even before the staring of the polymerization reaction. From a historical point of view, the *m*-APB was replaced by another, more-performant, amine-curing agent starting from 1998. The newly proposed aming-curing agent was the well-

Table 5.5 Flammability Results of the BPh/IM-7 Composites with those of other CF-reinforced Composites

Composites	T_{ig} (s)	PHR (kW/m^2)
BPh/CF	75	118
PMR-15/CF	55	85
BMI/CF	22	270
PEEK/CF	42	85
Epoxy/CF	28	232
Phenolic/CF	25	96

CF, carbon fibers; *PHR*, peak heat release; *PEEK*, poly(ether ether ketone).
Reprinted from Sastri SB, Armistead JP, Keller TM. Phthalonitrile-carbon fiber composites. Polym Compos 1996;17:816–22.

known bis[4-(4-aminophenoxy) phenyl]sulfone (*p*-BAPS) [35]. In contrast with the *m*-APB, the *p*-BAPS is a slower reacting agent with lower volatility which allows more control over the curing process. Moreover, considering its highest boiling point, the *p*-BAPS can be added to either the melted or the powdered form of the PN monomers. Considering the benefits of this curing agent, Dominguez et al., prepared BPh/CF composites cured by the *p*-BAPS and compared the resulting properties with those cured with *m*-APB [36].

To find the adequate amount of *p*-BAPS to be added to the BPh monomers, Dominguez and coworkers evaluated the viscosity changes over time at 280°C for several weight ratios of *p*-BAPS ranging between 2 and 4 wt.% as seen in Fig. 5.1. The investigations revealed that the increase in viscosity was closely related to the amount of the aming-curing agent. For example, higher amounts of *p*-BAPS (>2.8 wt.%) showed a quick and rapid viscosity increase, while for fewer amounts of *p*-BAPS (<2.4 wt.%) the viscosity slowly increased over a period of 2 h. Therefore, the authors chose the 2.4 wt.% *p*-BAPS as the best match for a proper control over the curing process. If compared with the *m*-APB, the amount of *p*-BAPS to be added to the PN monomers was slightly higher because of the lowest reactivity of the latter.

Taking into account the viscosity results, PN prepolymers containing 2.4 wt. of *p*-BAPS were deposed on Hercules unsized IM-7 12 K carbon by a solution impregnation process in which the amount of the resin was about 33 wt.%. Prior to the consolidation step, small amount of the

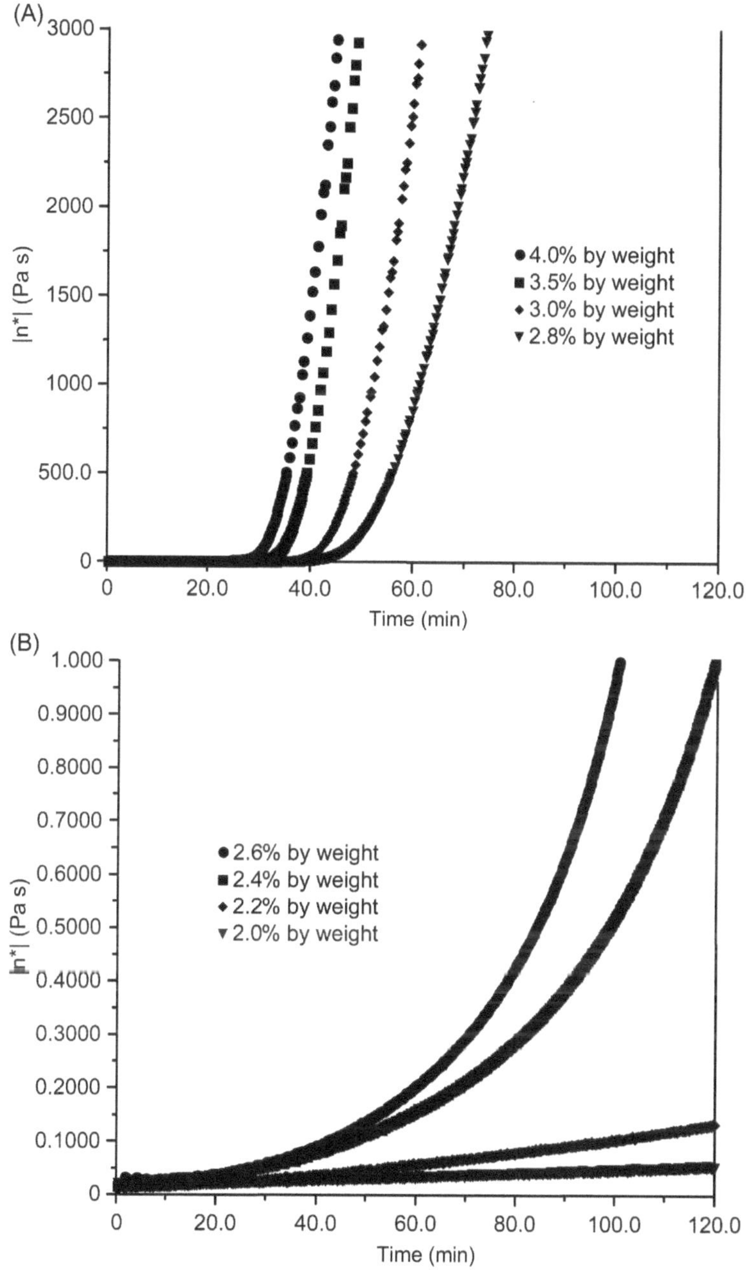

Figure 5.1 Complex viscosity variations of the BPh prepolymers as a function of the *p*-BAPS loading contents: from 2.8 to 4.0 wt.% (A) and from 2.0 to 2.6 (B). Reprinted from Dominguez DD, Jones HN, Keller TM. The effect of curing additive on the mechanical properties of phthalonitrile-carbon fiber composites. Polym Compos 2004;25:554–61.

PN prepolymers containing higher contents of *p*-BAPS (2.8 wt.%) were dispersed between alternative plies in order to improve the structural adhesion between the plies. For a comparative purpose, unidirectional, BPh/IM-7 composites containing 8 and 16 plies were fabricated by the conventional prepregs consolidation process as described by Sastri et al., and further detailed in the previous paragraphs. Subsequently, a four-step, curing process involving dwells at several temperatures, vacuum processing, and pressure application was employed to achieve the fully cured composites. The obtained composites had a thickness of about 1.3 and 2.5 mm according to the number of plies. The structural integrity of the composites was assessed about by visual and ultrasonic inspections and the BPh/IM-7 composites were confirmed to be void-free and high-quality panels.

It is noteworthy to point out that the cure schedule with a 1 h dwell at each temperature was initially chosen to reduce the cure time of the PN composites in the autoclave. However, to confirm whether this curing program allowed fully cured panels, the DMA were performed on the machinated samples to assess the extent of cure by monitoring the storage modulus and Tan Delta variations as a function of temperature. The obtained data, displayed in Fig. 5.2, revealed that the storage modulus at 40°C was about 10 GPa and this value was markedly reduced above 300°C. In the meantime, the Tan Delta showed the appearance of a peak at about 450°C implying that a viscoelastic transition occurred at this temperature. Generally, the fully cured PN polymers do not show any T_g even for temperatures in excess of 450°C. Further room- and elevated-temperature, short beam tests supported the conclusion that the PN composites were not fully cured using the initially adopted curing schedule.

In order to complete the cure of the PN composites, the panels were additionally postcured at 375°C. The previously discussed, PN composites cured with *m*-APB showed that the fully cured panels were obtained after postcuring at high temperature and extended time (8 h at 375°C). Thus, in a similar way, these PN composite cured with *p*-BAPS were postcured at 375°C for an additional 1–7 h and the obtained materials were once again tested by DMA. By analyzing the data, a gradual improvement was noticed in the overall performances. For instance, the PN composites postcured at 375°C for 7 h showed only a 30% reduction in the storage modulus at 450°C while the Tan Delta curve showed only minor changes up to the same temperature. As for other PN resins and their curing agents, increasing the curing temperature or performing a postcure practically always increases the crosslinking degree resulting

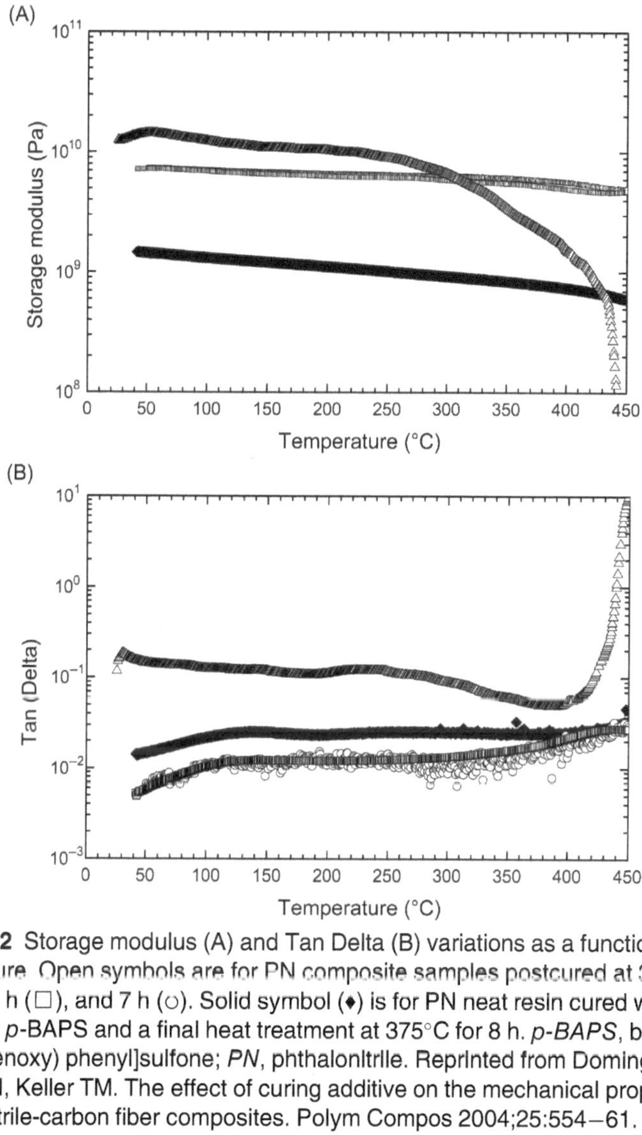

Figure 5.2 Storage modulus (A) and Tan Delta (B) variations as a function of temperature. Open symbols are for PN composite samples postcured at 375°C for 0 h (△), 1 h (□), and 7 h (○). Solid symbol (♦) is for PN neat resin cured with 2.5% by weight *p*-BAPS and a final heat treatment at 375°C for 8 h. *p-BAPS*, bis[4-(4-aminophenoxy) phenyl]sulfone; *PN*, phthalonitrile. Reprinted from Dominguez DD, Jones HN, Keller TM. The effect of curing additive on the mechanical properties of phthalonitrile-carbon fiber composites. Polym Compos 2004;25:554–61.

in enhancements in the thermal and thermomechanical properties. Meanwhile, the high temperature short beam tests for samples postcured for less than 7 h at 375°C revealed a softening and viscous behavior for these composites at 371°C confirming that the chosen postcure conditions were not satisfactory. In fact, based on the DMA and short beam tests only the PN composites postcured at 375°C for 7 h could be referred to as "fully cured."

Table 5.6 Effect of the Different Amine-Curing Agent on the Mechanical Properties of the BPh/IM-7 Composites

Property	*p*-BAPS Cured BPh/IM-7	*m*-APB Cured BPh/IM-7
0° Tension		
Strength (MPa)	1975	2500
Modulus (GPa)	159	146
0° Flexure		
Strength (MPa)	2278	2350
Modulus (GPa)	144	174
Short beam shear (MPa)	93	85

p-BAPS, bis[4-(4-aminophenoxy) phenyl]sulfone; *m-APB*, 1,3-bis(3-aminophenox)-benzene; *IM*, intermediate modulus.
Reprinted from Dominguez DD, Jones HN, Keller TM. The effect of curing additive on the mechanical properties of phthalonitrile-carbon fiber composites. Polym Compos 2004;25:554–61.

By adapting the initial, autoclave, curing program followed by an additional postcure at 375°C for 7 h, the fully cured, PN composites were investigated for their mechanical performances, including the tensile and bending properties, as well as the apparent interlaminar shear strength. The obtained results along with those obtained with the BPh/IM-7 composites fully cured in the presence of *m*-APB are summarized in Table 5.6. It should be understood that the apparent interlaminar shear strength was measured at room and elevated temperatures. At room temperature, the PN composites cured with *p*-BAPS exhibited the value of 93 MPa while those cured with *m*-APB displayed the shear strength value of 85 MPa. The tensile and bending modulus and strength for both types of composites only revealed a small advantage for the *m*-APB-based composites. Generally, the tensile largely depends on the fibers properties, while the bending and shear are dominated by the matrix properties. Therefore, the effect of the curing agent type should be extrapolated from these two latter tests. However, in this case, the differences were in the range of errors and no significant advantage for any curing agent was depicted. Therefore, it was concluded that *p*-BAPS and *m*-APB had quite an akin effect on the overall mechanical properties.

5.2.2.2 X-Functional Phthalonitrile/Carbon Fibers Composites

Similarly to the typical PN resins, functional PN resins were also used as matrices for CF-based composites. As previously discussed in Chapter 3, X-Functional Phthalonitrile Monomers and Polymers, the preparation of functional, PN monomers/polymers aims to achieve better performances as those obtained with their typical PN counterparts. Generally, the functional, PN monomers are prepared by incorporating either different monomers or simple functionalities containing active groups into the PN backbone. The as-such, prepared, PN materials can combine the performances of both systems to achieve targeted improvements in some desired properties. The field of functional, PN materials has been extensively studied and nowadays the number of available, modified, PN monomers surpasses that of the typical PN resins.

The first, functional, PN/CF composites were reported about by Augustine et al. in 2012 [37]. The authors prepared several, novolac-functional, PN oligomers by varying the substitution degree of the PN groups. Thus, 44%, 71%, 87%, and 99% of the hydroxyl groups were replaced by PN units to understand how the overall performances could be altered by these changes in the novolac backbone. The general idea was to take advantage from the active hydroxyl groups from the novolac to initiate the cyano addition reaction of the PN units. Readers are referred to Chapter 3, X-Functional Phthalonitrile Monomers and Polymers, where detailed discussions about the oligomers preparation, curing behavior, and the resulting polymers properties are provided. The novolac-functional, PN/CF prepregs were obtained by dipping Grade T-300 CF into a methyl ethyl ketone dissolved solution of novolac-functional, PN oligomers. Following the aforementioned process, several types of prepregs were fabricated by changing the substitution degree of the PN groups in the novolac-functional, PN oligomers. Indeed, studying the effect of the amount of hydroxyl groups in the PN oligomers was another objective of the study. The weight ratio of the PN to CF was about 30/70. Subsequently, the impregnated prepregs were dried in a vacuum oven at 60°C for 30 min to completely evacuate the solvents. Then, the panels were consolidated by curing under vacuum following the curing schedule of 100, 150, and 200°C for 30 min each; 250 and 300°C for 1 h each and 350°C for 3 h.

The performed mechanical tests were chosen in a way to highlight the contribution of the polymeric matrices in the CF-based composites. For instance, the bending and shear tests are well-known for their high sensitivity towards the polymeric matrices, while the tensile test is

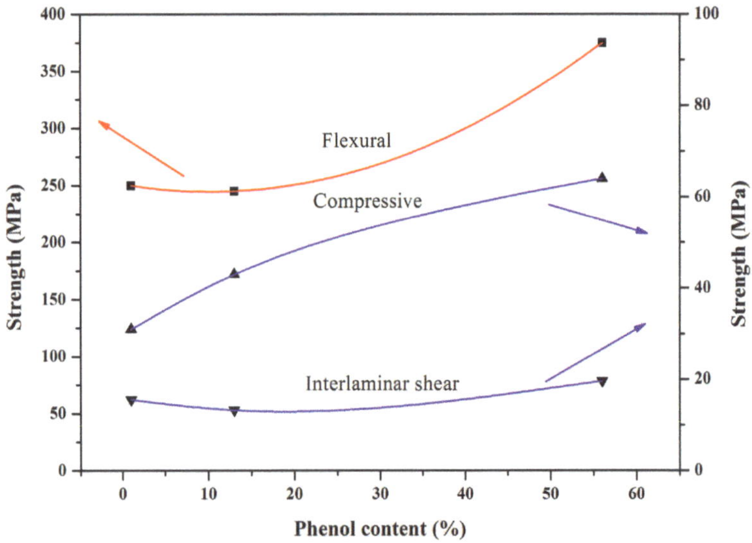

Figure 5.3 Effects of the hydroxyl groups contents in the novolac-functional, PN matrix on the bending, compression, and shear properties of the resulting CF-based composites. *PN*, phthalonitrile; *CF*, carbon fibers. Reprinted from Augustine D, Mathew D, Nair CPR. Phenol-containing phthalonitrile polymers—synthesis, cure characteristics and laminate properties. Polym Inter 2012;62:1068–76.

generally governed by the contribution of the fibers. Augustine et al., reported about the effect of the hydroxyls contents on the composites bending, compression, and shear properties. As seen in Fig. 5.3, the composites based on the novolac-functional, PN oligomers containing 99% of PN units exhibited flexural and compressive strengths of about 250 and 30 MPa, respectively, with a shear strength of less than 20 MPa. Surprisingly, if compared with the results obtained with the BPh/IM-7 composites prepared by either Sastri or Dominguez, the overall performances of these PN composites were catastrophically lower and seemed to result from a different kind of resin. One of the reasons behind these limited mechanical properties was the adopted short and low temperature, curing schedule. On the other hand, it was noticed that increasing the phenol contents induced slight improvements in all the studied mechanical properties.

Augustine et al., also investigated the fractured surfaces of the novolac-functional, PN-based composites by scanning electron microscopy (SEM) to understand the microstructural changes happening during the mechanical tests. The fractured surfaces of all the PN composites with various proportions of PN units are displayed in

Fig. 5.4. By analyzing the images, it appeared that the PN matrices containing higher percentage of PN units (>87%) showed a poor fiber wetting. In those cases, the bare fiber surfaces were clearly visible. Contrarily, the PN matrices containing higher hydroxyl groups allowed a good adhesion between the matrix and the CF. Although these observations effectively corroborated the mechanical results, the poor adhesion between the PN matrices and the CF cannot justify the weak overall performances of these composites. Nevertheless, as seen from Table 5.7, these novolac-functional PN/CF showed good flammability properties with limiting oxygen index values up to 70 at higher amounts of PN units.

Zong et al., also reported about a new kind of functional PN oligomers and their subsequent CF-based composites [38]. Initially, the authors aimed the preparation of new series of PN oligomers containing

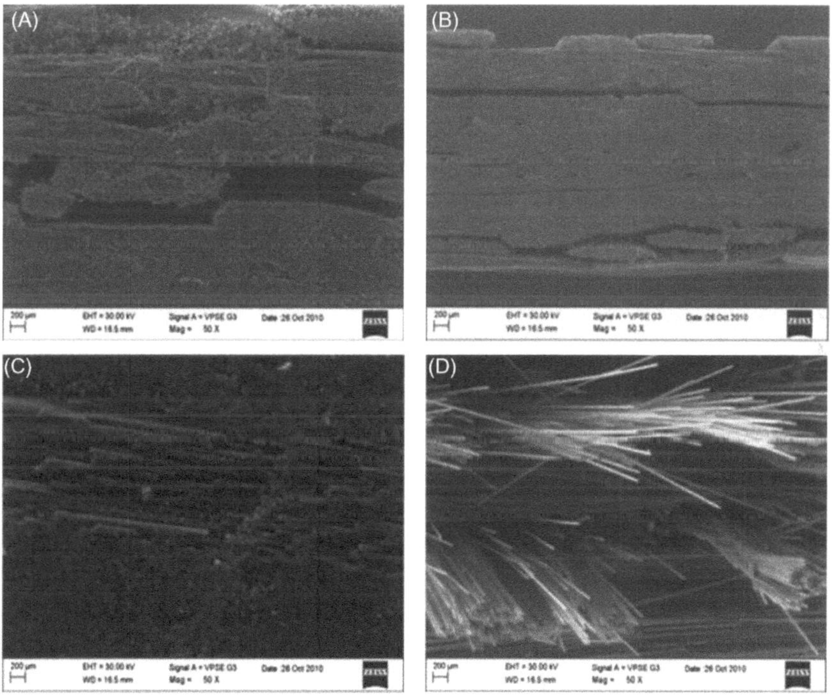

Figure 5.4 SEM images of the fractured surfaces of the CF composites containing various extent of PN units in the novolac-functional, PN matrix: 44% (A), 71% (B), 87% (C), and 99% (D). *PN*, phthalonitrile; *CF*, carbon fibers; *SEM*, scanning electron microscopy. Reprinted from Augustine D, Mathew D, Nair CPR. Phenol-containing phthalonitrile polymers—synthesis, cure characteristics and laminate properties. Polym Inter 2012;62:1068–76.

Table 5.7 Limiting Oxygen Index (LOI) Values of the Novolac-Functional, PN/CF Composites

Degree of PN Substitution	LOI
44	58
71	70
87	72
99	73

LOI, Limiting oxygen index; *PN*, phthalonitrile.
Data extracted from Fig. 12 Augustine D, Mathew D, Nair CPR. Phenol-containing phthalonitrile polymers—synthesis, cure characteristics and laminate properties. Polym Inter 2012;62:1068–76.

heteroaromatic phenyl-*s*-triazine moieties of several lengths. The obtained, PN oligomers were cured in the presence of 5 wt.% *p*-BAPS as the curing catalyst. DSC investigations revealed that the curing exotherm was somewhere between 254 and 278°C depending on the length of the repeating unit between the PN ends. Subsequently, rheometric measurements were also conducted to study the processability of the prepared PN oligomers. The isothermal viscosity measurements at 280°C over time showed that the gelation occurred quite rapidly after about 9.3 min for the shortest main-chain length and even less for longer chains. Moreover, viscosity changes over temperature highlighted the fact that the melt viscosity of the PN oligomers was lower than 100 Pa s at 218, 275, 316, and 325°C for the main chain-length of 1, 3, 5, and 10, respectively. Although the new, PN oligomers seems to have acceptable processability, if compared with the typical, BPh monomers cured either with *m*-APB of *p*-BAPS, the triazine-based, PN oligomers showed relatively lower processing advantages. For instance, at 280°C the BPh monomers cured with 4 wt.% *p*-BAPS only showed a sharp increase in the viscosity after about 30 min. Furthermore, the melt viscosity of the latter was as low as 0.05 Pa s. It should be reminded that for the PN resins or similar polymeric materials, a reduced reactivity along with a low melt viscosity are desired to meet the requirements of the industrial processing techniques. Back to the work of Zong and coworkers, it would have been better to study the effect of various amounts of *p*-BAPS on the curing behavior and processability and to choose fewer amounts since the chosen 5 wt.% of this amine-curing agent appears to greatly increase the rate of the cure regardless of the main-chain length. Nevertheless, the main advantage of these PN resins is their high thermal resistance, which was found to surpass that of the typical PN or even that of many functional, PN polymers. For instance, when cured following the curing schedule of 250°C for 3 h, 285°C for 1 h, 325°C for 3 h, 350°C for 2 h, and 375°C for 8 h, in the presence of 5 wt.% *p*-BAPS, the resulting polymer had a temperature of 5 wt.% weigh loss ($T_{5\%}$) ranging from 538 to 582 in nitrogen atmosphere and between 543 and 575°C under air atmosphere, with respect to the main chain length.

The triazine-linked, PN/CF composites were fabricated using the T-700 CF from TORAY. The general process of fabrication was a three-step procedure starting from the preparation of the impregnating solution. During this step, 50 g of triazine-based, PN oligomers was heated to 280°C and 5 wt.% of *p*-BAPS was added to the melt with vigorous stirring. The obtained mixture, known as the B-stage, was then cooled down to room temperature prior to dissolving it NMP solvent. In the second step, prepregs were afforded by dipping the T-700 CF in the previous solution and drying them under an infrared lamp. The prepregs were further dried in a vacuum oven at 180°C for 36 h and 200°C for 1 h. During the final step, 12 plies of prepregs were first unidirectionally stacked in a steel mold under a pressure of 2.5 MPa at 280°C for 2 h and then transferred to a muffle furnace to achieve the consolidation of the panels. The adopted curing schedule was akin to that presented in the previous paragraph. The weight percentage of the PN resin in the composite was between 62.7% and 69.2%. It should be understood that some minor modifications in the fabrication process took place according the chain length of the repeating unit in the PN oligomers.

The bending and short beam tests for the triazine-linked, PN/T-700 composites with various, main chain lengths of the PN oligomeric precursors were also investigated and the obtained results are listed in Table 5.8. The flexural strength of the PN composites ranged between 1722 and 1855 MPa when the main chain length moved from 1 to 5, respectively. These improvements were attributed to the enhancement in toughness upon increasing the main chain length. However, further increasing the length to 10 reduced the flexural strength to 1339 MPa. The authors attributed this failure to the processing difficulties of its oligomeric precursors. On the other hand, the flexural modulus decreased from 164 to 113 GPa as the chain length moved from 1 to 10, respectively. The shear beam strength showed values between 71.6 MPa for the chain length of 1 and 92.2 MPa for the chain length of 3. As previously discussed, the 16 plies of BPh/IM-7 composites cured

Table 5.8 Flexural and Short Beam Tests Results for the Triazine-Linked, PN/T-700 Composites with Various Main Chain Lengths of the PN Oligomeric Precursors

Chain Length	Flexural Strength (MPa)	Flexural Modulus (GPa)	Interlaminar Shear Strength (MPa)
1	1722	164	71.8
3	1772	153	92.2
5	1855	125	80.8
10	1339	113	80.5

Data extracted from Figs. 10 and 11 Zong L, Liu C, Guo Y, Wang J, Jian X. Thermally stable phthalonitrile resins based on multiple oligo (aryl ether)s with phenyl-s-triazine moieties in backbones. RSC Adv 2015;5:77027–36.

Figure 5.5 SEM images of the fractured surfaces of the triazine-linked PN/T-700 composites with various main chain lengths of the PN oligomeric precursors: 1 (A), 3 (B), 5 (C), and 10 (D). *PN*, phthalonitrile; *SEM*, scanning electron microscopy. Reprinted from Zong L, Liu C, Guo Y, Wang J, Jian X. Thermally stable phthalonitrile resins based on multiple oligo (aryl ether)s with phenyl-s-triazine moieties in backbones. RSC Adv 2015;5:77027–36.

with *m*-APB and the 8 plies BPh/IM-7 composites cured with *p*-BAPS had at least 500 MPa higher flexural strength with a quite comparable flexural modulus and shear beam strength.

The flexural fractured samples were analyzed by SEM to study the fracture mechanisms. As seen from Fig. 5.5, the brittle (blue pane) and coarse (red pane) fractures were depicted on each micrograph. The brittle fractures were found to decrease upon increasing the chain length which indicated that the toughness was enhanced during this process.

Additionally, evidences of the CF pull out were also depicted as indicated in the white pane. The interface between the PN matrices and the CF was smooth and featureless confirming the weak adhesion between both phases. Therefore, improving the PN/CF compatibility is the challenge to be addressed if higher mechanical performances are targeted.

In summary, although the functional, PN resins showed some better performances than their typical PN counterparts, the CF composites based on functional, PN resins did not achieved better or even comparable performances. Therefore, more efforts need to be dedicated on improving the processability and the interface compatibility between the PN and the CF.

5.2.2.3 Phthalonitrile Blends/Carbon Fibers Composites

Blending the PN monomers with other polymeric systems has been extensively explored since the appearance of the first-known, PN monomers. In contrast with the functional, PN monomers, the ease of preparation and the reduced cost of the blends, favorited the development of several hybrid polymeric materials. In the PN field, the blends and copolymers are essentially studied to either overcome the sluggish curing behavior and brittleness of the typical, PN resins or to improve the thermal and thermomechanical properties of the non-PN phase. In either case, the variety of possible combinations allows the preparation of HP polymers with tailored properties.

As an extension of their previously discussed attempt to prepare CF composites based on functional, PN matrices, Augustine and coworkers opted to improve the toughness and the overall mechanical properties of the novolac-functional, PN resin by blending it with the poly(ether ether ketone) (PEEK) thermoplastic terminated with either PN (PEEKPN) or hydroxyl (PEEKOH) groups [39]. It should be reminded that the neat novolac-functional, PN polymers possess an excellent thermal stability along with good mechanical properties and high flame resistance. However, the curing of the novolac-functional, PN monomers was also found to be difficult and required high temperatures and extended time. Thus, the PEEKOH were expected to improve the curing behavior of the monomers and the PEEKPN were supposed to improve the integration of the soft PEEK moieties in the PN network for toughness improvements. The novolac-functional, PN blends with PEEKOH or PEEKPN were prepared by mixing the two phases in various weight ratios in tetrahydrofuran (THF) solvent. To the mixtures was also added 1 wt.% of 4,4′-diaminodiphenyl sulfone (DDS) curing agent to catalyze

Table 5.9 Mechanical Properties of Some CF-Reinforced Composites Issued from Various PN Precursors

Blend Matrix	Flexural Strength (MPa)	Interlaminar Shear Strength (MPa)	Impact Strength (kJ/m^2)
Neat novolac-functional PN	147	15	75
Novolac-Functional PN/PEEKOH by Weight Ratio			
90/10	324	23	128
80/20	483	40	186
70/30	257	22	100
Novolac-Functional PN/PEEKPN by Weight Ratio			
80/20	284	19	119

PN, phthalonitrile.

Reprinted from Augustine D, Mathew D, Reghunadhan Nair CP. End-functionalized thermoplastic-toughened phthalonitrile composites: influence on cure reaction and mechanical and thermal properties. Polym Inter 2015;64:146–53.

the polymerization reaction. The authors opted for the T-300 8H CF from TORAY to prepare the composites. The prepregs were obtained by impregnating the CF in the PN oligomeric solutions. Then, the prepregs were subsequently dried under atmospheric and vacuum conditions to completely remove the solvent. Finally, the cured composites were obtained by stacking and curing the prepregs following the cure schedule of 120°C for 30 min, 150 and 200°C for 1 h each, 250°C for 2 h, 300°C for 3 h, and 330°C for 4 h. The afforded composites were also further postcured at 330°C for 10 h.

The mechanical properties of the resulting composites were investigated and the obtained results are listed in Table 5.9. For comparison purpose, the neat, novolac-functional, PN/T-300 composites were also prepared and tested. The obtained data indicated that in all instances the blend-based composites showed better mechanical performances than those prepared from the neat resin. Meanwhile, increasing the amount of the PEEKOH in the blend to 20 wt.% resulted in a composite displaying the highest performances. For instance, the flexural and shear beam strengths reached up to 483 and 40 MPa, respectively, and the apparent impact strength was about 186 kJ/m^2. Further increasing the PEEKOH loading resulted in reductions in these values. Nevertheless, it

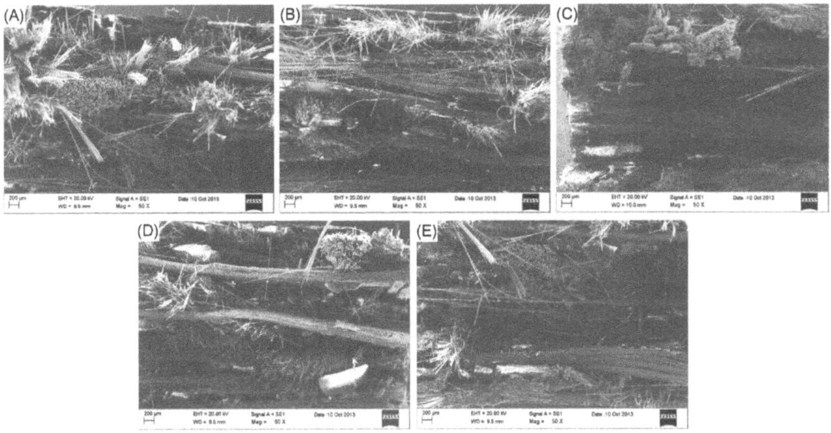

Figure 5.6 SEM images of the fractured surfaces of the composites prepared from CF and various polymeric matrices: novolac-functional PN (A), 90: 10 of novolac-functional PN and PEEKOH (B), 80:20 of novolac-functional PN and PEEKOH (C), 70:30 of novolac-functional PN and PEEKOH (D), and 80:20 of novolac-functional PN and PEEKCN (E). *PN*, phthalonitrile; *CF*, carbon fibers; *SEM*, scanning electron microscopy. Reprinted from Augustine D, Mathew D, Reghunadhan Nair CP. End-functionalized thermoplastic-toughened phthalonitrile composites: influence on cure reaction and mechanical and thermal properties. Polym Inter 2015;64:146–53.

appeared that the PEEKOH successfully improved the chain flexibility, and as a result, enhanced the toughness of the neat PN resin. On the other hand, the composite made from PEEKPN had a less toughening effect, which may be attributed to the PN ends that participated in the polymerization reaction and subsequently increased the crosslinking degree of the resulting polymeric structure.

The fractured samples issued from the bending test were analyzed by SEM to visualize the toughening effect of the different PEEKs. As seen from Fig. 5.6, the neat novolac-functional, PN/T-300 composites interphase was weak and evidences of breakages and fibers pull out were also depicted. The addition of the PEEK and especially the PEEOH improved the fiber wetting and enhanced the adhesion between the matrix and the fibers resulting in a toughened structure. As clearly seen in Fig. 5.6C, the PN matrix containing 20 wt.% of PEEKOH afforded the best interphase which further corroborated the mechanical properties findings. In summary, Augustine et al., succeeded in improving the overall mechanical performances of the novolac-functional, PN/CF composites by introducing the PEEK flexible segments. However, from a wider point of view these results remained a way below those afforded by the BPh/IM-7 composites.

In another recently published work by Bulgakov et al., the VIP was used to prepare HP, PN/CF composites [16]. The PN matrix was carefully chosen to meet the viscosity requirements of this process. Indeed, the VIP method imposes a melt viscosity lower than 1 Pa s for a longer period of time at a specific processing temperature. Thus, to meet this exigent requirement, the authors chose three, PN monomers and performed a step-by-step viscosity and performance investigations in order to find the suitable percentage of each component in the blend. The studied PN monomers were the bis(3-(3,4-dicyanophenoxy)phenyl) phenyl phosphate (RPPhPN), BPh, and 4-[3-(prop-2-yn-1-yloxy)phenoxy] benzene-1,2-dicarbonitrile (PPN).

The first PN monomer, referred to as RPPhPN, was synthetized by incorporating a phosphate-bridge between the PN ends. This monomer possesses a low T_g (about 21°C) and its cured form showed excellent thermal and thermomechanical properties similar to those obtained with the typical BPh monomer. The latter was also chosen as the second PN monomer of the blend. The third PN monomer was prepared by incorporating a propargyl functionality in the free end of a mono-functional PN monomer. The as such obtained PN monomer (PPN) was introduced considering its lower melt viscosity (around 220 mPa s at 120°C). The appealing low melt viscosity of the PPN is related to the high mobility of the propargyl ether group.

To avoid any misleading, it is noteworthy to point out that the authors fixed the processing temperature at 150°C and the amount of *m*-APB curing agent at 4 mol.%. By knowing these facts, it should be acknowledged that any change in these parameters would result in different blends' properties and therefore the following findings cannot be extrapolated to other processing temperatures or different types and amounts of curing agents. To find the proper blend to meet the VIP viscosity requirements, the authors first prepared a 30/70 weight ratio of RPPhPN and BPh. However, the blend crystallized during the rheometric measurements and was deemed nonsuitable for further investigations. By lowering the BPh ratio to 50 wt.%, the crystallization was avoided and the blend exhibited a 600 mPa s of viscosity at 150°C while its pot life was over 10 h. To achieve even lower melt viscosity, the PPN was introduced and ternary blends were formulated by varying the weight ratio of BPh and PPN while that of RPPhPN remained constant. Thus, three blends containing 30/60/10, 30/50/20, and 30/40/30 of RPPhPN, BPh, and PPN, respectively, were studied by rheometric analyses to find the best match in terms of processability and pot life. Except for the first composition, the remaining blends displayed low viscosity plateaus (250–300 mPa s) starting below 150°C. Although the third blend had the lowest starting plateau temperature, the high enthalpy released during its curing was judged inappropriate. As a consequence, the second blend was chosen as the most appropriate matrix for the preparation of CF composites by VIP process. Moreover, the pot life of more than 24 h at 150°C for the second blend was another processing advantage.

The three blends were also cured and investigated for their mechanical properties. The copolymers were obtained after curing the blends at 250°C over night and at 315°C for 5 h plus an additional postcure at 375°C for 8 h. The bending test revealed that all the copolymers had relatively lower performances than the typical, BPh polymer with respect to the type and amount of the used curing agent and to the adopted curing schedule. Fortunately, these copolymers showed major improvements in the toughness. For instance, the copolymer 2 and 3 had up to 173 and 202 J/m^2 of strain energy release rate in crack growth (G_{IC}) with a stress intensity factor (K_{IC}) of about 0.64 and 0.72 MPa, respectively. On the other hand, the BPh and PMR-15 had a G_{IC} of around 120 and 87 J/m^2. Considering the appealing mechanical properties of the copolymers 2 and 3, the authors chose to use their respective blends as matrices for the CF composites.

The PN/CF composites were manufactured using the VIP technique. Basically, 12 plies of 3k HTA40 TohoTenax CF from Teijin Company were impregnated for about 30 min in a programmable oven equipped with outlets for a vacuum pump. For practical limitations, the samples were first cured at 200°C for 24 h inside the vacuum bag. Afterwards, the panels were removed from the bag and further postcured at 375°C for 8 h. The average volume ratio of the resin in the composite was about 37 vol.%.

The mechanical properties of the composite panels were compared in terms of their tensile, compressive, and shear strengths at room temperature as well as at 300°C. The obtained results are resumed in Table 5.10. Regardless of the PN matrix composition, all the studied mechanical properties showed quite similar performances at room and high temperature (300°C). Interestingly, the tensile strength was further enhanced at 300°C due to the increase in the ductility upon heating. Moreover, increasing the proportion of the PPN in the matrix significantly improved all the tested mechanical properties. In contrast with the PN/CF composites from the 30/50/20 composition, the PN/CF composites from the 30/40/30 composition showed improvements of about 37.4%, 25.8%, and

Table 5.10 Mechanical Properties of the PN/CF Composites Prepared from Two Different Blends of PN Monomers and Manufactured by VIP

Blend Matrix	Temperature (°C)	Tensile Strength (MPa)	Compressive Strength (MPa)	Interlaminar Shear Strength (MPa)	In-plane Shear Strength (MPa)
Composition 2	25	487	341	31	66.3
	300	535	315	30	64.5
Composition 3	25	669	429	37	78.1
	300	691	382	35	72.6

Reprinted from Bulgakov B, Sulimov A, Babkin A, Timoshkin I, Solopchenko A, Kepman A, et al. Phthalonitrile-carbon fiber composites produced by vacuum infusion process. J Compos Mater 2017. doi: 10.1177/0021998317699452.

19.4% in the tensile, compressive, and interlaminar shear strengths, respectively. The authors attributed these improvements to the enhanced wettability of the CF in the presence of more PPN contents.

Since the authors did not report about the flexural properties, the comparison with other PN/CF composites is limited to the tensile and shear strengths values. First of all, the mechanical properties BPh/IM-7 composites remain unchallenged since 1996. The interlaminar shear strength of these composites was slightly better than that obtained with the neat novolac-functional PN/CF composites. Moreover, the overall performances of the triazine-linked, PN/CF composites were also superior. Nonetheless, the work performed by Bulgakov et al., tailored the processing conditions of the PN to meet for the first time the exigent requirements of the VIP process, rendering possible the manufacturing of complex-shaped, PN-based composites. Meanwhile, room for future improvements is also available by changing the initial processing conditions and by modifying the blends compositions.

5.3 Glass Fibers/Phthalonitrile Composites

5.3.1 Properties of Glass Fibers

Glass fibers (GF) are the most common among all reinforcing fibers in the field of polymer composites. Since their first conception in the 1930s by Owens Corning Textile Products, the GF rapidly initiated a revolution in reinforced composite materials. The main advantages of the GF are their exceptional, high tensile strength, low cost of fabrication, and superior chemical resistance. Owing to their many interesting features, the GF have been extensively used in common and exigent applications. Thus, the global annual GF consumption in 2000 was about 2.6 million tons and this value exponentially increased since then.

A large number of commercially available GF are produced by varying the proportion of the raw materials in the batch. The principal ingredient in all the GF is the silica (SiO_2) while the introduction of other oxides such as Al_2O_3, B_2O_3, CaO, and so on, allows the modification of the GF final properties [21]. Table 5.11 provides the oxide designations and their respective weight ranges for the main types of the commercially available GF. The most commonly used GF are known as the E- and S-glass. The E-glass GF have the lowest manufacturing cost and are designated for GP where strength and electrical resistivity are required. The S-glass GF, also known as the HT GF, were first developed for military applications, such as for missile casings and military aircraft components. Later another version (S-2-glass), prepared with a lower cost of fabrication and quite similar performances, was commercialized for civilian applications. Other types of GF were also designed for different application purposes. For instance, the C-glass GF allow greater resistance to acid-induced corrosion and the AR-glass are used in cement substrates and concrete.

The general manufacturing process of the GF can be resumed as follows: first the raw materials are dry-mixed and melted in a furnace at about 1400°C. The melted glass is then transformed into filaments by passing through a number of orifices to achieve filaments of about 10 μm in diameter. At this step, each filament is coated with a protective layer made from lubricants, antistatic agents, binders, and sometime with coupling agents for a promoted adhesion with the polymeric matrices. The as-such, obtained filaments are collected and stacked together to form continuous strands. Another form of GF, known as roving, is obtained by wounding a group of untwisted parallel strands on a cylindrical forming package. GF with reduced length can also be achieved by cutting the continuous stands into short lengths. Moreover, GF with bidirectional properties can be afforded by the preparation of woven roving or woven cloth. All these GF forms can be used for the preparation of polymeric composites by the conventional fabrication processes.

The properties of the commercially available GF are summarized in Table 5.12. As previously stated, the high tensile strength of the GF is

Table 5.11 Oxide Designations and their Respective Weight Ranges for the Main Types of Commercially Available GF

GF Type Oxide	A-Glass %	C-Glass %	D-Glass %	E-Glass %	ECR-Glass %	AR-Glass %	R-Glass %	S-2-Glass %
SiO_2	63–72	64–68	72–75	52–56	54–62	55–75	55–60	64–66
Al_2O_3	0–6	3–5	0–1	12–16	9–15	0–5	23–28	24–25
B_2O_3	0–6	4–6	21–24	5–10		0–8	0–0.35	
CaO	6–10	11–15	0–1	16–25	17–25	1–10	8–15	0–0.2
MgO	0–4	2–4		0–5	0–4		4–7	9.5–10
ZnO					2–5			
BaO		0–1						
Li_2O						0–1.5		
$Na_2O + K_2O$	14–16	7–10	0–4	0–2	0–2	11–21	0–1	0–0.2
TiO_2	0–0.6			0–1.5	0–4	0–12		
ZrO_2						1–18		
Fe_2O_3	0–0.5	0–0.8	0–0.3	0–0.8	0–0.8	0–5	0–0.5	0–0.1
F_2	0–0.4			0–1		0–5	0–0.3	

GF, glass fibers.

Table 5.12 **Some of The Physical Properties of the Commercially Available GF**

	A-Glass	C-Glass	D-Glass	E-Glass	ECR-Glass	AR-Glass	R-Glass	S-2-Glass
Density (g/cm^3)	2.44	2.52	2.11−2.14	2.58	2.72	2.70	2.54	2.46
Tensile strength (MPa)	3310	3310	2415	3445	3445	3241	4135	4890
Tensile modulus (GPa)	68.9	68.9	51.7	72.3	80.3	73.1	85.5	86.9
Elongation (%)	4.8	4.8	4.6	4.8	4.8	4.4	4.8	5.7

the main advantage of these fibers. For instance, the E-glass tensile modulus can reach up to 3.4 Gpa, while that of the S-glass is around 4.5 GPa. These high tensile values largely surpass those of the typical available CF. Moreover, the GF are also characterized by a high ductility, which is up to 5.2% for the S-glass. On the other hand, the GF suffer from a reduced tensile modulus (less than 100 GPa), which is one of the lowest among all the fibers and whiskers. In addition, the high density of the GF (more than 2.4 g/cm^3) hindered their uses in many applications where light structures are needed. Other GF shortcomings include a high sensitivity to abrasion during handling, low fatigue resistance, and high hardness.

5.3.2 Glass Fibers Reinforced Phthalonitrile Composites

According to the type of the PN resin matrix, three (03) categories can be distinguished. The first category englobes the PN matrices prepared from typical or common, PN monomers/oligomers, while the second category includes PN matrices prepared from X-functional, PN monomers/oligomers. The last category regroups the PN matrices made from physical blends of typical or functional PN blended with other polymeric systems including thermosets and thermoplastics.

5.3.2.1 Typical Phthalonitrile/Glass Fibers Composites

The first research concerned about the PN/GF composites was published by Sastri et al., in 1997 [40]. The authors used the typical BPh resin as the matrix for the E-glass GF (#7781 with high temperature 497a finish from BFG Corporation). The BPh/GF composite panels were obtained by the conventional prepregs consolidation technique and by the RIM process.

The prepregs consolidation technique used to prepare the BPh/GF composites was quite similar to that reported by the same authors for the fabrication of the BPh/CF panels. Briefly, a B-stage resin was first prepared by adding to the melt of PN monomers 1.8 wt.% of *m*-APB curing agent. The mixture was vigorously stirred for a short period of time before being quenched to room temperature to form the prepolymer. Subsequently, the BPh/GF prepregs, with about 32% of PN weight, were afforded by the solution impregnation process using a mixture of solvents. The impregnated panels were then stacked together (16 plies) between 3-mil-thick DuPont Kapton release films. Prior to the consolidation step, viscosity measurements were performed on the prepregs to find the suitable autoclave cure cycle. The rheometric measurements for the BPh/GF prepregs revealed that the viscosity remained as low as 0.02 Pa s for 1 h at 260°C. Therefore, an initial dwell at 260°C was first applied and then the autoclave temperature was increased to 325°C for 4 h. Meanwhile, 200 psi pressure was applied soon after the resin melted to ensure an adequate flow. The as-such, cured panels were confirmed to be void-free and high-quality composites. Physical characterizations showed that the 16 plies-based BPh/GF composites had a density of about 2.1 g/cm^3 with an average thickness of 2.9–3.0 mm.

In addition to the autoclave cure, some samples were postcured under inert atmosphere either at 8 h for 325°C or at 4 h at 350 and 4 h at 375°C. The effects of the postcure conditions were studied by comparing the DMA, bending, and shear test results. The DMA results confirmed the ability of the postcured samples to maintain higher percentages of their initial storage, modulus values at 400°C. For instance, the samples postcured at 4 h for 350°C and 4 h at 375°C showed a storage modulus retention at 400°C of 79%. The T_g was also markedly improved by extending the postcuring time and temperature. In fact, the T_g moved from 320°C for the non-post-cured samples to more than 450°C for the postcured samples.

As clearly seen in Table 5.13, the BPh/GP composites exhibited excellent mechanical performances with a flexural strength and modulus that reached 824 MPa and 31 GPa, respectively, for the samples cured up to 8 h at 325°C. In addition, the bending tests also revealed that the flexural strength was slightly affected by the postcure. In the meantime,

Table 5.13 Effect of The Postcure Conditions on the Flexural Properties of the BPh/GF Composites

Postcure Conditions	Strength (MPa)	Strain (%)	Modulus (GPa)
None	773	3.2	30.7
325°C for 8 h	824	3.1	31.0
350°C for 4 h and 375°C for 4 h	819	2.6	32.9

Reprinted from Sastri SB, Armistead JP, Keller TM. Phthalonitrile-glass fabric composites. Polym Compos 1997;18:48–54.

the strain decreased and the modulus increased by extending the post-curing conditions. Based on the above observations, it was confirmed the formation of a stiffer matrix upon postcuring the samples. Further mechanical properties investigations, showed that the BPh/GP composite panels postcured up to 375°C had a short beam shear strength of about 54 MPa. The shear tests also highlighted the fact that the post-cured samples behaved better in the compressive mode, while the non-post-cured sample had a better behavior in the interlaminar mode. In contrast with other HP polymer composites such as those prepared from epoxy, benzoxazine, and cyanate ester, the BPh/GP composites proved to be highly superior. For instance, the neat epoxy/GF composites exhibited flexural strength up to 367 Mpa, while the GF composites from the typical bisphenol-A (BPA) benzoxazine had a flexural strength of about 420 MPa [41]. The BPh/GP composites also surpassed the BPA cyanate ester/GF composites which exhibited a flexural strength and modulus up to 691 MPa and 36 GPa, respectively [42].

The same BPh/GF composites were also fabricated via the RIM process and the obtained composites performances were compared with those obtained from the conventional prepregs consolidation technique. The RIM is particularly appealing since it combines the impregnation and consolidation processes into one step. In this case, the resin viscosity is the most important factor that determines the properties of the cured panels. Similarly to the VIP technique discussed in previous sections, the RIM is attractive for resins with low viscosity values at high temperatures. The initial melt viscosity of the BPh when cured in presence of 1.7−1.8 wt.% of *m*-APB ranges between 0.4 and 0.6 Pa s at 260°C. Therefore, as far as the viscosity is concerned these PN prepolymers are suitable to be used in the RIM process. Nevertheless, in order to determine the proper amount of the *m*-APB curing agent to be added to the PN monomers, the authors tried various amounts and studied their effect of the physical properties of the panels. First, the PN prepolymers containing 1.8 wt.% of *m*-APB afforded good quality panels with 12 layers of GF. However, when 16 plies were used, the PN prepolymers did not achieve the complete wettability of all fibers. Therefore, it was predicted that reducing the contents of the amine-curing agent could increase the fibers wetting. Indeed, when 1.68 wt.% *m*-APB was used, the wettability was improved and high-quality, PN/GF panels with homogenous thickness were achieved. Table 5.14 lists some of the properties of the BPh/GF composites obtained by both prepregs consolidation and RIM techniques. Regardless of the minor differences in the fiber volume fraction, density, and thickness, both manufacturing

Table 5.14 Properties Comparison of the BPh/GF Composites Manufactured by the Conventional Prepregs Consolidation Technique and by RIM

Property	RIM	Prepregs Consolidation Technique
Fiber volume fraction	54	60
Density (g/cm^3)	2.04	2.10
16 plies panel thickness (mm)	3.2	2.9
T_g after 4 h at 350°C and 4 h at 375°C (°C)	>450	>450
Short beam shear strength (MPa)	52	54

RIM, resin infusion molding.
Reprinted from Sastri SB, Armistead JP, Keller TM. Phthalonitrile-glass fabric composites. Polym Compos 1997;18:48–54.

techniques afforded composites with quite comparable properties. However, if the cost of production is taken into account, the RIM process seems to be a best alternative to the conventional prepregs consolidation technique.

Some of the many advantages of the PN resins are their low water uptake and superior flame resistance. As previously discussed in the PN/CF composites, the BPh/IM-7 composites had less than 1% of water uptake after 16 months of immersion and their flammability performances met the US Navy flammability MIL-STD-2031 requirements. Considering the low cost of the GF, the BPh/GF can be used in marine applications where their relatively higher density can be tolerated. Therefore, the water uptake and flammability properties of the BPh/GF composites were explored to see whether these materials can also meet the US Navy requirements. As expected, the water uptake test revealed that the BPh/GF composites had also less than 1% of water uptake after 5 months of immersion. The flammability tests included specific optical measurements during a smoke density test and determinations of the heat release and ignitability data of the BPh/GF composites. As seen in Table 5.15 where are presented the smoke and combustion gas data, the BPh/GF composites displayed the lowest specific optical density at 300 s (D_s) and maximum specific optical density among a large number of polymeric composites. In addition, the concentration of carbon monoxide evolving during the combustion was minimal for the BPh/GF composites. Moreover, in terms of T_{ig} and PHR, the BPh/GF composites

Table 5.15 Smoke and Combustion Data of Several GF-Reinforced Composites

Composite	D_s (300 s)	D_{max}	CO (ppm)	CO_2 (%V)
BPh/GF	1	1	40	0.5
Vinyl ester/GF	463	576	230	0.3
Epoxy/GF	56	165	283	1.5
BMI/GF	40	130	300	0.1
Phenolic/GF	7	20	300	1.0
Polyimide/GF	3	15	200	1.0

GF, glass fibers.
Reprinted from Sastri SB, Armistead JP, Keller TM. Phthalonitrile-glass fabric composites. Polym Compos 1997;18:48–54.

also clearly met the US Navy flammability MIL-STD-2031 requirements. In light of these facts, the BPh/GF composites appeared to be a good alternative to many polymeric composites in use in marine and fire resistant applications.

Another attempt to prepare HP, PN/GF composites was reported by Zhou et al. [43]. The authors dedicated their efforts on studying the effect of the amino position on the aminophenoxy PN (APN) monomers curing behavior and processability. In the margin of their work, the authors also used the conventional prepregs consolidation technique to study the thermomechanical properties of the APN/GF composites. It should be pointed out that the APN monomers have played various roles in the field of the PN materials. In fact, the APN monomers were used as stand-alone, PN monomers and also as advanced curing agents for other, PN and non-PN monomers. In their work, Zhou et al. used the APN monomers as matrices for the preparation of GF-reinforced composites. The fabrication process involved the dissolution of the monomers in a proper solvent at a concentration of 0.2 g/L. Subsequently, 8 plies of GF were dipped into the solution to achieve the impregnation. The prepregs were then properly dried to remove any trace of solvent. The consolidation was achieved by curing the prepregs in a closed steel mold following the curing schedule of 70°C for 1 h, 200°C for 1 h, and 250°C for 5 h. An additional heat treatment was applied at 315°C for 5 h and 375°C for 5 h in an open-air oven. The weight ratio of the PN matrices in the composites ranged from 37 to

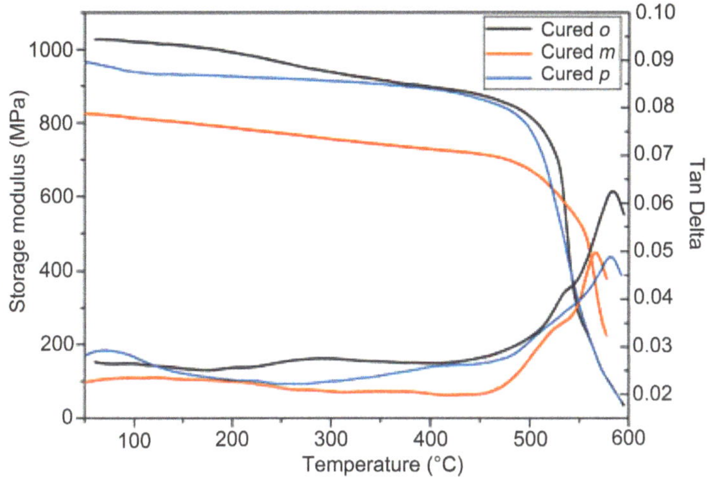

Figure 5.7 Storage modulus and Tan Delta variations of the GF-composites issued from APN matrices bearing an amino group in different positions: *ortho* (*o*), *meta* (*m*), and *para* (*p*). *APN*, aminophenoxy PN; *GF*, glass fibers. Reprinted from Zhou H, Badashah A, Luo Z, Liu F, Zhao T. Preparation and property comparison of ortho, meta, and para autocatalytic phthalonitrile compounds with amino group. Polym Adv Technol 2011;22:1459–65.

40 wt.% according to the position of the amino group in the APN monomer.

The thermomechanical properties of the APN/GF composites were investigated by DMA and the obtained results are presented in Fig. 5.7. Regardless of the amine position, all the composites showed T_g values over 550°C. The storage modulus was also found to maintain a large percentage of its initial value up to 490°C. After 500°C, the storage modulus showed catastrophic decreases because the polymeric matrices started to decompose at these temperatures. Although the authors did not report about other mechanical properties, the APN/GF composites have the potential of the BPh/GF composites and are expected to exhibit even better properties.

In a different report, Chen et al. used the well-known BPA-based PN (BAPh) resin to prepare HP, PN/GF composites [44]. The BAPh monomers were cured in the presence of various amount of 3-APN (with the amine functionality in the *meta* position) to study the effect of the latter on the thermal and mechanical properties of the resulting composites. Prior to the composites fabrications, the authors investigated the impact of the 3-APN on the melt viscosity of the BAPh monomers. Thus, the BAPh monomers in the presence of 10–70 wt.% 3-APN exhibited initial melt viscosities at 200°C between 0.01 and 0.1 Pa s. Meanwhile, the isothermal viscosity measurements for the 50/50 3-APN/BAPh prepolymers at 200°C revealed that a sharp increase in the viscosity started after 37 min. Clearly, the 3-APN allows good processing conditions favorable for composites fabrication. Moreover, by changing the amount of 3-APN and the processing temperature, a proper control on the manufacturing process can be easily achieved.

The BAPh/GF composites were fabricated by the conventional prepregs consolidation technique. First, the mixtures of 3-APN and BAPh monomers with various ratios (100/0, 70/30, 50/50, and 30/70) were dissolved in dimethyl formamide. Then, the impregnation process took place by dipping the GF into the solutions. After a proper drying process, 20 plies of impregnated GF were stacked together for each composition and hot-press cured following the curing program of 200°C for 2 h, 220°C for 2 h, 250°C for 4 h, and 280°C for 4 h. The obtained composites were machinated to meet the size specifications of each test.

The thermal and thermo-oxidative stability of the fabricated, BAPh/GF composites were studied by TGA. From the obtained curves, it appeared that all the composites achieved an excellent thermal resistance with $T_{5\%}$ ranging from 501 to 546°C under nitrogen atmosphere and from 479 to 536°C under air atmosphere. In contrast with the thermal performances of the neat matrices, the introduction of highly thermally stable fibers further improved the overall thermal stability. Interestingly, the 50/50 composition exceled over the other matrices and afforded the best match in term of thermal resistance. This may be ascribed to the high compatibility between the 3-APN and BAPh monomers at this ratio.

The mechanical performances of the composites were studied in terms of flexural strength and modulus. Globally, the composites showed good flexural strengths ranging from 500 to 766 MPa and the flexural modulus varied between 21 and 27 GPa. Once more, the 50/50 matrix composition was found to impart the best match with a flexural strength of about 766 MPa. This observation corroborated the viscosity and thermal stability findings. With respect to the curing and processing conditions, these BAPh/GF composites showed quite comparable performances to that of the previously discussed BPh/GF composites. Further SEM investigations of the fractured surfaces of the composite from 50/50 composition revealed that a good state of adhesion between the matrix and the fibers was achieved. Indeed, at lower magnification, the GF appeared to be attached by the polymeric matrix, while at higher magnification, the surfaces of the GF also appeared to be clearly covered by the matrix indicating an ameliorated degree of impregnation.

Chen et al., also explored the effect of another curing agent, namely the BPA, on the thermal and mechanical properties of the BAPh/GF composites [45]. Viscosity measurements for the BAPh/BPA compositions with various molar ratios (2/0.5, 2/1, 2/1.5, and 2/2) confirmed that all the compositions had initial melt viscosities between 0.1 and 1 Pa s. The BAPh/GF composites cured in the presence of BPA were manufactured in a similar procedure to that described for the BAPh/GF composites cured in the presence of APN with some minor changes. The modifications included the reduction of the number of plies to 15 and the last curing temperature was risen to 300°C.

The TGA investigations were carried out in both nitrogen and air atmospheres and the obtained results are listed in Table 5.16. The thermal stability parameters were found to be improved by more BPA contents. For instance, the 2:0.5 BAPh/BPA composition afforded a composite with a $T_{5\%}$ of 509°C and a char yield at 800°C of 84.5% under nitrogen atmosphere, these values were improved by 17°C and 5.6%, respectively, for the composite prepared from the 2/2 composition. In fact, with larger amounts of BPA, the crosslinking degree was increased resulting in a more thermally stable polymeric structure.

The bending and tensile tests were used to assess the mechanical properties of these BAPh/GF composites. The flexural strength and modulus values, listed in Table 5.17 revealed the presence of a close relationship between these properties and the amount of BPA in the matrix. Typically, the flexural strength and modulus moved from

Table 5.16 Thermal and Thermo-Oxidative Properties of the BAPh/GF Composites Cured in the Presence of Various Amounts of BPA

	Nitrogen		Air	
BAPh/BPA Matrix (Molar Ratio)	$T_{5\%}$ (°C)	Char Yield at 800°C (%)	$T_{5\%}$ (°C)	Char Yield at 800°C (%)
2:0.5	509	84.5	508	79.3
2:1	514	86.6	518	86.2
2:1.5	529	89.4	543	89.9
2:2	526	90.1	523	84.6

BPA, bisphenol-A.

Reprinted from Chen Z, Guo H, Tang H, Yang X, Xu M, Liu X. Preparation and properties of bisphenol A-based bis-phthalonitrile composite laminates. J Appl Polym Sci 2013;129:2621–8.

Table 5.17 Flexural Strength and Modulus of the BAPh/GF Composites Cured in the Presence of Various Amounts of BPA

BAPh/BPA Matrix (Molar Ratio)	Flexural Strength (MPa)	Flexural Modulus (GPa)
2:0.5	525	24.9
2:1	554	25.8
2:1.5	633	30.1
2:2	571	26.4

BPA, bisphenol-A.
Data extracted from Figs. 10 and 11 Chen Z, Guo H, Tang H, Yang X, Xu M, Liu X. Preparation and properties of bisphenol A-based bis-phthalonitrile composite laminates. J Appl Polym Sci 2013;129:2621–8.

525 MPa and 24.9 GPa to 633 MPa and 30.1 GPa, respectively. Nevertheless, for an equivalent molar ratio of BAPh monomers and BPA, the flexural strength and modulus were reduced to 571 MPa and 26.4 GPa, respectively. The same trend was also observed for the tensile strength which reached its maximum value of 415 MPa at the 2/1.5 BAPh/BPA composition. Globally, the studied composites showed good mechanical performances but slightly lower than those obtained with the 4-APN/BAPh matrices. Moreover, the initial improvements in these properties by increasing the BPA contents were attributed to the development of a highly crosslinked network essentially afforded by the presence of a sufficient number of hydroxyl groups. However, when more than necessary amount of BPA was added, the unreacted molecules started to volatilize at the high curing temperatures causing voids and defects that lowered the mechanical properties of the composites.

5.3.2.2 X-Functional Phthalonitrile/Glass Fibers Composites

Although many functional PN monomers/oligomers were developed, only few of them were used as matrices for GF composites. The well-known, benzoxazine-functional, PN (BZPN) is one of the most studied functional, PN materials. The BZPN monomers are known for their self-curing behavior and good processability at lower temperature while their thermal and thermomechanical properties are comparable to those of the typical, PN polymers. Moreover, extensive investigations about the BZPN revealed that a large number of nitriles remained unreacted at the end of the polymerization process. Therefore, in order to achieve better performances Xu et al., claimed that the addition of small amounts of BPA could effectively provide the required hydroxyl groups to advance the polymerization degree of the nitrile groups [46].

As previously stated, the fabrications of the PN/GF composites were mainly performed at relatively high temperature (>200°C) to achieve an adequate flow of the resin for impregnation purposes. On the other hand, the BZPN monomers can be easily processed at relatively lower temperatures and allow a better control over the polymerization reaction. Meanwhile, it was found that adding increasing amounts of BPA resulted in gradual increases in the reactivity. Although the increase in the reactivity is generally not appreciated, small sacrifices were tolerated considering the overall improvements in the thermal and mechanical properties. In summary, when 2/1, 2/2, and 2/3 molar ratios of BZPN to BPA were studied by rheometric analyzes; all the compositions showed reduced viscosity values with minimum melt viscosities of 1−6 Pa s. Moreover, isothermal viscosity measurements at 160°C revealed that the low melt viscosity of 3 Pa s was maintained for all the compositions for at least 30 min. Subsequently, the viscosity increased at various rates according the BPA contents in each composition.

In the course of their work, Xu et al. also reported about several PN/GF composites prepared from the E-glass GF cloth-7628, provided by Jiangxi Changjia glass fiber Co. Ltd., Jiangxi, China and PN matrices composed of BZPN and BPA in various proportions [46]. The composites were fabricated by the conventional impregnation process in which the BZPN and BPA were first physically mixed and then dissolved in butanone at 80°C for 2 h. Afterwards, the obtained viscous mixtures were used to impregnate the GF clothes. Next, 10 plies of dried prepregs were stacked together and hot-press cured at 160°C for 4 h. The as-such, obtained composites panels were first cooled down to room temperature and then further cured in an oven at 200°C for 4 h, 240°C for 2 h, and 280°C for 2 h.

The flexural properties of the neat- and BPA-based BZPN/GF composites at various treated temperatures are displayed in Fig. 5.8. The neat BZPN/GF composites treated at 160°C had a flexural strength and modulus of about 370 MPa and 12 GPa, respectively. These values were gradually ameliorated to reach 542 MPa and 25.8 GPa at 280°C, respectively. On the other hand, all the BPA-based, BZPN/GF composites exhibited gradual improvements in the flexural properties to a certain extent of heat treatment. Indeed, the highest flexural strengths were obtained at the heat treatment of 200°C. For instance, the 2/1, 2/2, and

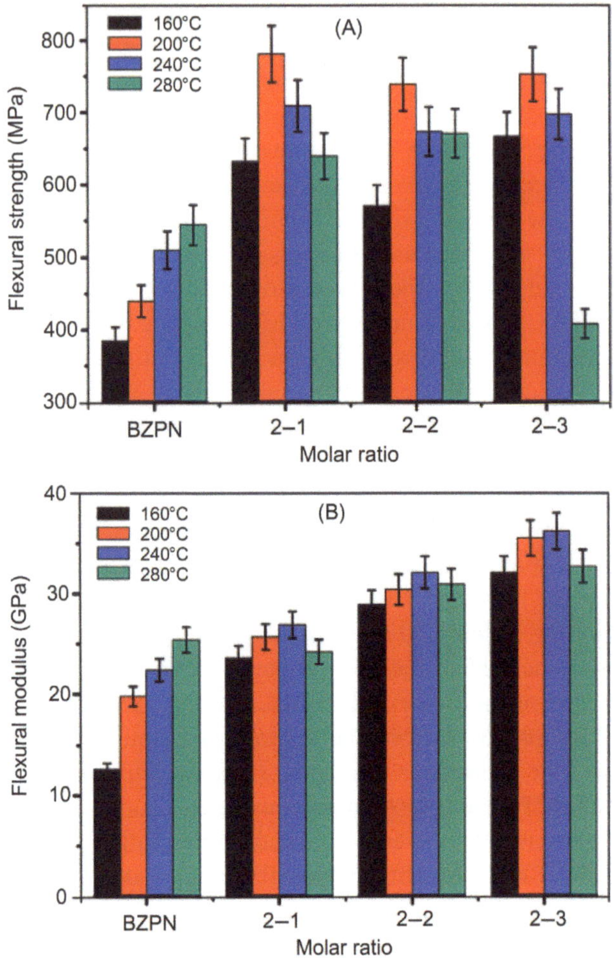

Figure 5.8 Flexural properties of the neat- and BPA-based BZPN/GF composites at various treated temperatures. *BPA*, bisphenol-A; *BZPN*, benzoxazine-functional; *GF*, glass fibers. Reprinted from Xu M, Jia K, Liu X. Effect of bisphenol-A on the structures and properties of phthalonitrile-based resin containing benzoxazine. Exp Polym Lett 2015;9.

2/3 BZPN/BPA compositions reached their maximum flexural strength value of 786, 737, and 753 MPa at that temperature, respectively. More importantly, it was proven that in all instances the flexural properties of the BPA-based, BZPN/GF composites were better than those of the neat, BZPN/GF composites. This was due to the fact that with more BPA contents, the extent of polymerization of the nitriles was increased resulting a highly crosslinked network. Furthermore, the reductions in the flexural properties of the BPA-based BZPN/GF composites upon increasing the heat treatment temperature over 200°C were attributed to the volatilization of the unreacted BPA causing structural failures in the composites. In summary, although these composites were processed and manufactured at lower temperature and reduced number of plies, they exhibited similar flexural properties to those of the neat BPh/GF composites cured at higher temperatures and extended period of time. Therefore the BPA-based BZPN/GF composites have the potential to be used in extremely exigent applications if they pass the flammability and water uptake requirements.

Encouraged by the promising results of the BPA-based, BZPN/GF composites, Luo et al., studied the effect of another bisphenol, namely the diallyl BPA (DBPA), on the overall properties of the BZPN and BZPN/GF composites [47]. The presence of the additional allyl groups on the DBPA was believed to improve the crosslinking density of the resulting polymer. The rheometric investigations confirmed that similarly to the BZPN/BPA composition, the BZPN/DBPA can be used to manufacture PN/GF composites using the conventional preparation techniques with low processing temperatures. The DBPA-based, BZPN/GF composites were fabricated in a similar way to that used for the preparation of the BPA-based, BZPN/GF composites with some minor changes in the curing cycle.

The bending test showed that the flexural strength and modulus were both ameliorated by increasing the DBPA contents up to an equivalent molar ratio with the BZPN. For instance, at that molar ratio the flexural strength and modulus reached 560 MPa and 37 GPa, respectively, which is about 40 MPa and 12 GPa higher than those of the neat BZPN/GF composites. These improvements in the bending properties were attributed to the increase in the crosslinking density of the polymeric matrix upon adding the DBPA. In fact, during the polymerization reaction several mechanism were involved; first the hydroxyl groups of the DBPA accelerated the ring opening of the oxazine rings and catalyzed the cyano addition reaction of the PN units. Then, the free hydroxyl groups generated from the oxazine rings also participated in activating the cyano addition reaction. Finally, the self-polymerization of the allyl groups generated supplementary hydroxyl groups that subsequently increased the extent of nitriles polymerization. Conversely, with such a high number of curing paths the mechanical properties were not better than those obtained with the BPA-based, BZPN/GF composites. Indeed, with the occurring of several polymerization reactions, phases separations probably occurred which may explain the relatively lower enhancements in the mechanical properties. In addition, at high DBPA contents, volatilization of the unreacted molecules may also have weakened the polymeric matrix resulting in negative impact of the overall performances. In summary, the functional PN matrices proved to afford HP, GF-based composites with lower processing temperatures and easy control over the final composites properties. Considering the large panel of available, functional, PN resins, the next few years may show an exponential increase in the number of studies on their use as HP matrices for GF-reinforced composites.

5.3.2.3 Phthalonitrile Blends/Glass Fibers Composites

Several PN blends were studied and evaluated for their possible use as HP matrices for FRC. The ease of preparation, cheap cost, and possibilities of tailoring the properties are the main advantages of the PN blends-based, FRC. All the PN blends discussed in this part were previously reviewed in Chapter 4, Phthalonitrile-Based Blends and Copolymers, and herein brief explanations were introduced where necessary.

Xu et al., reported about PN/GF composites prepared from several blends of BAPh oligomers and BZPN monomers [48]. The authors studied the effect of each phase amount on the thermal and mechanical properties of the resulting GF-based composites. As far as the processability of the blends was concerned, it was confirmed a low, initial melt viscosity for all the blends (between 0.1 and 1 Pa s). Moreover, with higher amounts of BZPN, the blends were more reactive. Meanwhile, isothermal rheometric measurements for an equivalent weight ratio of BZPN and BAPh confirmed that the pot life at 180°C was about 60 min and at 200°C around 25 min. The BAPh/BZPN/GF composites, containing 20 plies of E-glass GF EW170-100, provided by Shenyang Gaote glass fiber Co. Ltd., Shenyang, China, were fabricated following the conventional prepregs consolidation technique with a curing schedule of 200°C for 8 h under 20 MPa of pressure.

The flexural properties of the BAPh/BZPN/GF composites are summarized in Table 5.18. As clearly seen, increasing the percentage of the BAPh in the blends resulted in gradual increases in the flexural properties. In contrast with the neat, BAPh/GF composite which displayed a flexural strength of 494 MPa and a flexural modulus of 20 GPa, the BAPh/BZPN/GF composites containing 70 wt.% of BAPh exhibited improvements of about 6.4% and 25% in the flexural strength and modulus, respectively. These findings are quite similar to those obtained

Table 5.18 Flexural Strength and Modulus of the BAPh/BZPN/GF Composites

Loading Content of BZPN (wt.%)	Flexural Strength (MPa)	Flexural Modulus (GPa)
0	494	20
30	512	21
50	521	23
70	530	25

BZPN, benzoxazine-functional PN.
Data extracted from Fig. 8 and 9 Xu M, Liu M, Dong S, Liu X. Design of low temperature self-cured phthalonitrile-based polymers for advanced glass fiber composite laminates. J Mater Sci 2013;48:8108–16.

with the BPA- and DBPA-based BZPN/GF composites. Regardless of the fibers properties, the good flexural performances were attributed to both the high crosslinking density of the matrices and to the good interfacial adhesion between them and the GF.

Xu et al., also used the TGA to assess the thermal and thermo-oxidative stabilities of the BAPh/BZPN/GF composites. Globally, all the studied samples showed a high thermal resistance in either nitrogen or air atmospheres. For example, the $T_{5\%}$ was in all instances superior to 430°C and the best performances were depicted for the composite containing the highest amount of BAPh. Considering the low curing temperatures and time used during the fabrication process, the BAPh/BZPN/GF composites displayed a good cost/performance ratio in comparison with other PN/GF composites.

Following the same fabrication process, various other PN blends were used as matrices for GF-reinforced composites. Considering the similarities in the manufacturing processes and with respect to the curing behavior and processability of each system, Table 5.19 regroups the flexural properties of all the remaining, blend-based, PN/GF composites in accordance with their respective number of plies and curing schedule. The mechanical performances were limited to the bending properties in order to highlight the contribution of the polymeric matrices in the GF composites. Indeed, it has been established that the bending and shear tests are highly sensitive towards the polymeric matrices while the tensile test is generally governed by the contribution of the fibers. Other information about the monomers preparation and overall properties of

Table 5.19 Flexural Strength and Modulus of Several Blend-Based PN/GF Composites in Accordance with their Respective Number of Plies and Curing Schedule

Matrix Composition	Flexural Strength (MPa)	Flexural Modulus (GPa)	Number of Plies	Curing Schedule	Reference
PN/PN (BZPN/DBPA-Ph) by Weight Ratio)					
10/90	187	17.2	10	200°C for 2 h, and 240°C for 4 h, under a pressure of 20 MPa	[49]
20/80	355	18.5			
30/70	345	22.5			
40/60	430	22.5			
50/50	491	23.8			
BMI/PN (BMI/3-APN by Molar Ratio)					
100/0	380	16.0	10	180°C for 2 h, 220°C for 2 h, 260°C for 2 h, and 300°C for 2 h, under a pressure of 20 MPa	[50]
50/50	435	25.3			
33/67	590	36.6			
25/75	634	38.7			
0/100	693	39.1			
BMI/PN (BMI/BZPN by Molar Ratio)					
0/100	542	25.8	10	160°C for 2 h, 180°C for 2 h, 200°C for 2 h, 240°C for 2 h, and 280°C for 2 h, under a pressure of 20 MPa	[51]
40/60	678	36.3			
50/50	484	15.3			
60/40	464	10.0			
BMI/PN (BMI/DBPA-Ph) by Molar Ratio)					
50/50	498	33.1	10	180°C for 2 h under a pressure of 20 MPa followed by 220°C for 3 h, 260°C for 3 h, 300°C for 2 h, and 320°C for 3 h	[52]

Table 5.19 Flexural Strength and Modulus of Several Blend-Based PN/GF Composites in Accordance with their Respective Number of Plies and Curing Schedule—cont'd

Matrix Composition	Flexural Strength (MPa)	Flexural Modulus (GPa)	Number of Plies	Curing Schedule	Reference
PEN/PN (PEN-OH/BAPh by Weight Ratio)					
0/100	507	24.0	10	200°C for 4 h, 240°C for 4 h; 280°C for 4 h, and 300°C for 4 h, under a pressure of 20 MPa	[53]
4/96	521	32.1			
6/94	561	35.8			
8/92	445	30.4			
CTBN/PN (CTBN/BAPh by Weight Ratio)					
0/100	528	20	12	200°C for 2 h, 250°C for 2 h, and 300°C for 3 h, under a pressure of 20 MPa	[54]
10/90	534	32			
20/80	507	24			
30/70	473	27			
40/60	466	21			

APN, aminophenoxy PN; *BZPN*, benzoxazine-functional, PN; *DBPA*, diallyl bisphenol-A; *PN*, phthalonitrile.

each blend were already discussed in Chapter 4, Phthalonitrile-Based Blends and Copolymers.

The first PN/GF composites presented in Table 5.19 were reported about by Zou et al., [49]. The polymeric matrices were prepared by blending two different PN monomers, namely the BZPN and the DBPA-based PN (DBPA-Ph) monomers. According to the relatively low curing temperature and time, the resulting composites exhibited interesting flexural properties which further increased by more BZPN contents in the matrix. For instance, the highest flexural strength and modulus of 491 MPa and 23.8 GPa, respectively, were obtained at the maximum weight ratio of 50/50. On the other hand, it appeared that the neat DBPA-Ph/GF composites had the lowest flexural properties among all the studied compositions. To understand why these advanced PN monomers failed in delivering HP composites, a quick look back in the polymerization reaction is needed. Although the polymerization of the allyl functionality in the DBPA-Ph can provide free active hydroxyls, their concentration was not sufficient to fully catalyze the cyano addition reaction of the PN units. In such case, the crosslinking density was low and the polymeric structure failed more likely under small amounts of stress. In the meantime, adding an additional source of hydroxyl groups from the BZPN increased the extent of the nitriles transformation resulting in better flexural properties. Ultimately, the DBPA-Ph/GF composites could have exhibited improved properties if cured in the presence of an advanced curing agent such BAPS, APB, or APN and for extended curing temperatures and time.

In another report, Huang and coworkers first blended the well-known APN with the amino group of the *meta* position (3-APN) with various molar ratios of bismaleimide (BMI), the resulting mixtures were then used as matrices for GF composites [50]. The composites, cured up to 300°C, showed flexural strengths and modulus over 320 MPa and 16 GPa, respectively. The BMI/GF composite had the weakest performances while the neat 3-APN/GF composite showed the best match with a flexural strength of 693 MPa and a flexural modulus of 39.1 GPa. The composites issued from various molar ratios of 3-APN and BMI exhibited flexural performances ranging between those of the composites prepared from the neat precursors. Through this work, it has been confirmed that the PN monomers can effectively improve the properties of the BMI by sample physical blends.

In the same context, Jiang et al., achieved PN/GF composites with improved flexural properties by blending the BZPN with BMI [51]. Indeed, the resulting composites cured only to 280°C showed a flexural strength of about 678 MPa for the 60/40 molar ratio of BZPN/BMI which was 44 MPa higher than that obtained with the composite from 75/25 molar ratio of 3-APN/BMI cured up to 300°C. Once more, it has been confirmed that the PN resins can easily improve the flexural properties of the BMI. In a different report, Zou et al., tried to circumvent the initially low flexural properties of the DBPA-Ph/GF composites by changing the matrix to a blend of an equivalent molar ratio of DBPA-Ph and BMI [52]. The afforded prepregs were also cured at higher temperatures and extended time. The bending test carried out on the resulting composite revealed consequent improvements in both flexural strength and modulus which reached about 498 MPa and 33.1 GPa, respectively. Thereby, the addition of BMI along with the changes in the curing schedule successfully improved the flexural performances of the DBPA-Ph-based composites.

Other PN/GF composites were also afforded through the preparation of blends from PN monomers and thermoplastics. For instance, Chen et al., first functionalized PEN precursors with hydroxyl groups (PEN-OH) and subsequently physically blended them with the BAPh monomers in the presence of minim quantity of DDS curing agent [53]. The introduction of small amounts of PEN-OH was found to positively affect the overall properties of the composites. For instance, the flexural strength and modulus moved from 507 MPa and 24 GPa for the neat BAPh/GF composite to about 561 MPa and 35.8 GPa for the BAPh/GF composite containing 6 wt.% of PEN-OH. Finally, the blends from BAPh and CTBN rubber were also used by Pan et al. to prepare PN/GF composites [54]. The fabricated composites showed some slight improvements in the flexural properties of the neat BAPh/GF composite (cured in the presence of 2 wt.% DDS) when 10 wt.% of CTBN was introduced. To sum up, all the studied PN/GF composites, regardless of the composition of their blends, proved to achieve HP, GF-reinforced composites for use in applications requiring low cost of fabrication and advanced mechanical and thermal properties.

5.4 Conclusions

In the course in this chapter all the fiber-reinforced, PN composites prepared from either typical, functional, or blend of PN precursors were discussed. In summary, it was confirmed that all the PN resins presented advanced processing conditions favorable for use in various manufacturing techniques. The cured composites' thermal and mechanical properties also proved to be superior to those obtained from other HP polymers. Hence, the fiber-reinforced PN composites can be used in many applications ranging from the common to the most exigent ones. In addition, this chapter not only reviewed all the finding in the field, but also highlighted the areas that need further investigation for future studies.

References

[1] Ting RY, Keller TM, Price TR, Poranski JCF. Characterization of the cure of diether-linked phthalonitrile resins. ACS Sym Ser 1982;195:337–50.

[2] Keller T.M., Price T.R. Polymer of dithioether-linked phthalonitrile. United States Patent US 5004801. 1991 Apr 2.

[3] Keller T.M., Roland C.M. High temperature adhesive. United States Patent US 5242755. 1993 Sep 7.
[4] Keller TM. Synthesis and polymerization of multiple aromatic ether phthalonitriles. Chem Mater 1994;6:302–5.
[5] Sastri SB, Keller TM. Phthalonitrile polymers: cure behavior and properties. J Polym Sci: Part A: Polym Chem 1999;37:2105–11.
[6] Sastri S.B., Keller T.M. Fiber-reinforced phthalonitrile composite cured with low-reactivity aromatic amine curing agent. United States Patent US 6001926. 1999 Dec 14.
[7] Keller TM, Dominguez DD. High temperature resorcinol-based phthalonitrile polymer. Polymer 2005;46:4614–18.
[8] Laskoski M, Dominguez DD, Keller TM. Processable phthalonitrile resins with high-thermal and oxidative stability. ACS Symp Ser 2005;922:378–88.
[9] Derradji M, Ramdani N, Zhang T, Wang J, Gong LD, Xu XD, et al. Effect of silane surface modified titania nanoparticles on the thermal, mechanical, and corrosion protective properties of a bisphenol-A based phthalonitrile resin. Prog Org Coat 2016;90:34–43.
[10] Derradji M, Ramdani N, Gong LD, Wang J, Xu XD, Lin ZW, et al. Mechanical, thermal, and UV-shielding behavior of silane surface modified ZnO-reinforced phthalonitrile nanocomposites. Polym Adv Technol 2016;27:882–6.
[11] Sastri SB, Armistead JP, Keller TM. Phthalonitrile-carbon fiber composites. Polym Compos 1996;17:816–22.
[12] Fowler G. Proceedings of the 34th international SAMPE technical conference. Soc Adv Mat Proc Eng 2002;34:1123.
[13] Duck DE. Proceedings of the 44th international SAMPE symposium and exhibition. Soc Adv Mat Proc Eng Ser 1999;44:705.
[14] Van Dine P, Spencer BE. Proceedings of the 46th international SAMPE symposium and exhibition. Soc Adv Mat Proc Eng Ser 2011;46:1095.
[15] Smith MD, Spencer BE, Van Dine P. Proceedings of the 30th international SAMPE technical conference. Soc Adv Mat Proc Eng 1998;30:138.
[16] Bulgakov B, Sulimov A, Babkin A, Timoshkin I, Solopchenko A, Kepman A, et al. Phthalonitrile-carbon fiber composites produced by vacuum infusion process. J Compos Mater 2017. Available from: https://doi.org/10.1177/0021998317699452.
[17] Miks M.W., Shigley J.K. Addition-polymerization resin systems for fiber-reinforced nozzle ablative components. United States Patent US 5645219. 1997 Jul 8.
[18] Keller T.M. Phthalonitrile prepolymer as high temperature sizing material for composite fibers. United States Patent US 5389441. 1995 Feb 14.
[19] Chung D. Carbon fiber composites. Oxford: Butterworth-Heinemann; 2012.
[20] Morgan P. Carbon fibers and their composites. Boca Raton, Florida: CRC Press; 2005.

[21] Mallick PK. Fiber-reinforced composites: materials, manufacturing, and design. CRC Press; 2007.
[22] Ting R, Nash H. The development of C-10 phthalocyanine for composite matrix applications. Polym Eng Sci 1981;21:441–8.
[23] Griffith JR, O'Rear JG, Walton TR. Phthalonitrile resins. Adv Chem; Am Chem Soc 1975;142:458–64.
[24] Keller TM, Price TR. Amine-cured bisphenol-linked phthalonitrile resins. J Macromol Sci Chem 1982;18:931–7.
[25] Keller T.M. Phthalonitrile resin from diphthalonitrile monomer and amine. United States Patent US 4408035. 1983 Oct 4.
[26] Keller T.M. Synthesis and polymerization of phthalonitrile monomers containing multiple phenoxy and sulfone linkages. United States Patent US 4409382. 1983 Oct 11.
[27] Keller T.M. Processable and stable conductive polymers from diether-linked bisorthodinitrile monomers. United States Patent US 4587325. 1986 May 6.
[28] Keller TM. Phthalonitrile-based conductive polymer. J Polym Sci: Part A: Polym Chem 1987;25:2569–76.
[29] Keller TM. Phthalonitrile-based high temperature resin. J Polym Sci: Part A: Polym Chem 1988;26:3199–212.
[30] Keller T.M. Amino phenyl containing curing agent for high performance phthalonitrile resin. United States Patent US 5003039. 1991 Mar 26.
[31] Keller TM. Imide-containing phthalonitrile resin. Polymer 1993;34:952–5.
[32] Keller T.M. Phthalonitrile monomers containing imide and/or phenoxy linkages, and polymers thereof. United States Patent US 5132396. 1992 Jul 21.
[33] Warzel M, Keller T. Tensile and fracture properties of a phthalonitrile polymer. Polymer 1993;34:663–6.
[34] Burchill PJ. On the formation and properties of high-temperature resin from a bisphthalonitrile. J Polym Sci: Part A: Polym Chem 1994;32:1–8.
[35] Sastri SB, Keller TM. Phthalonitrile cure reaction with aromatic diamines. J Polym Sci: Part A: Polym Chem 1998;36:1885–90.
[36] Dominguez DD, Jones HN, Keller TM. The effect of curing additive on the mechanical properties of phthalonitrile-carbon fiber composites. Polym Compos 2004;25:554–61.
[37] Augustine D, Mathew D, Nair CPR. Phenol-containing phthalonitrile polymers—synthesis, cure characteristics and laminate properties. Polym Inter 2012;62:1068–76.
[38] Zong L, Liu C, Guo Y, Wang J, Jian X. Thermally stable phthalonitrile resins based on multiple oligo (aryl ether)s with phenyl-*s*-triazine moieties in backbones. RSC Adv 2015;5:77027–36.
[39] Augustine D, Mathew D, Reghunadhan Nair CP. End-functionalized thermoplastic-toughened phthalonitrile composites: influence on cure reaction and mechanical and thermal properties. Polym Inter 2015;64:146–53.

[40] Sastri SB, Armistead JP, Keller TM. Phthalonitrile-glass fabric composites. Polym Compos 1997;18:48–54.

[41] Kimura H, Matsumoto A, Ohtsuka K. Glass fiber-reinforced composite based on benzoxazine resin. J Appl Polym Sci 2009;114:1256–63.

[42] Suman JN, Kathi J, Tammishetti S. Thermoplastic modification of monomeric and partially polymerized Bisphenol A dicyanate ester. Eur Polym J 2005;41:2963–72.

[43] Zhou H, Badashah A, Luo Z, Liu F, Zhao T. Preparation and property comparison of *ortho*, *meta*, and *para* autocatalytic phthalonitrile compounds with amino group. Polym Adv Technol 2011;22:1459–65.

[44] Chen Z, Guo H, Yang J, Zhao R, Liu X. Manufacturing and thermal and mechanical properties of advanced 3-aminophenoxyphthalonitrile/bisphthalonitrile composite laminates. High Perform Polym 2012;25:214–24.

[45] Chen Z, Guo H, Tang H, Yang X, Xu M, Liu X. Preparation and properties of bisphenol A-based bis-phthalonitrile composite laminates. J Appl Polym Sci 2013;129:2621–8.

[46] Xu M, Jia K, Liu X. Effect of bisphenol-A on the structures and properties of phthalonitrile-based resin containing benzoxazine. Exp Polym Lett 2015;9:567–81.

[47] Luo Y, Xu M, Pan H, Jia K, Liu X. Effect of ortho-diallyl bisphenol A on the processability of phthalonitrile-based resin and their fiber-reinforced laminates. Polym Eng Sci 2016;56:150–7.

[48] Xu M, Liu M, Dong S, Liu X. Design of low temperature self-cured phthalonitrile-based polymers for advanced glass fiber composite laminates. J Mater Sci 2013;48:8108–16.

[49] Zou X, Yang X, Xu M, Jia K, Liu X. Curing behaviors and properties of allyl-and benzoxazine-functional phthalonitrile with improved processability. J Polym Res 2016;23:2.

[50] Huang Y, Jiang M, Xu M, Zou X, Luo Y, Lei Y, et al. Curing behavior and processability of BMI/3-APN system for advanced glass fiber composite laminates. J Appl Polym Sci 2016;133:43640–7. Available from: http://doi.org/10.1002/app.43640.

[51] Jiang M, Xu M, Jia K, Liu X. Copolymerization of self-catalyzed phthalonitrile with bismaleimide toward high-temperature-resistant polymers with improved processability. High Perform Polym 2016;28:895–907.

[52] Zou X, Xu M, Jia K, Liu X. Copolymerizing behavior and processability of allyl-functional bisphthalonitrile/bismaleimide system. Polym Compos 2015. Available from: https://doi.org/10.1002/pc.23726.

[53] Chen Z, Lei Y, Tang H, Wei J, Liu X. Mechanical, thermal, electrical, and interfacial properties of high-performance bisphthalonitrile/polyarylene ether nitrile/glass fiber composite laminates. Polym Compos 2013;34:2160–8.

[54] Pan H, Xu M, Lei X, Jia K, Liu X. Curing behaviors and performance of a carboxyl-terminated butadiene acrylonitrile rubber/bisphthalonitrile resin system. High Perform Polym 2016;28:581–90.

6 Ceramic-Based Phthalonitrile Micro- and Nanocomposites

6.1 Introduction

Technological advances require either new materials or ameliorations in the already existing ones. Hybrid composites prepared from ceramics and polymers are a class of materials that combines the benefits of both phases to produce materials with enhanced properties not achieved by either phase alone. The resulting hybrids are known as micro- or nanocomposites according to the ceramic fillers' sizes. Considering the large panel of existing ceramics and polymers, numerous possible combinations can be envisaged and the resulting hybrid properties can be tailored to meet the needs of different industries. Some interesting features of these hybrids such as their light weights, low cost of fabrication, and improved thermal and mechanical properties are particularly appealing for electronic, automotive, and aviation industries [1–6]. Indeed, the past few decades showed a growing interest in this kind of materials in both academic and industrial fields. The characteristics of these hybrid materials depend on several factors such as the properties of the polymeric matrices and the ceramic fillers, including their size and amount as well as the type of interaction between the two phases [7–9]. Thus any improvement in these parameters will certainly result in ameliorations in the hybrids final properties. For instance, with the recent advances in the polymer design and the emergence of high-performance polymers, a way for further progress in this field was opened.

High-performance polymers include, but are not limited to, epoxies, cyanate esters, polyimides, polybenzoxazines, bismaleimides, and phthalonitrile (PN) resins. From the above, the PN resins present the best match in terms of combination of properties. The PN resins are recognized for their outstanding thermal resistance with starting decomposition temperatures over 450°C for the typical PN polymers. The good thermal stability of the PN resins is accompanied by an ameliorated processability, superior flame resistance, absence of a glass transition temperature (T_g) before their thermal decomposition, low water absorption, advanced UV shielding, and excellent corrosion protective properties [10–21]. Such a combination of properties promotes the PN resins

Phthalonitrile Resins and Composites.
DOI: https://doi.org/10.1016/B978-0-12-812966-1.00006-8

over the traditional, high-performance thermosets and makes them suitable to be used as advanced polymeric matrices for ceramic-based composites. On the other hand, the PN resins, as any other thermosets, are brittle in nature and their mechanical properties are lower than those of ceramics or metals. Although the thermal stability of the PN resins is superior to that of other, high-performance polymers, it still remains incomparable to that of the ceramic materials. Attempting to ameliorate the properties of the PN resins by molecular design or by blending with other polymers can be envisaged; however, the preparation techniques are time consuming and from an economic point of view quite expensive. On the other hand, reinforcing the PN resins with ceramic particles is a cost-effective method that allows better enhancements and more possibilities of tailoring the properties of the final products.

The ceramic fillers for possible use as reinforcement can be classified into carbides, nitrides, and oxides. The most commonly used ceramics are silicon carbide (SiC), boron carbide, silicon nitride (Si_3N_4), boron nitride (BN), aluminum nitride (AlN), alumina or aluminum oxide (Al_2O_3), titania or titanium oxide (TiO_2), zinc oxide (ZnO), chromium oxide (Cr_2O_3), and zirconia or zirconium oxide (ZrO_2). These materials can be referred as the first generation of ceramic reinforcements for the polymeric matrices including the PN resins. In contrast with the polymeric matrices, these fillers possess ameliorated hardness, strength, modulus, thermal stability, corrosion resistance, and thermal conductivity. Inclusion of such materials within any polymer, if properly achieved, will result in crucial enhancements in the properties of the matrix [22]. A quick look in the open literature confirms that these ceramic fillers have been successfully incorporated within a substantial number of polymeric matrices such as poly (ether ether-ketone) [23,24], polyamide [25], polypropylene [26,27], poly(methyl methacrylate) [28–30], epoxy [31–35], cyanate ester [36–39], polyimide [40–43], polybenzoxazine [44–47], and bismaleimide [48–50]. The PN resins have also been reinforced with many of these ceramics and high-performance, polymeric composites were obtained in each case. For example, the inclusion of SiC microparticles in a bisphenol A (BPA)-based PN (BAPh) resin resulted in ameliorations in the hardness, flexural, and tensile properties [51]. The thermal stability and the stiffness of the neat resin were also highly ameliorated. The Si_3N_4 nanoparticles, added at various amounts, resulted in exceptional improvements in thermomechanical properties of the neat BAPh matrix [52]. The BAPh resin toughness, hardness, and thermal stability were also significantly enhanced with alumina nanoparticles [53]. Moreover, Derradji et al., reported about the use of titania nanoparticles to improve the corrosion protective properties of the BAPh resin [54]. The second generation of PN resins consisting on oligomeric containing aliphatic moieties were also reinforced with various amounts of zirconia nanoparticles and global improvements in the thermal and mechanical properties were achieved [55]. Furthermore, the nanosized ZnO particles not only improved the thermomechanical properties of the neat resin but also markedly enhanced its UV shielding behavior [56]. Derradji et al., also achieved excellent thermal conductivity with the highly-filled, BN-based, PN nanocomposites [57].

The recent advances in the field of ceramics led to the appearance of a new class of ceramic particles commonly known as the "MAX phases." MAX phases are the second generation of ceramics especially designed to be incorporated in polymeric, metallic, and ceramic matrices aiming the preparation of high-performance composites. MAX phases are a family of ternary carbides or nitrides with the general formula $M_{N+1}AX_n$, where M is a class of early transition metal, A is a IIIA or IVA group element and X is C or N element [58–61]. The dual behavior resulting from the combination of the advantageous attributes of both metals and ceramics is what gives the MAX phase materials their outstanding electrical, mechanical and thermal properties. These interesting features include good thermal and electrical conductivities, easy machinability, low density, excellent wear resistance, high hardness and good oxidation resistance [62]. These attractive properties make the MAX phases highly-promising candidates to address today's manufacturing challenges. Moreover, the MAX phases possess ameliorated ductility and their inclusion in the polymeric matrices can enhance the toughness of the resulting composites to a certain extent. Different MAX phases can be used such as Ti_2AlC, Ti_3AlC_2, Ti_3SiC_2, and Ti_4SiC_3. Although the concept of the MAX phases is relatively new in the field of polymer composites, many researchers are starting to explore this interesting area. For instance, Mahesh et al., investigated the properties of polyaryletherketone polymer reinforced with Ti_3SiC_2 [62]. Zhang et al., also investigated the mechanical and antifriction performances of Ti_3AlC_2/ultrahigh molecular weight polyethylene nanocomposites [63]. Wang et al. reported on the electrical, dielectric, and mechanical properties of novel Ti_3AlC_2/epoxy composites [64]. The PN resins were also reinforced with two types of MAX phases (Ti_3AlC_2 and Ti_3SiC_2) and remarkable improvements in the toughness and thermal properties were noticed [65,66].

Another important parameter in developing hybrid materials is the fillers' size. The past few decades showed an extensive use of microsized particles to reinforce polymeric matrices. In those cases, the properties of the composites involved compromises; stiffness for toughness, or thermal stability at the cost of mechanical properties. Moreover, macroscopic defects caused by nonuniform dispersion often resulted in poor mechanical properties. These engineering limitations were gradually overcome by the use of nanoscaled fillers. Indeed, the polymer nanocomposites showed ameliorated properties without compromises or limitations. The advantage of the nanoscaled fillers is obviously their size. For instance, the light scattering is reduced with small nanoparticles, making it possible to prepare composites with altered thermal and mechanical properties with a high degree of optical clarity. In addition, the reduced particles size does not cause large stress concentrations and therefore the ductility of the polymer is not greatly affected. Moreover, the problems related to the particles deposition at the bottom of the structure, caused by the difference in density between the phases, are not seen in the nanocomposites mainly due to the nanoparticles large interfacial area, which allows more compatibility with the matrix. Generally, the ceramic nanoparticles are preferred to their microsized counterpart; however, for some applications requiring high thermal conductivity some controversial reports can be found on the preferred fillers' size. The fillers amount is also a determining factor of the composites final properties. The amount of particles to be added depends on the targeted applications and the engineering reliability of the product. For example, fewer amounts of nanoparticles can be enough to enhance the mechanical properties of the polymeric matrices. Meanwhile, lager amounts may result in the formation of agglomerations that can have a negative effect on the properties. Conversely, for thermally conductive applications, higher amounts of particles are needed to form a conductive path, which may result in high thermal conductivity values for the composites. Indeed, the particles size and amount are two other parameters that allow more flexibility in tailoring the properties of the hybrid materials. The PN resins have been reinforced with both micro- and nanosized ceramic particles at different loadings and a detailed comparison will be provided later.

Avoiding the ceramic particles agglomerations during the composite preparation is also an important challenge to be addressed. The problem of agglomerations is more accentuated in polymer nanocomposites due to the nanoparticles high specific surface area and volume effects. This problem can be solved or minimized by a proper surface modification of the fillers. One solution involves using silane coupling agents to ameliorate the compatibility between the two phases. Various silane coupling agents are commercially available and excellent results were achieved in most of the cases. Another solution consists on grafting monomer or oligomeric molecules on the outer surface of the particles. The advantages of this procedure lie on the great flexibility in the choice of the grafted species. Thus, the properties of the resulting hybrids can also be tailored according to the specific needs.

The present chapter reports on the preparation and evaluation of properties of the newly-developed, high-performance, polymeric micro and nanocomposites prepared from multiple ceramic fillers and various PN resins. The effects of various ceramic fillers' size, amount, and state of surface on the PN resins are scrutinized in terms of impact on the morphology, curing behavior, mechanical properties, thermal properties, UV-shielding behavior, and corrosion resistance. The obtained improvements are also compared to some similar types of polymer–ceramic based composites.

6.2 Properties of the Studied Ceramic Fillers

As previously stated the ceramics can be categorized into carbides, nitrides, oxides, borides, and MAX phases. The PN resins were reinforced with a variety of ceramics; some of those ceramic fillers physical properties are listed in Table 6.1. By comparing these ceramics in terms of their properties, it appears that the nitrides have high melting points around 2000°C, which means high thermal stability in the window of application of the PN-based composites. The resistivity of the nitrides varies between 10 and 20 μΩ cm with some exceptions [67,68]. For instance, the Si_3N_4 has a decomposition temperature around 1900°C with a resistivity in excess of 1019 μΩ cm. The Si_3N_4 also presents the relatively modest microhardness value of about 16.9 GPa. Globally, the nitrides have high dielectric constant about 8.3 for Si_3N_4 and 5 for AlN. In contrast with other ceramics, the nitrides possess the highest thermal conductivity values of 80, 120, and 400 W/m K for Si_3N_4, AlN, and BN, respectively. Therefore, for the preparation of highly thermal conductive, PN-based composites, the BN particles were found to be the best reinforcing fillers' choice. The ceramic carbides, such SiC, present densities similar to those of nitrides ranging between 2.5 and 15 g/cm^3. On the other hand, the melting points of these ceramics are in excess of 2500°C. Although the carbides are thermally more stable than the

Table 6.1 Thermal and Mechanical Properties of Various Ceramic Fillers

Ceramic Code	Modulus (GPa)	Hardness (GPa)	Thermal Conductivity (W/m K)	CTE (10^{-6}/°C)	Dielectric Constant	Density (g/cm^3)
SiC	410	28.2	40	4.7	—	3.1
Al_2O_3	300	11.8	30	9.8	8.2	3.97
SiO_2	73	6.0	1	0.55	3.82	2.2
TiO_2	230	9.3	12	8.4	10	3.97
ZrO_2	200	11.0	2	5–10	—	5.5
ZnO	112	2.2	60	2	—	5.1
BN	46.9	—	65	4.3	4.5	2.27
AlN	330	11.1	140–180	4.8	4.5–5	3.26
Si_3N_4	303	16.9	80	33	8.3	3.2
Ti_3AlC_2	306	3.5	48	—	—	4.0
Ti_3SiC_2	320	4.0	60	—	—	4.4

nitrides, the impact of their introduction on the thermal stability of the polymeric matrices is quite the same since most of the high-performance, polymer composites operate at temperatures below 500°C. The specific resistivity of the carbides is also quite higher than the nitrides with values ranging between 10^9 and 10^{11} μΩ cm for SiC. As a result, the thermal conductivity of the carbides is limited around 40 W/m K for SiC. In contrast with other ceramics, the carbides exhibit the highest microhardness values. For example, SiC has the remarkable microhardness of about 28.2 GPa. Thus, the ceramic carbides seem to be good candidates to ameliorate both the wear and thermal resistances of the PN resins. The oxides and borides have densities ranging from 4 to 10 g/cm^3. Their melting points are lower than both nitrides and carbides and their specific resistance can reach up to 10^{22} μΩ cm. Therefore, the oxides and borides have low thermal conductivities. For instance, the thermal conductivities of alumina, titania, and zirconia are 30, 12, and 2 W/m K, respectively. Furthermore, the microhardness of the oxides and borides are lower than that of the carbides and similar to that of the nitrides. The MAX phase ceramics have densities from 3 to 7 g/cm^3 and their hardness is lower than all the other ceramics. In addition, the MAX phases also present ameliorated thermal conductivity values ranging between 40 and 60 W/m K. It is also important to point out that the MAX phases are ductile ceramics especially advantageous in enhancing the strength while retaining some of the initial ductility of the matrix. Thus, the MAX phases are particularly appealing for the new generation of polymeric nanocomposites.

6.3 Preparation Methods of the Ceramic-Based Phthalonitrile Hybrid Materials

The challenges that rise when preparing ceramic-based, PN micro- or nanocomposites are the particles dispersion and adhesion with the matrix. Several preparation techniques can be envisaged such as the sol–gel processing, in situ polymerization, and solution blending.

6.3.1 Sol–Gel Processing

This method allows the preparation of hybrid materials through the formation of a sol starting from a solution of precursor previously hydrolyzed, such as tetraethyloxysilane, tetrabutyl titanate, or aluminum iso-propoxide and organic monomers or oligomers. Then, the sol is transformed into gel by condensation reaction and, finally, the used solvents are removed [69–71]. An example of the sol–gel synthetic route is presented in Scheme 6.1. The sol–gel technique was not applied in the field of PN composites because of the specific curing character of the PN monomers. Indeed, the curing of the PN monomers is a sluggish process requiring elevated temperatures and extended time. Therefore, the sol–gel technique, which is generally performed at ambient temperature, is not appealing for the preparation of the PN-based hybrids. Moreover, this technique is quite expensive and not environment friendly, since acids and solvents are extensively used in this process. Thus at the time this book was written no records were found on any attempt to prepare any PN-based composite using the sol–gel technique. It is also important to point out that in terms of particle dispersion the sol–gel method is preferred over the other synthetic routes. With the rapid development in the field of PN materials, the use of the sol–gel process to prepare ceramic-based, PN hybrids will certainly be scrutinized. The sol–gel process described herein is presented as a research opportunity for the specialists in the field of PN and for any researcher interested in high-performance, polymeric composites-based on PN.

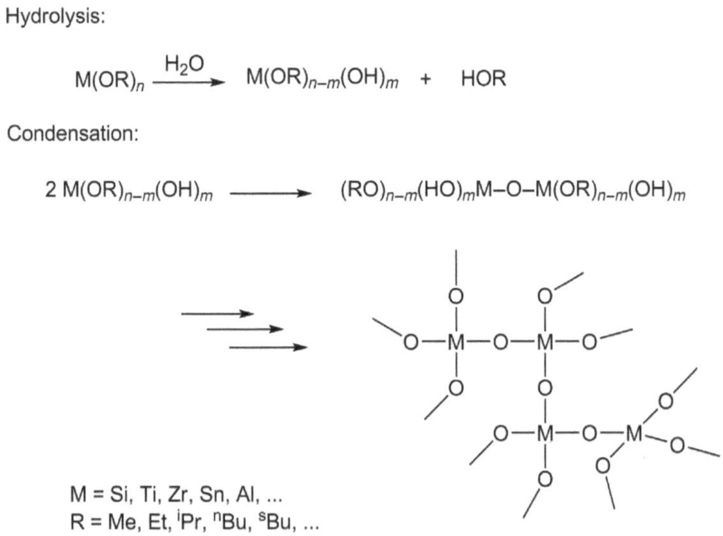

Scheme 6.1 Sol–gel synthetic route of hybrid materials. Reprinted from Kango S, Kalia S, Celli A, Njuguna J, Habibi Y, Kumar R. Surface modification of inorganic nanoparticles for development of organic–inorganic nanocomposites—a review. Prog Polym Sci 2013;38:1232–61.

6.3.2 In Situ Polymerization

In this process, the ceramic particles are first dispersed in a powdered monomers or oligomers and the mixture is polymerized in a way similar to bulk polymerization [72,73]. During this process no solvents are needed, and as a result, the prepared composites are void free and cost effective. Various ceramic fillers were introduced within the PN matrices using this technique. For instance, Derradji et al., prepared series of SiC-based, PN microcomposites containing various amounts of SiC ranging between 0 and 20 wt.% [51]. The composites were introduced by direct mixing of the BAPh monomers, aminophenoxy PN (APN) curing agent, and a proper amount of SiC particles. The mixtures were first thoroughly mixed, melted, and then quenched to room temperature. The mixtures were once again grinded to powder and hand mixed before polymerization. At that point, the mixtures can be stored indefinitely at room temperature; this step is known as the B-stage. After that, the mixtures were first placed in a vacuum oven heated to 120°C for 5 h and then transferred to the appropriate molds according to the specific test standards. Derradji et al., also prepared a new kind of nanocomposites from the typical BAPh monomers and Si_3N_4 nanoparticles [52]. The nanocomposites were also achieved by the in situ polymerization process. Typically, the nanoparticles were vigorously mixed with the BAPh monomers and their curing agent to assure a good dispersion of the nanofillers. The mixtures were then melted, mixed at their melting stage, and quenched to room temperature. Afterwards, the powdered mixtures were again mixed to further improve the nanoparticles dispersion, dried in a vacuum oven to remove any adsorbed moisture, and then placed in steel molds according to the mechanical and thermomechanical test shape requirements. Shun et al. also reported on nanocomposites prepared from alumina nanoparticles and resorcinol-based PN (RPh) monomers [74]. The preparation technique involved mixing the RPh monomers with 4,4′-diamino-diphenyl ether as curing agent at a weight ratio of 100:5. The alumina nanoparticles were added at different weight ratios ranging from 1 to 5 wt.% with an increment of 2 wt.%. The mixtures were melted in a 500 mL reaction kettle and mechanically stirred for about 3 min before being quenched to room temperature. The obtained prepolymers were placed in an aluminum mold, degassed in a vacuum oven, and then cured in a muffle furnace. The alumina nanoparticles were used in the same way by Derradji et al., to prepare high-performance nanocomposites from the BAPh monomers [53]. These monomers were first mixed with APN at a weight ratio of 90:10, and then the Al_2O_3 nanoparticles were added at various amounts ranging from 3 to 15 wt.% at increments of 3 wt.%. The mixture compounds were melted, vigorously stirred and then cooled to room temperature. In order to obtain void-free thermosets, the mixtures were first degassed in a vacuum oven at 180°C for 5 h, and then transferred to the appropriate steel molds according to the dynamic mechanical analysis (DMA), and tensile test shape requirements. A hot compression technique was used to cure these nanocomposites.

6.3.3 Solution Blending Technique

This method is the most conventional and most preferred way of preparing polymer/inorganic composites. As indicated by its name, this method is a liquid state processing that allows a relatively good level of dispersion between the two phases. This process overcomes the limitations of the melt blending by dissolving both monomers and particles in a solution. The additional cost of this method is therefore dependent on the cost of the solvent and its recovery. This composites route of preparation was largely applied in the field of PN materials. For instance, Derradji et al., used the solution blending technique to prepare TiO_2-reinforced, high-performance, PN nanocomposites with ameliorated thermal, mechanical, and anticorrosion properties [54]. The PN monomers and their curing agent were first dissolved in acetone prior to adding the nanoparticles at various weight ratios. Aiming to improve the dispersion of the nanofillers, the resulting mixture compounds were mechanically stirred for 4 h and sonicated at the rate of 4000 rpm for 4 h. For the mechanical and thermal tests, the uncured nanocomposites were first placed in different steel molds to fit the shape requirements for the DMA and tensile tests and then finally cured with a hot compression molding technique. For the corrosion protective properties, the authors chose as substrate the carbon steel panels. The substrates, with a dimension of 50 $\times$ 50 $\times$ 1 mm, were previously cleaned and degreased in acetone. The deposition of the nanocomposites on the substrate was achieved by a dip coating process. Practically, the substrates were six times immersed into the prepared mixtures and each immersion time was about 1 min with no waiting time between consecutive immersions. Prior to the curing, the plates were dried in a vacuum oven at 180°C for 5 h and finally cured in a muffle furnace.

The solution blending technique was also used by Liu et al., to prepare PN-terminated poly(arylene ether nitriles) composites with hybrid-functionalized, Al_2O_3 particles [75]. The route of preparation can be resumed as follows. A certain amount of alumina was weighted, dissolved in *N*-methyl-2-pyrrolidone (NMP) solvent, and violently sonicated to form a suspension. The same procedure was applied for the PN monomers. After that, the solution containing alumina was slowly added to the monomers mixture. The resulting solution was refluxed at 200°C with continuous mechanical stirring for 30 min. Moreover, the mixture was once again sonicated for 30 min to enhance the particles state of dispersion. Finally, the mixture was casted, dried in a vacuum oven, and cured according to a specific curing schedule chosen by the authors.

Derradji et al., also reported on the successful dispersion of zirconia nanoparticles within two oligomeric containing aliphatic moieties PN resins using the solution blending technique [55]. Moreover, the same group of researchers prepared BN- and ZnO-based PN nanocomposites with enhanced nanoparticles dispersions, largely attributed to the route of preparation and state of surface of the reinforcing phase [56,57]. The MAX phases, as the second-generation of ceramic fillers, were also introduced by a solution blending technique within a BAPh matrix and an ameliorated state of dispersion was clearly achieved [76,77].

6.4 Surface Modification of Ceramic Particles

Although the routes of preparation previously discussed can ensure a relatively good state of dispersion, if high-filled. polymeric composites are targeted, the use of raw ceramic fillers will certainly result in agglomerations and therefore reduced performances. In fact, the problem of agglomerations is more accentuated in polymer nanocomposites due to the nanoparticles highly specific, surface area and volume effects. Thus, it is highly advised to treat the surface of the fillers by chemical treatment or by grafting of synthetic polymers.

6.4.1 Chemical Treatments

The chemical treatment of the fillers is the most preferred technique to improve the dispersion of raw ceramics in various liquid media. The most preferred chemical treatment in the field of particles-reinforced, polymer composites is the direct adsorption of a silane coupling agent on the outer surfaces of the fillers. The silane coupling agents have the general formula of X_3SiRY, as seen in Scheme 6.2, where Y is an organofunctional group and X represents the hydrolysable group [78,79]. The organofunctional group (Y) is chosen for its reactivity and compatibility with the polymer. The hydrolysable group (X), typically alkoxy, acyloxy, halogen, or amine, is the intermediate possessing the ability to bond with the inorganic surfaces. The silane coupling agents can also be defined as the intermediates between the organic and inorganic materials. The surface modification occurs between the hydrolysable group and the substrate in a four steps reaction. First, the three labile groups are hydrolyzed and condensed to form oligomers, and then the presence of the hydroxyl groups on the surface of the substrate allows the formation of hydrogen bonds. Finally, after drying or curing, covalent linkages are formed with the substrate accompanied by water removal. It is also important to note that these steps can eventually occur at the same time, after the first hydrolysis reaction. On the other hand, the organofunctional group (Y) remains free for possible interaction with the polymeric matrix. An example showing the four steps of the hydrolytic depositions of silanes is presented in Scheme 6.3. Nowadays, with the exceptional development in the molecular design, any conceivable organofunctional silane can be envisaged. Some of the most famous silanes are based on vinyl, chloropropyl, epoxy, methacrylate, primary and secondary amines, mercapto, and cationic styryl. The chemical structure of these silanes can be depicted in Scheme 6.4. The necessary amount of silane to be added depends on the fillers' size, and rarely exceeds 8 wt.% of the reinforcing phase. The PN matrices have been extensively reinforced with various, silane, surface-modified, ceramic fillers. For instance, SiC, Si_3N_4, Al_2O_3, TiO_2, ZnO, ZrO_2, BN, Ti_3AlC_2, and Ti_3SiC_2 particles were all previously subjected to an amino-based, silane, surface modification to enhance their dispersion and adhesion with the PN matrices [51−57]. The choice of the amine as the organofunctional group is mainly attributed to the characteristics of the PN materials. As seen in Chapter 1, Introduction to the Phthalonitrile Resins and Their Curing Behavior, the PN monomers are very reactive in the presence of amines. Indeed, the amines are the most preferred curing agents in the field of PN resins and composites. Thus, the amino-based, silane, coupling agents can easily react with the cyano groups to assure a good adhesion with the matrix. Until now, no other organofunctional group has been proven to react as effectively with the PN resins as the amine ones.

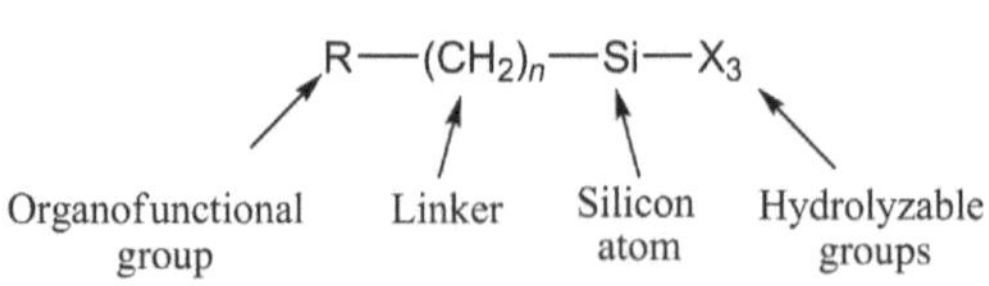

Scheme 6.2 General formula of the silane coupling agents.

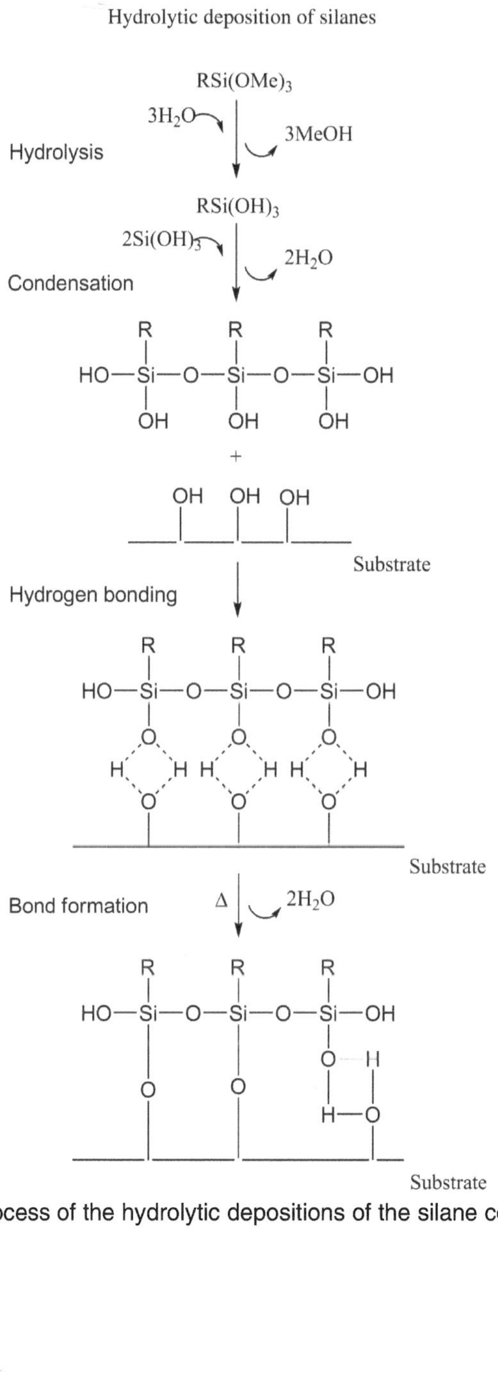

Scheme 6.3 Process of the hydrolytic depositions of the silane coupling agents.

Organofunctional group	Chemical structure
A. Vinyl	$CH_2{=}CHSi(OCH_3)_3$
B. Chlorogropyl	$ClCH_2CH_2CH_2Si(OCH_3)_3$
C. Epoxy	$\overset{O}{\widehat{CH_2CH}}CH_2OCH_2CH_2CH_2Si(OCH_3)_3$
D. Methacrylate	$CH_2{=}\overset{CH_3}{\overset{\vert}{C}}{-}COOCH_2CH_2CH_2Si(OCH_3)_3$
E. Primary amine	$H_2NCH_2CH_2CH_2Si(OCH_3)_3$
F. Diamine	$H_2NCH_2CH_2NHCH_2CH_2CH_2Si(OCH_3)_3$
G. Mercapto	$HSCH_2CH_2CH_2Si(OCH_3)_3$
H. Cationic styryl	$CH_2{=}CHC_6H_4CH_2NHCH_2CH_2NH(CH_2)_3Si(OCH_3)_3$ +HCl

Scheme 6.4 Chemical structures of several commercially available, silane coupling agents.

Derradji et al., resumed the procedure of the silane surface modification for ceramics designated to reinforce the PN matrices as follows. Prior to silane grafting, pristine ceramic particles have to be dried at 110°C in a vacuum oven for a specific period of time to remove any moisture adsorbed on the surface. The dried particles can be mixed with the appropriate silane coupling agent (4–8 wt.% of the fillers) in ethanol and mechanically stirred for 6 h. The particles, collected by filtration, should be first washed several times with toluene to remove any unreacted siloxane moieties, and then dried overnight under vacuum at 110°C to remove the residual solvent [21]. The impact of the silane surface modification of the morphology and properties of the PN matrices will be discussed in the sections related to the properties of the ceramic-based, PN composites.

6.4.2 Grafting of Synthetic Polymers

This technique takes advantage of the low molecular weight of the monomers, which can easily occupy the interstitial spaces between the particles and react with the activated sites on the particle surfaces. Thus the distance between the particles further increases, leading to a more enhanced state of dispersion and adhesion of the fillers. Moreover, the grafting of monomers or eventually oligomers renders the surfaces of the particles more hydrophobic, resulting in an ameliorated miscibility between the two phases. Scheme 6.5 shows the dispersion behavior of inorganic particles with and without the grafting [80].

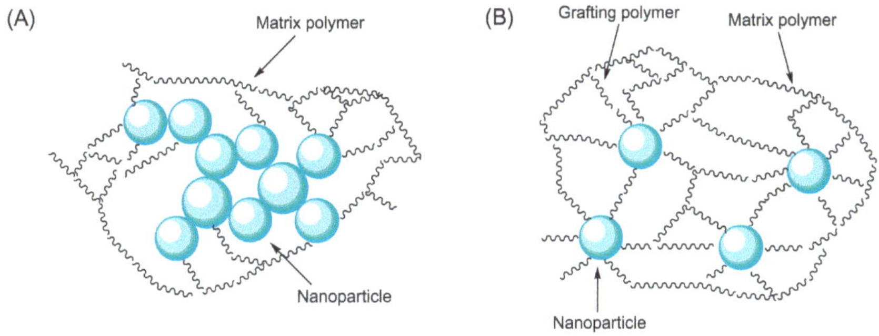

Scheme 6.5 Dispersion behavior of inorganic particles with and without the grafting. Reprinted from Rong MZ, Zhang MQ, Zheng YX, Zeng HM, Walter R, Friedrich K. Structure–property relationships of irradiation grafted nano-inorganic particle filled polypropylene composites. Polymer 2001;42:167–83.

Practically, two ways can be envisaged to effectively graft polymeric chains on the ceramic particles' surfaces. The first one is known as the "grafting to" method and consists on reacting end-functionalized polymers or monomers with the substrate. The second one is the "grafting from" technique where polymeric chains are grown at the surface of the substrate using an anchor linked to an initiating specie [9,81]. Scheme 6.6 summarizes the grafting procedure of both approaches. In the field of the PN-based composites, the grafting of synthetic polymers on ceramic fillers is limited to the "grafting to" approach. For instance, Liu et al. reported on the use of the PN monomers as organic modifiers to prepare surface modified Al_2O_3. The treated particles were used as new kind of reinforcement for a PN-terminated poly(arylene ether nitriles) matrix [75]. The grafting was carried out as follows. A certain amount of native alumina was mixed with NMP solvent and violently sonicated for 30 min. The PN monomers were also dissolved in NMP solvent and further heated for 30 min to assure the complete dissolution of the monomers. After that, the alumina suspension and the PN curing agent were slowly added to the previous mixture. The resulting mixtures were sequentially refluxed at 200°C for 6 h, centrifuged, washed with distilled water, and finally dried at 80°C overnight.

Another interesting work has been reported by Shan et al., concerning the surface modification of alumina nanoparticles by both silane coupling agent and polymer grafting [74]. In this procedure, the nanoparticles were first treated with aminopropyltrimethoxy silane in similar way to that described by Derradji et al. Then, the silane, surface-modified nanoparticles were further reacted with 4-(4-carboxylphenoxy) PN (CPN) monomers in the presence of the dimethylformamide solvent

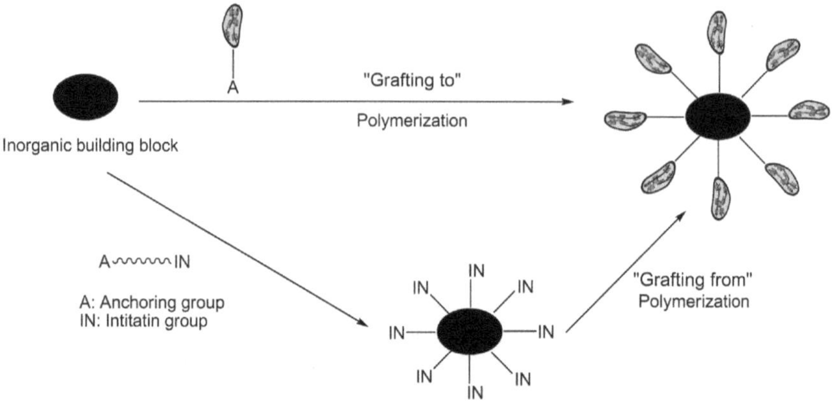

Scheme 6.6 Schematic description of grafting-to and grafting-from approaches. Reprinted from Kickelbick G. Concepts for the incorporation of inorganic building blocks into organic polymers on a nanoscale. Prog Polym Sci 2003;28:83–114.

and $SOCl_2$. After that, the effects of the treated, alumina nanoparticles on the thermomechanical properties and toughness of an RPh resin were investigated.

6.5 Properties of Ceramic-Based, Phthalonitrile Micro- and Nanocomposites

6.5.1 Properties of Phthalonitrile/SiC Microcomposites

Derradji et al., investigated the effects of various amounts and state of surfaces of SiC microparticles (average particles diameter of 10 μm) on the mechanical and thermal properties of a BAPh resin [51]. The aim of the work was to synergistically combine the properties of both phases to prepare high-performance, polymeric composites with improved thermal resistance and mechanical properties. The SiC particles were previously modified by an aminopropyl trimethoxy silane coupling agent, and the treated and raw particles were compared by Fourier infrared spectroscopy (FTIR) as seen in Fig. 6.1. As shown in this figure, the native SiC particles only have few absorption bands. The raw SiC particles also revealed the presence of a large absorption band at around 3410 cm^{-1} attributed to the vibration of the hydroxyl group (−OH). The presence of such a large amount of hydroxyl group is advantageous for grafting a silane coupling agent. As previously discussed, the silane coupling agent hydrolysable groups react these

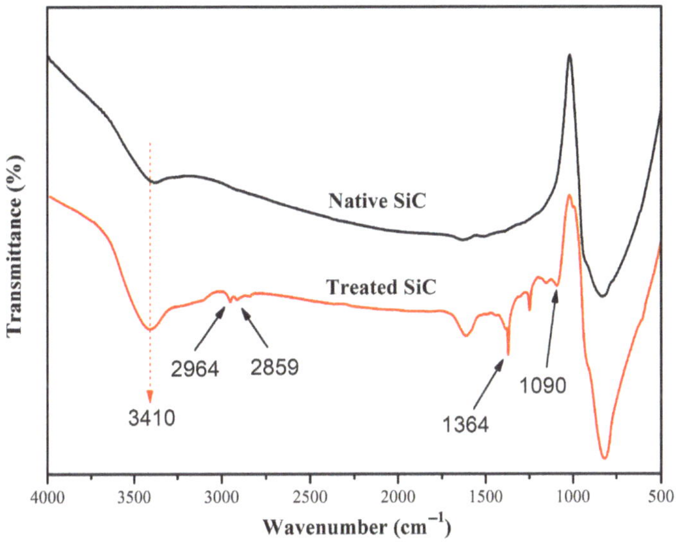

Figure 6.1 FTIR spectra of native and silane-treated SiC particles. *FTIR*, Fourier infrared spectroscopy; *SiC*, silicon carbide. Reprinted from Derradji M, Ramdani N, Zhang T, Wang J, Feng TT, Wang H, et al. Mechanical and thermal properties of phthalonitrile resin reinforced with silicon carbide particles. Mater Des 2015;71:48–55.

hydroxyls present at the surfaces of the particles to covalently bond with the substrate. In contrast with the native SiC particles, the silane surface modified ones showed an FTIR spectrum containing several new bands. For instance, the absorption depicted at 1090 cm^{-1} was attributed to the (Si–O) groups and the bands at 1630 and 1364 cm^{-1} were ascribed to (N–H) shear vibration and bending mode of the hydroxyl groups, respectively. Moreover, the bands appearing at 2964 and 2859 cm^{-1} were attributed to the asymmetric and symmetric ($-CH_2$) stretching vibrations, respectively. Interestingly, the large band at around 3410 cm^{-1} was broadened; the authors attributed this to the presence of more polar groups.

FTIR analyses were also carried out for the neat BAPh monomers and cured polymers and the results are depicted in Fig. 6.2. Clearly, both spectra revealed the presence of the band at 1380 cm^{-1} attributed to the triazine structure. This is related to the type of curing agent used; in this study, the authors opted for the APN curing agent, which initiates the polymerization reaction by the amino function attacking the cyano groups of the monomers. This kind of activated reaction generally results in the formation of triazine structure even at the early stages

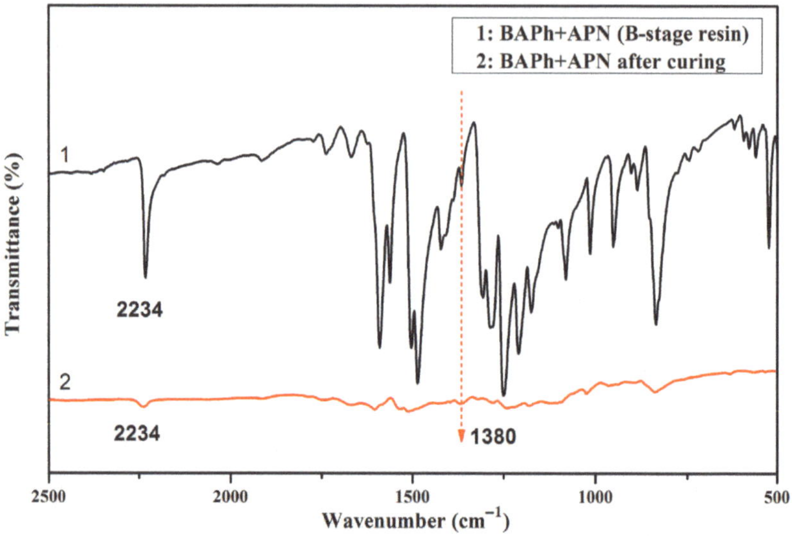

Figure 6.2 FTIR spectra of the neat BAPh resin before and after the curing. *FTIR*, Fourier infrared spectroscopy; *BAPh*, bisphenol A-based PN. Reprinted from Derradji M, Ramdani N, Zhang T, Wang J, Feng TT, Wang H, et al. Mechanical and thermal properties of phthalonitrile resin reinforced with silicon carbide particles. Mater Des 2015;71:48–55.

of the curing reaction. Meanwhile, the absorption band at 2234 cm^{-1} related to the CN groups became weaker after the curing proving the ability of the chosen curing agent to catalyze the cure of the BAPh monomers. The small nitrile absorption remaining after the curing further confirmed the triazine structure formation.

The curing behavior of the neat BAPh resin and its SiC-based microcomposites was studied by dynamic scanning calorimetry (DSC) and the results are plotted in Fig. 6.3. All the compositions showed two endothermic peaks related to the melting of the curing agent and the BAPh monomers and an exothermic peak at around 240°C, confirming the ability of the curing agent to catalyze the cure of these PN monomers. The impact of the SiC particles can be seen through the measurement of the heat polymerization. As seen in Fig. 6.3, increasing the amount of the reinforcing phase to 20 wt.% reduced the heat of polymerization by 33.6%. It is important to point out that a reduced heat of polymerization is pursued in the industry especially during the preparation of thick, composite sections.

Before moving forward with the detailed thermal and mechanical properties, it should be understood that these properties are closely related to the curing procedure and amount of the curing agent used.

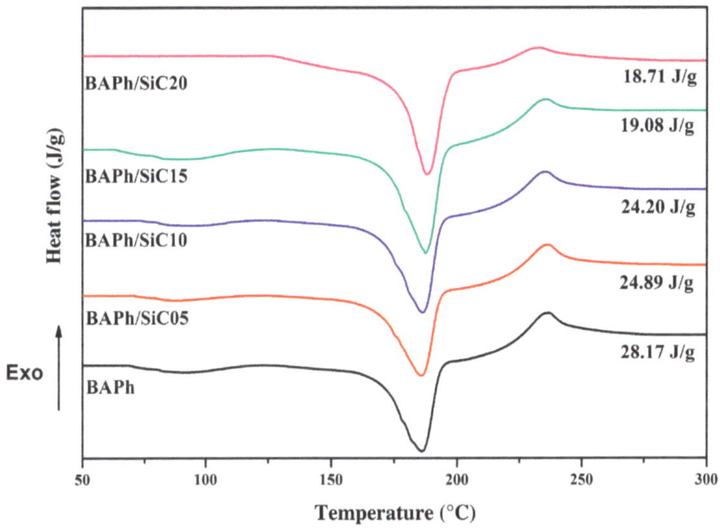

Figure 6.3 DSC thermograms of the BAPh/SiC composites at the heating rate of 20°C/min. *DSC*, dynamic scanning calorimetry; *BAPh*, bisphenol A-based PN; *SiC*, silicon carbide. Reprinted from Derradji M, Ramdani N, Zhang T, Wang J, Feng TT, Wang H, et al. Mechanical and thermal properties of phthalonitrile resin reinforced with silicon carbide particles. Mater Des 2015;71:48–55.

Derradji et al., reported that the BAPh/SiC composites were cured in the presence of 10 wt.% of APN curing agent following the curing procedure of 240°C for 2 h, 260°C for 5 h, and 280°C for 8 h. Thus, any result presented herein can be ameliorated or reduced by changing either the curing agent or the curing temperatures.

The high hardness of the SiC was expected to result in PN composites with improved microhardness. Thus the authors investigated the effect of the native and silane treated SiC particles on this important mechanical property. The microhardness experimentally recorded values are shown in Fig. 6.4. According to the obtained results, the neat BAPh resin exhibited a microhardness value of 380 MPa. The addition of both native and treated SiC particles gradually improved the neat resin microhardness to 477 and 536 MPa, respectively, at 20 wt.% fillers loading. The high hardness of the SiC (28.2 GPa) is the main reason behind these excellent improvements. Moreover, the native SiC particles showed lower enhancements, in contrast with the silane modified ones, because of their poor state of dispersion and adhesion with the PN matrix. The importance of the silane treatment can be clearly seen through the obtained results. For instance, at 20 wt.% micro-SiC, the microhardness of the BAPh resin, reinforced with the silane treated SiC

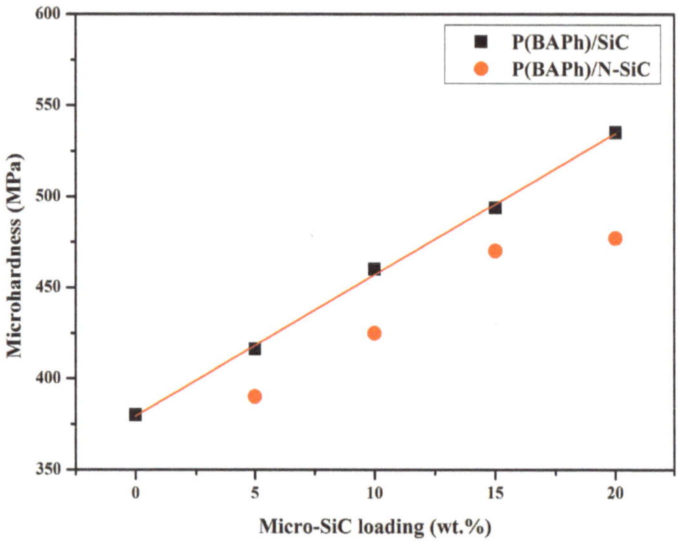

Figure 6.4 Microhardness of the native and silane-treated SiC-based PN composites. *SiC*, silicon carbide. Reprinted from Derradji M, Ramdani N, Zhang T, Wang J, Feng TT, Wang H, et al. Mechanical and thermal properties of phthalonitrile resin reinforced with silicon carbide particles. Mater Des 2015;71:48–55.

particles, is 59 MPa higher than that of the BAPh resin, reinforced with raw SiC particles. Furthermore, if compared with another, high-performance resin, such as the BPA-based polybenzoxazine reinforced with the same type and amount of fillers, the PN matrix exhibited an ameliorated hardness before and after the reinforcement. For example, the microhardness of the BPA-based polybenzoxazine was measured to be 354 MPa, and at 20 wt.% of silane treated micro-SiC loading the microhardness was about 450 MPa [68].

The impact of the native and silane treated SiC particles on the flexural properties of the BAPh resin was also investigated. The flexural strength and modulus of the pristine BAPh matrix and its SiC-based composites are presented Fig. 6.5. Clearly, the neat BAPh resin exhibited the lowest, flexural strength and modulus with 60 MPa and 1.9 GPa, respectively. By contrast, as the contents of the native or silane surface modified SiC increased in the resin, both the flexural strength and modulus of their resulting composites increased to reach 99 and 136 MPa, and 2.8 and 3.5 GPa, respectively, at their maximum microfillers contents. The ability of the SiC particles to effectively transfer the load was behind these exceptional improvements in the flexural properties of the PN matrix. Meanwhile, the extraincrease in these properties

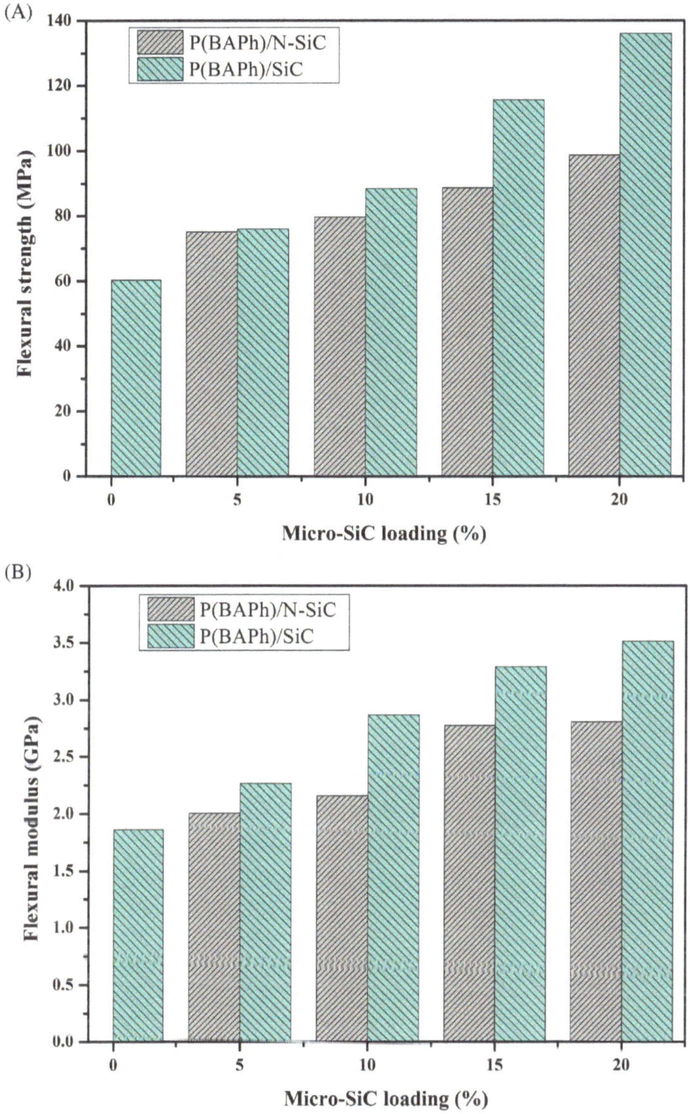

Figure 6.5 Flexural strength (A) and modulus (B) of the cured BAPh/SiC composites. *BAPh*, bisphenol A-based PN; *SiC*, silicon carbide. Reprinted from Derradji M, Ramdani N, Zhang T, Wang J, Feng TT, Wang H, et al. Mechanical and thermal properties of phthalonitrile resin reinforced with silicon carbide particles. Mater Des 2015;71:48–55.

obtained with the silane-treated, SiC particles was due to their better contacts and uniform dispersion with the BAPh matrix.

Aiming the judge the toughness and strength of the neat BAPh resin and its related silane treated SiC composites, the tensile test was performed and the toughness values were measured by integrating the surface under the stress–strain curves. Both the stress–strain curves and toughness values are presented in Fig. 6.6. The strength, defined as the value of stress before breaking occurs, was measured to be 27 MPa

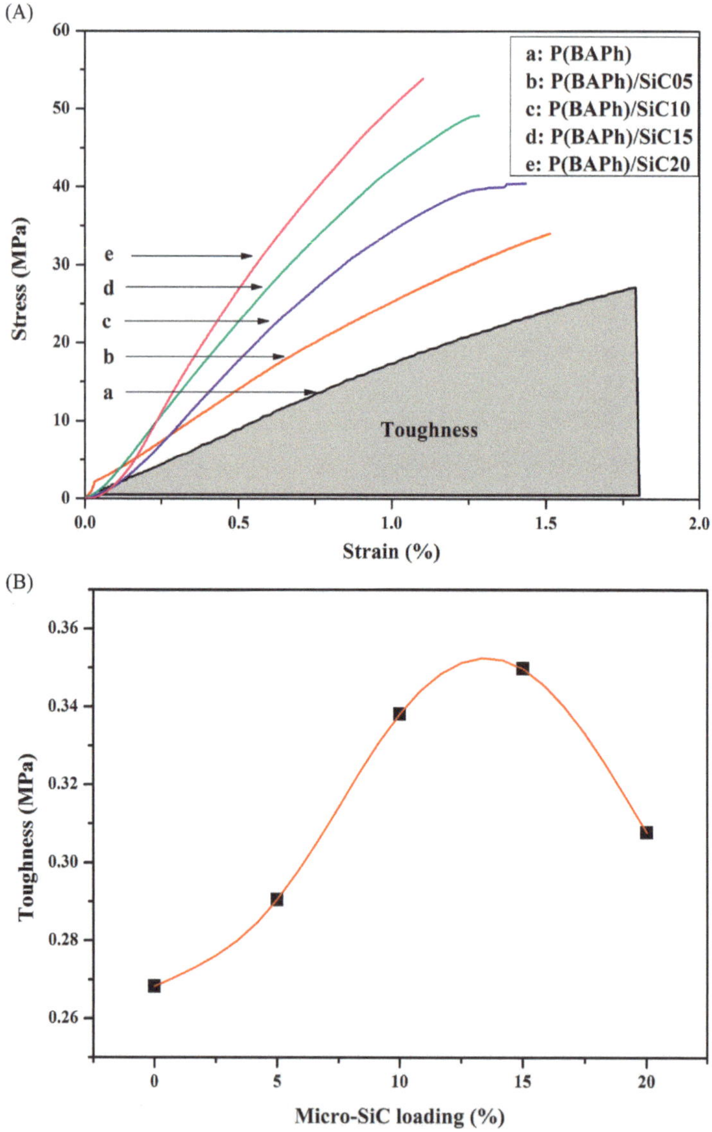

Figure 6.6 Stress–strain curves (A) and toughness (B) of the cured BAPh/SiC composites. *BAPh*, bisphenol A-based PN; *SiC*, silicon carbide. Reprinted from Derradji M, Ramdani N, Zhang T, Wang J, Feng TT, Wang H, et al. Mechanical and thermal properties of phthalonitrile resin reinforced with silicon carbide particles. Mater Des 2015;71:48–55.

for the neat, BAPh resin and this value was further enhanced by adding more silane-treated SiC to reach 54 MPa at 20 wt.% fillers loading. On the other hand, the toughness reached its maximum value at 15 wt.% micro-SiC loading and adding more contents seemed to negatively affect the toughness state. It is also important to point out that even at 20 wt.% of fillers contents, the toughness was better than that of the neat BAPh resin. The previous observations confirmed the ability of the SiC particles to enhance the strength and toughness of the PN matrix to a certain extent. Moreover, the particles treatment also played a key role in improving the tensile properties of the pristine PN resin.

The dispersion of the SiC particles in the PN matrix was also investigated by a polarization optical microscope (POM). A comparison between the neat BAPh resin and its composite of 20 wt.% treated micro-SiC are presented in Fig. 6.7. Clearly, the silane surface modification of the fillers allowed an ameliorated state of dispersion with no apparent agglomerations, even at relatively high fillers amount. The POM investigations also further confirmed the enhancements seen in the mechanical properties.

After the tensile test was carried out, the sample fractured surfaces were analyzed by a scanning electron microscope (SEM) to study the deformation mechanisms. The micrographs shown in Fig. 6.8 are those of the SiC fillers, the neat BAPh matrix, and its SiC-based composites at 5, 10, 15, and 20 wt.% of fillers loading, respectively. Clearly, the fractured surfaces of neat BAPh resin showed long fracture lines confirming the brittle character of this kind of thermosets. Meanwhile, the

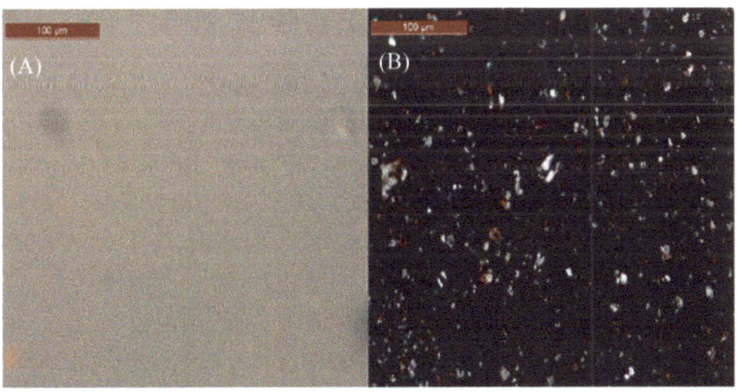

Figure 6.7 POM micrographs of the cured unfilled resin (A) and BAPh/SiC at 20 wt.% treated micro-SiC (B). *POM*, polarization optical microscope; *BAPh*, bisphenol A-based PN; *SiC*, silicon carbide. Reprinted from Derradji M, Ramdani N, Zhang T, Wang J, Feng TT, Wang H, et al. Mechanical and thermal properties of phthalonitrile resin reinforced with silicon carbide particles. Mater Des 2015;71:48–55.

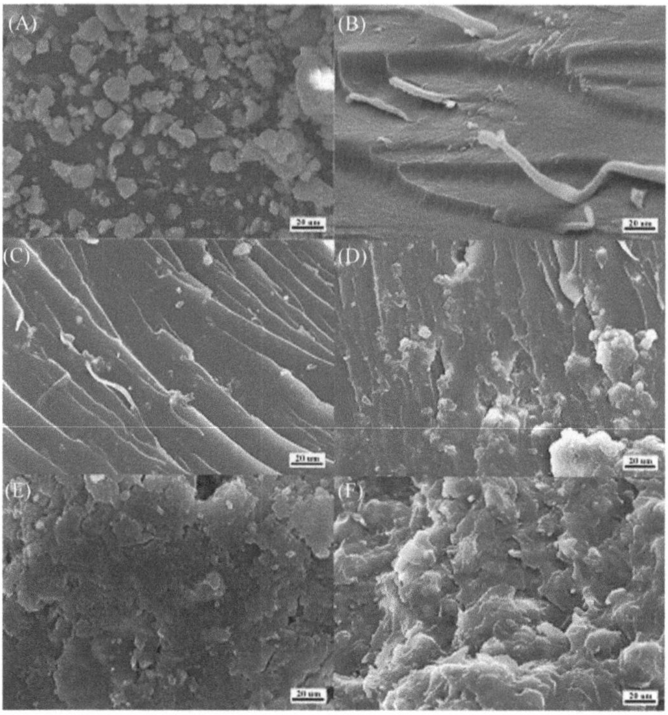

Figure 6.8 SEM micrographs of the fractured surface of the cured BAPh/SiC composites at various SiC contents: micro-SiC powder (A), pure resin (B), 5 wt.% (C), 10 wt.% (D), 15 wt.% (E), and 20 wt.% (F). *SEM*, scanning electron microscope; *BAPh*, bisphenol A-based PN; *SiC*, silicon carbide. Reprinted from Derradji M, Ramdani N, Zhang T, Wang J, Feng TT, Wang H, et al. Mechanical and thermal properties of phthalonitrile resin reinforced with silicon carbide particles. Mater Des 2015;71:48–55.

introduction of the reinforcing phase completely changed the deformation mechanism. Indeed, the long fracture lines were no longer seen and were replaced by small-sized cracks as the amount of the fillers' increased. The disappearance of the long fracture lines was mainly caused by the presence of the microparticles, which deviated the initial fractures from its path, resulting in smaller cracks. Moreover, the composites' fracture surfaces appeared rougher than those of the neat resin and evidences of large amounts of absorbed energy can be easily seen in the micrographs. Other fracture mechanisms were also depicted, such as particles debonding from the matrix, voids, and formation of cavities, on the rough surfaces of the BAPh/SiC composites.

The thermal stability investigations were performed by thermogravimetric analyses (TGA) under nitrogen atmosphere. The comparisons

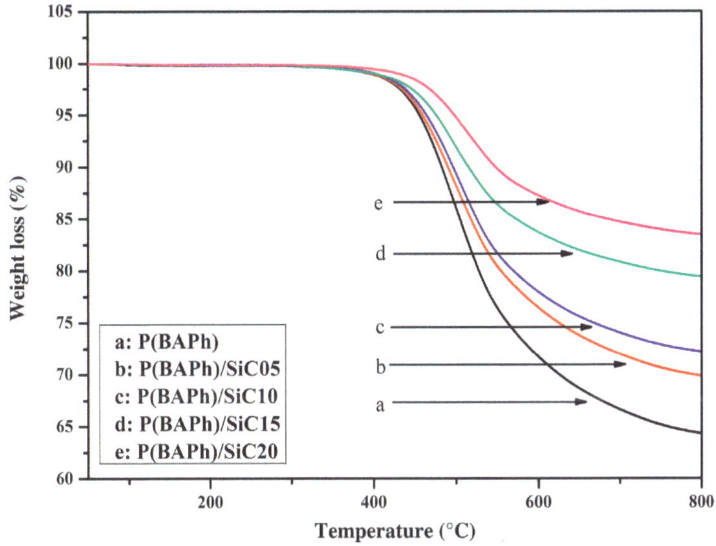

Figure 6.9 Thermal stability of the cured BAPh/SiC composites at various micro-SiC contents. *BAPh*, bisphenol A-based PN; *SiC*, silicon carbide. Reprinted from Derradji M, Ramdani N, Zhang T, Wang J, Feng TT, Wang H, et al. Mechanical and thermal properties of phthalonitrile resin reinforced with silicon carbide particles. Mater Des 2015;71:48–55.

between the thermal stability of the neat resin and its SiC-based composites were done after measuring the starting decomposition temperature at 5 and 10 wt.% ($T_{5\%}$ and $T_{10\%}$) as well as the char yield at 800°C (Y_c). The TGA curves are plotted in Fig. 6.9 and Table 6.2 lists all the thermal stability parameters. The outstanding, thermal stability of the PN resins was confirmed by a $T_{5\%}$ and $T_{10\%}$ of 456 and 483°C, respectively, with a char yield of 64.3%. The addition of the treated micro-SiC further improved the thermal stability of the neat resin. For instance, as the amount of the fillers increased, all the thermal stability parameters were gradually ameliorated to finally reach the values of 498 and 548°C for the $T_{5\%}$ and $T_{10\%}$, respectively, and 83.5% for the char yield. The improvement in the thermal stability of these composites can be attributed to the good dispersion of the SiC microfillers in the resin and the better matrix/filler adhesion, due to the presence of polar groups on the external surface of the particles afforded by the silane surface modification. Also, the high, thermal stability of the SiC ceramic particles allowed them to act as a thermal insulator protecting the PN matrix. Another factor is that the used micro-SiC particles can restrain the movement of the chain inside the PN matrix resulting in an

Table 6.2 Thermal Stability Parameters of the P(BAPh)/SiC Composites

Specimen Code	$T_{5\%}$ (°C)	$T_{10\%}$ (°C)	Y_c (%, 800°C)
P(BAPh)	455.5	482.6	64.3
P(BAPh)/SiC05	459.4	490.2	69.8
P(BAPh)/SiC10	464.4	496.6	72.2
P(BAPh)/SiC15	476.2	515.4	79.5
P(BAPh)/SiC20	497.8	548.2	83.5

Reprinted from Derradji M, Ramdani N, Zhang T, Wang J, Feng TT, Wang H, et al. Mechanical and thermal properties of phthalonitrile resin reinforced with silicon carbide particles. Mater Des 2015;71:48–55.

ameliorated thermal stability. The excellent thermal stability obtained with the BAPh/SiC composites has not been obtained with any other kind of high-performance resin reinforced with the same type and amount of fillers.

The DMA was also used to assess the variations in the storage modulus and Tan delta of the neat BAPh resin and its related, SiC-based composites. Both the storage modulus and Tan delta vary with temperature and allow the determination of two other important thermomechanical properties, known as the stiffness and T_g. The stiffness can be defined herein as the value of storage modulus at 50°C and the T_g is obtained from the maximum of Tan delta peak. The variations of the storage modulus and Tan delta are illustrated in Fig. 6.10 and Table 6.3 lists the stiffness and T_g values for all the tested samples. The stiffness and T_g of the neat resin was measured to be 1.1 GPa and 300°C, respectively. It is also important to point out that the obtained, thermomechanical properties of the neat, BAPh resin can be further enhanced by several means such as increasing the amount or changing the curing agent, extending the temperature and time of curing, or postcuring the samples at elevated temperatures. Indeed, the thermomechanical properties of the neat, PN resins can be tailored according to the required application. Interestingly, increasing the amount of the silane treated SiC particles positively affected the thermomechanical properties. For instance, the improvements reached up to 182% and 12.6% for the stiffness and T_g at the maximum fillers loading of 20 wt.%. The remarkable improvements in the stiffness of the neat BAPh resin were, to a large extent, attributed to the high modulus of the SiC fillers, and the T_g values increases also reflected an improvement in the BAPh resin

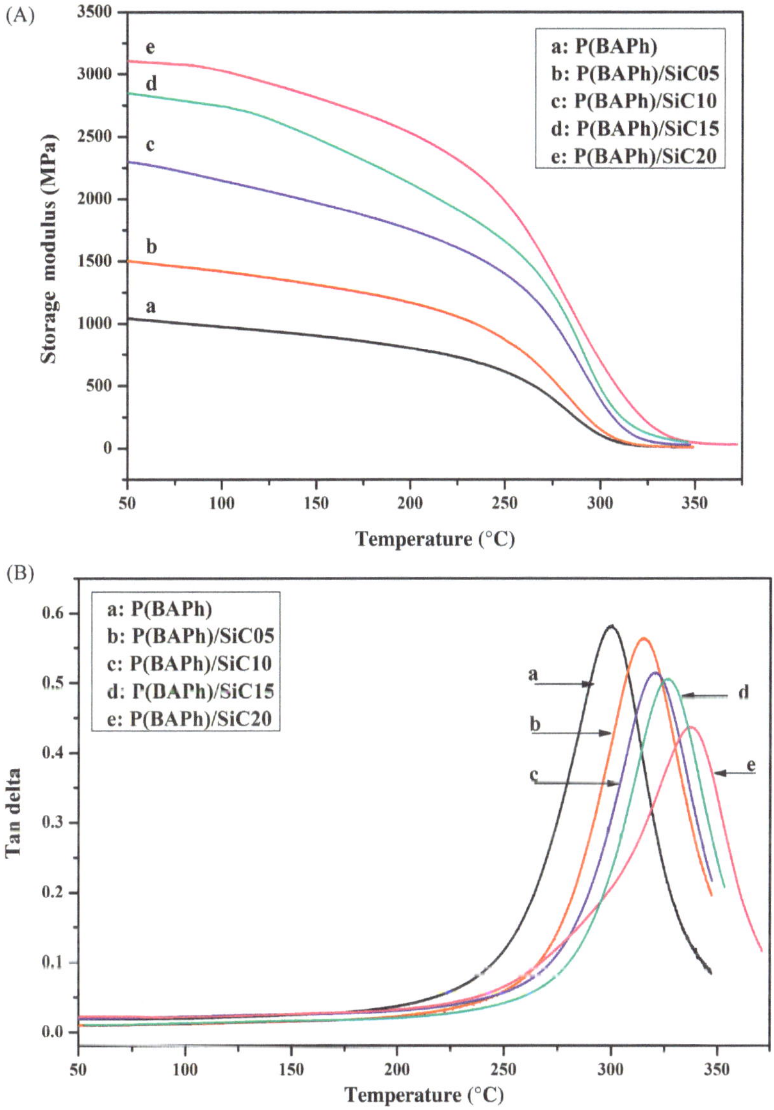

Figure 6.10 Evolution of storage modulus (A) and Tan delta (B) of the cured BAPh/SiC composites. *BAPh*, bisphenol A-based PN; *SiC*, silicon carbide. Reprinted from Derradji M, Ramdani N, Zhang T, Wang J, Feng TT, Wang H, et al. Mechanical and thermal properties of phthalonitrile resin reinforced with silicon carbide particles. Mater Des 2015;71:48–55.

rigidity. Furthermore, the Tan delta peak was reduced in height and broadened by the presence of more fillers contents. These phenomena are known as the dilution and relaxation process commonly seen in the field of hybrid materials.

Table 6.3 Thermomechanical Properties of the P(BAPh)/SiC Composites

Specimen Code	G' at 50°C (GPa)	The Gain in G' (%)	T_g (°C)
P(BAPh)	1.1	–	300
P(BAPh)/SiC05	1.5	36.4	315
P(BAPh)/SiC10	2.3	109.1	320
P(BAPh)/SiC15	2.9	163.6	328
P(BAPh)/SiC20	3.1	181.8	338

Reprinted from Derradji M, Ramdani N, Zhang T, Wang J, Feng TT, Wang H, et al. Mechanical and thermal properties of phthalonitrile resin reinforced with silicon carbide particles. Mater Des 2015;71:48–55.

The synergistic combination between the PN matrix and the SiC microparticles led to the preparation of high-performance, hybrid materials with better properties than those obtained with other polymeric systems. Thus it was concluded to further extend the researches to other ceramics and other PN resins.

6.5.2 Properties of Phthalonitrile/Si_3N_4 Nanocomposites

After the promising results found with the SiC microparticles, Derradji et al., extended their researches to other, PN-based nanocomposites and the first reported study was on the use of the Si_3N_4 (average nanoparticles diameter of 50 nm) nanoparticles as a new kind of reinforcement for the BAPh matrix [52]. The Si_3N_4 nanoparticles nanofillers have numerous interesting features, such as a high melting point, good hardness, superior thermal conductivity, relatively low density and good corrosion resistance. These nanofillers have the potential to improve the fracture toughness and lower the brittleness of various matrices including the polymeric ones. Moreover, due to their advantageous characteristics, Si_3N_4 nanoparticles were recognized as one of the most important ceramics for high temperature and mechanical strength applications. Indeed, various, high-performance thermosets, such as epoxy, polyimide, bismaleimide, cyanate ester, and polybenzoxazine have already been reinforced with different sizes and amounts of Si_3N_4 nanoparticles and interesting results were found [44,82–85].

The preparation of the BAPh/Si_3N_4 nanocomposites was preceded by a chemical treatment of the nanofillers by an amino silane, coupling agent. The FTIR was used to acknowledge about the effectiveness of the surface modification by comparing the spectra of both the native

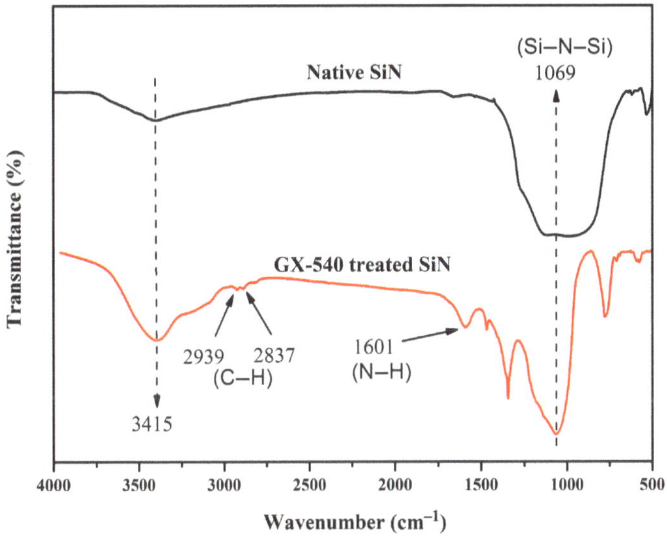

Figure 6.11 FTIR spectra of native and silane-treated Si_3N_4 nanoparticles. *FTIR*, Fourier infrared spectroscopy. *Si_3N_4*, boron carbide, silicon nitride. Reprinted from Derradji M, Ramdani N, Zhang T, Wang J, Lin Z-w, Yang M, et al. High thermal and thermomechanical properties obtained by reinforcing a bisphenol-A based phthalonitrile resin with silicon nitride nanoparticles. Mater Lett 2015;149:81–4.

and silane-treated, Si_3N_4 nanoparticles. The FTIR spectra are presented in Fig. 6.11. The native Si_3N_4 spectrum only showed few absorption bands such as the large absorption around 1000 cm^{-1} characteristic of the Si_3N_4 ceramic. A second absorption was depicted at 3415 cm^{-1} and was ascribed to the vibration of the hydroxyl group, probably resulting from the water absorption of the Si_3N_4 nanoparticles. The presence of the hydroxyl groups is appreciated because they allow the formation of hydrogen bonds with the silane coupling agent. The reaction with the silane coupling agent brought several changes in the Si_3N_4 spectrum. For instance, the bands appearing at 2964 and 2859 cm^{-1} were attributed to the asymmetric and symmetric ($-CH_2$) stretching vibrations, respectively. The band at 1601 cm^{-1} was ascribed to the (N–H) vibration group and the peak at 3415 cm^{-1} was broadened indicating the presence of more polar groups.

Before curing the samples, the authors studied the curing behavior of the PN prepolymers (BAPh monomers and APN curing agent in a 90:10 weight ratio) containing different proportions of the treated, Si_3N_4 nanoparticles ranging from 3 to 15 wt.%, under DSC analyzes and the obtained results are illustrated in Fig. 6.12. As clearly seen in

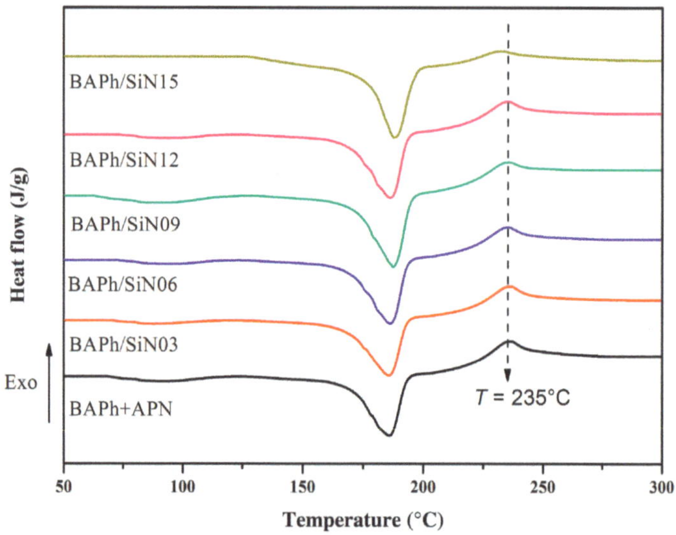

Figure 6.12 DSC thermograms of the BAPh/Si_3N_4 nanocomposites at the heating rate of 20°C/min. *DSC*, dynamic scanning calorimetry; *BAPh*, bisphenol A-based PN; *Si_3N_4*, boron carbide, silicon nitride. Reprinted from Derradji M, Ramdani N, Zhang T, Wang J, Lin Z-w, Yang M, et al. High thermal and thermomechanical properties obtained by reinforcing a bisphenol-A based phthalonitrile resin with silicon nitride nanoparticles. Mater Lett 2015;149:81–4.

this figure, all the samples showed two endothermic peaks related to the melting points of the BAPh monomers and the APN curing agent. Interestingly, the endothermic related to the curing agent was weaker than that of the BAPh monomers. The reason can be found from the preparation technique where the BAPh monomers and the curing agent were solid-state blended with the ceramic nanoparticles, and the resulting mixtures were melted and mechanically stirred before quenched to room temperature and grinded to powders. The samples analyzed under DSC were those resulting powders. Therefore, at this step, the reaction between the BAPh monomers and the APN curing agent could have been initiated resulting in a weak endothermic transition for the APN curing agent. All the samples also presented an exothermic peak at around 235°C, confirming the ability of the APN to cure the neat, BAPh monomers and its related nanocomposites. Moreover, the introduction of the silane-treated, Si_3N_4 nanoparticles did not affect the position of the exothermic transition. However, a reduction in the heat of polymerization was noticed upon adding more nanofillers contents. Thus similarly to the micro-SiC particles, the Si_3N_4 nanoparticles are appealing for the preparation of thick composites sections and can

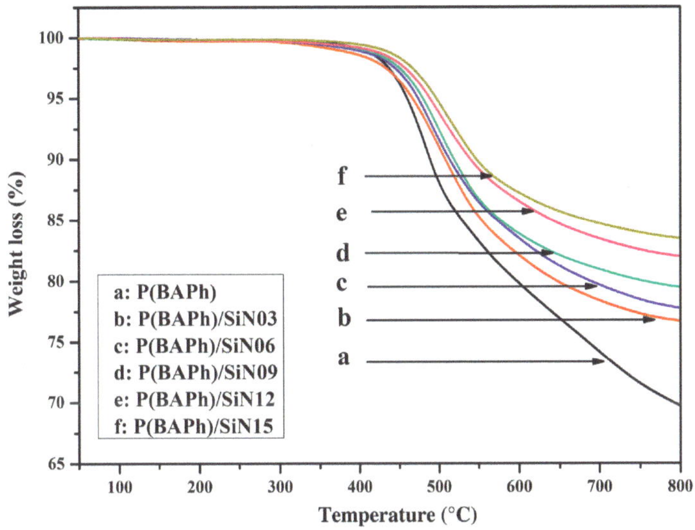

Figure 6.13 TGA curves of the cured BAPh/Si_3N_4 nanocomposites under nitrogen atmosphere. *TGA*, thermogravimetric analyses; *BAPh*, bisphenol A-based PN; *Si_3N_4*, boron carbide, silicon nitride. Reprinted from Derradji M, Ramdani N, Zhang T, Wang J, Lin Z-w, Yang M, et al. High thermal and thermomechanical properties obtained by reinforcing a bisphenol-A based phthalonitrile resin with silicon nitride nanoparticles. Mater Lett 2015;149:81–4.

easily be proceeded by the conventional industrial techniques, such as resin transfer molding (RTM) and vacuum-assisted RTM.

The BAPh/Si_3N_4 nanocomposites were prepared by the in situ polymerization process and cured by a hot press following the curing procedure of 240°C for 2 h, 260°C for 3 h, 280°C for 6 h, and 300°C for 6 h. The cured BAPh resin and its related Si_3N_4 nanocomposites were investigated by TGA and DMA for their thermal stability and thermomechanical properties, respectively. The TGA curves are presented in Fig. 6.13 and the thermal stability parameters such as the $T_{5\%}$, $T_{10\%}$, and Y_c are listed in Table 6.4. The neat PN resin, cured following the curing procedure presented above, showed an excellent thermal stability confirmed with a $T_{5\%}$ and $T_{10\%}$ of 460 and 489°C, respectively, and a char yield at 800°C of 69.7%. These values were higher than those obtained with the BAPh matrix used for the SiC-based composites because of the extended time and curing temperatures, as well as the use of the hot press curing technique. Moreover, the excellent thermal stability of the Si_3N_4 nanoparticles resulted in nanocomposites with an outstanding thermal resistance. For example, the addition of 15 wt.% Si_3N_4 nanoparticles ensued in a gain of 37 and 59°C in $T_{5\%}$, $T_{10\%}$,

Table 6.4 Thermal and Thermomechanical Properties of the P(BAPh)/ Si_3N_4 Nanocomposites

Specimen Code	G' at 50°C (GPa)	T_g (°C)	$T_{5\%}$ (°C)	$T_{10\%}$ (°C)	Y_c (%, 800°C)
P(BAPh)	1.7	300	459.8	488.7	69.7
P(BAPh)/SiN03	2.3	318	467.0	508.1	76.6
P(BAPh)/SiN06	2.8	338	474.1	513.6	77.7
P(BAPh)/SiN09	3.1	341	480.2	519.4	79.5
P(BAPh)/SiN12	3.4	349	488.7	539.7	82.0
P(BAPh)/SiN15	4.0	360	496.8	547.2	83.5

Reprinted from Derradji M, Ramdani N, Zhang T, Wang J, Lin Z-w, Yang M, et al. High thermal and thermomechanical properties obtained by reinforcing a bisphenol-A based phthalonitrile resin with silicon nitride nanoparticles. Mater Lett 2015;149:81–4.

respectively, and a 13.8% increase in the char yield. These remarkable enhancements in the thermal stability of the neat resin are related to both the shielding effect of the Si_3N_4 nanoparticles, which reduces the motion and insulates the polymeric matric from the heat source, and to the ameliorated state of dispersion and adhesion obtained by the appropriate particles silane surface modification.

The stiffness and T_g of the PN resin reinforced with different amounts of Si_3N_4 nanoparticles are also listed in Table 6.4 and the storage modulus and Tan delta variations are plotted in Fig. 6.14. The stiffness of the BAPh resin was 1.7 GPa and the T_g was 300°C. These values were markedly enhanced as the amount of the nanofillers increased. Indeed, at the maximum nanocontents of 15 wt.%, the stiffness and T_g reached 4 GPa and 360°C, respectively. As for the SiC microparticles, the relaxation and dilution processes were also observed in the Tan delta peaks. However, the nanoparticles brought better gains in the thermomechanical properties with fewer reinforcing amounts.

The SEM investigations were also conducted on the DMA fractured surfaces to get clear images about the mechanism of enhancements as well as to acknowledge about the nanoparticles dispersion and adhesion with the PN matrix. Fig. 6.15 shows the micrographs of the unfilled resin and its related nanocomposites at 6, 9 and 15 wt.% nano-Si_3N_4 loadings. The brittle nature of the BAPh matrix was confirmed by the presence of long fracture lines. On the other hand, adding the reinforcing phase resulted in a gradual increase in the roughness and the long fractures line were replaced by smaller microcracks indicating that the system can

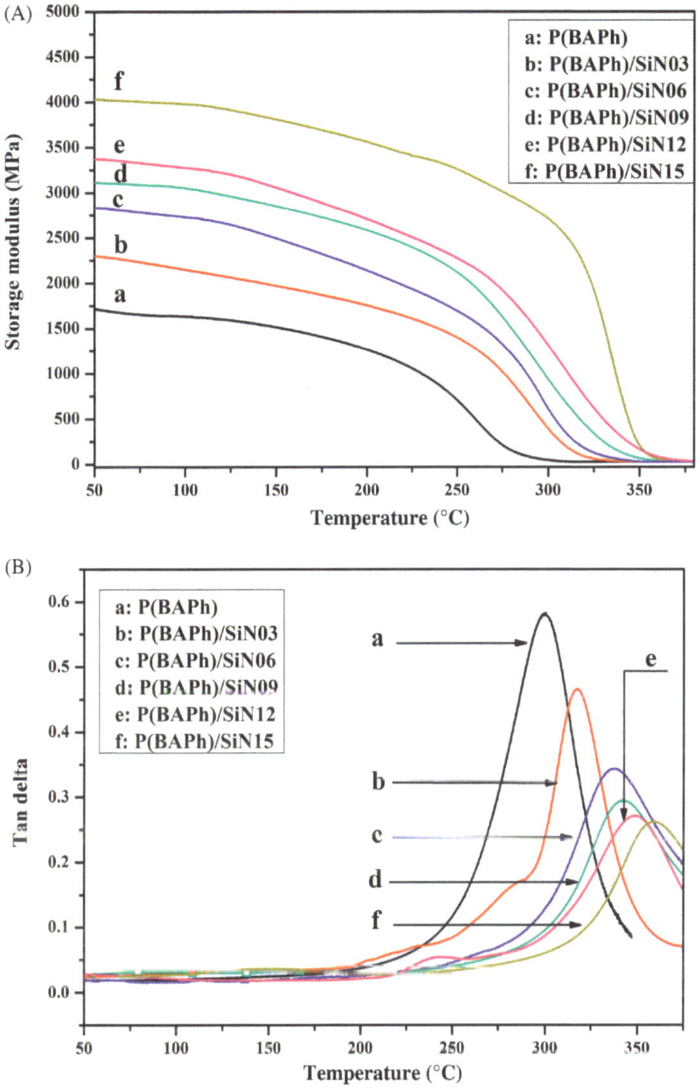

Figure 6.14 Evolution of storage modulus (A) and Tan delta (B) of the cured BAPh/Si_3N_4 nanocomposites. *BAPh*, bisphenol A-based PN; *Si_3N_4*, boron carbide, silicon nitride. Reprinted from Derradji M, Ramdani N, Zhang T, Wang J, Lin Z-w, Yang M, et al. High thermal and thermomechanical properties obtained by reinforcing a bisphenol-A based phthalonitrile resin with silicon nitride nanoparticles. Mater Lett 2015;149:81–4.

withstand elevated amount of stress before breaking occurs. Interestingly, the debonding phenomenon was not observed for the nanoparticles which may be attributed to the highly specific surface of these nanofillers, as well as the good adhesion afforded by the silane coupling agent.

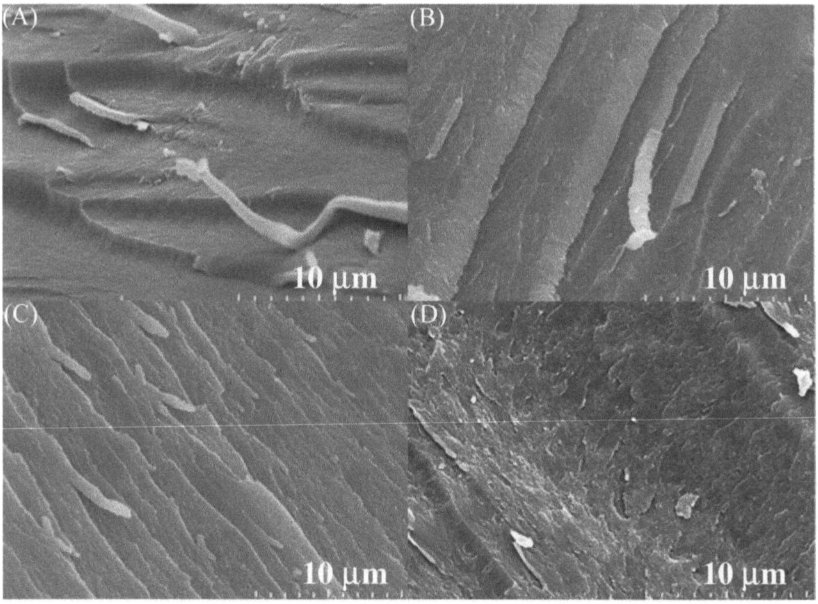

Figure 6.15 SEM micrographs of fracture surface of the cured BAPh/Si_3N_4 nanocomposites at various Si_3N_4 contents: (A) 0%, (B) 6%, (C) 9%, and (D) 15%. *SEM*, scanning electron microscope; *BAPh*, bisphenol A-based PN; *Si_3N_4*, boron carbide, silicon nitride. Reprinted from Derradji M, Ramdani N, Zhang T, Wang J, Lin Z-w, Yang M, et al. High thermal and thermomechanical properties obtained by reinforcing a bisphenol-A based phthalonitrile resin with silicon nitride nanoparticles. Mater Lett 2015;149:81–4.

6.5.3 Properties of Phthalonitrile/Al_2O_3 Nanocomposites

Al_2O_3 or alumina is a widely used ceramic known for its combination of interesting properties such as high modulus, good hardness, enhanced thermal conductivity, and good dielectric properties. The introduction of alumina nanoparticles into the PN matrices can bring global enhancements in the thermal and mechanical properties of the neat resins. Derradji et al., reported on the use of silane surface-modified, alumina nanoparticles (average nanoparticles diameter of 60 nm) as a new kind of reinforcement for a typical BAPh resin [53]. The authors investigated the effect of different nanofillers amounts on the microhardness, tensile, morphology, thermal, and thermomechanical properties of the pristine, PN resin. The FTIR technique was carried out for the native and silane treated alumina nanoparticles to judge about the effectiveness of the surface modification. The FTIR spectra of both

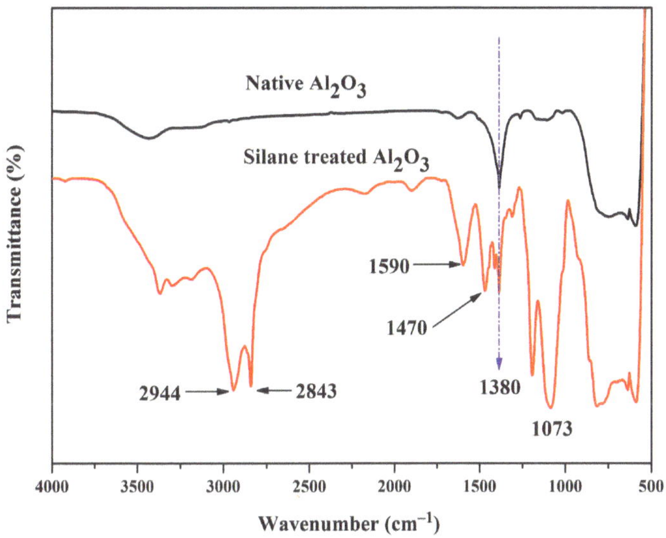

Figure 6.16 FTIR spectra of native and silane-treated Al_2O_3 nanoparticles. *FTIR*, Fourier infrared spectroscopy; *Al_2O_3*, alumina or aluminum oxide. Reprinted from Derradji M, Ramdani N, Zhang T, Wang J, Gong L-d, Xu X-d, et al. Thermal and mechanical properties enhancements obtained by reinforcing a bisphenol-A based phthalonitrile resin with silane surface-modified alumina nanoparticles. Polym Compos 2017;38:1546–58.

native and silane treated alumina nanoparticles are presented in Fig. 6.16. The silane surface modification allowed crucial changes in the FTIR spectrum of the native alumina. For example, the silane-treated alumina spectrum revealed the presence of new absorption bands at 2944 and 2843 cm^{-1} ascribed to the asymmetric and symmetric ($-CH_2$) stretching vibrations, respectively. In addition, the new bands depicted at 1590, 1470, and 1073 cm^{-1} are related to the (N–H) bending (scissoring), ($-CH_2$) bending (scissoring), and ($Si-O-CH_3$) stretching, respectively. Moreover, the broad band around 3435 cm^{-1} indicated the presence of a large number of polar groups mainly attributed to the effect of the surface silanization. Thus, the previous observations confirmed that the alumina nanoparticles were successfully treated with the amino-silane, coupling agent.

The authors also chose the APN as curing agent for the BAPh monomers. The extensive use of the APN to prepare the PN-based composites is closely related to the great advantages obtained from this combination. Indeed, the APN allows lower curing temperatures and better mechanical and thermal properties, and since this curing agent is a class of PN monomers, all the molecules are incorporated into the

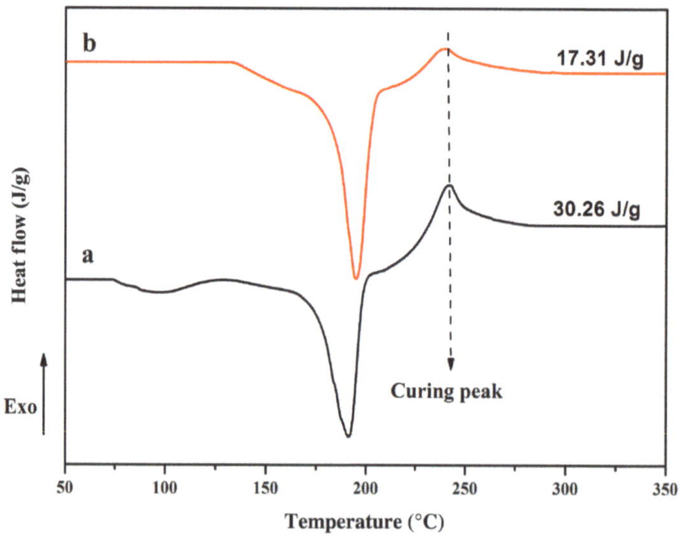

Figure 6.17 DSC thermograms of the unfilled BAPh matrix and the nanocomposites of 15 wt.% nano-Al_2O_3. *DSC*, dynamic scanning calorimetry; *BAPh*, bisphenol A-based PN; *Al_2O_3*, alumina or aluminum oxide. Reprinted from Derradji M, Ramdani N, Zhang T, Wang J, Gong L-d, Xu X-d, et al. Thermal and mechanical properties enhancements obtained by reinforcing a bisphenol-A based phthalonitrile resin with silane surface-modified alumina nanoparticles. Polym Compos 2017;38:1546–58.

final structure. The curing behavior of the neat BAPh monomers with APN in a ratio of 90:10, respectively, along with the alumina-based nanocomposites were studied by DSC. The DSC thermograms of the unfilled matrix and the BAPh/Al_2O_3 nanocomposite at 15 wt.% nano-contents are plotted in Figs. 6.17 and 6.18 shows the evolution of the heat polymerization energy for the BAPh/Al_2O_3 nanocomposites. As seen from Fig. 6.17, the APN was effectively able to cure the BAPh monomers with a relatively low curing temperatures and a large processing window. Subsequently, the addition of the nanofillers only resulted in diminutions in the heat of polymerization. For instance, the heat of polymerization was reduced by 75% when moving from the neat BAPh resin to the nanocomposites of 15 wt.% nano-Al_2O_3 contents.

The BAPh/Al_2O_3 nanocomposites were prepared following the in situ process and cured by a hot press according to the curing schedule of 240°C for 5 h, 260°C for 5 h, 280°C for 5 h, 300°C for 5 h, and 320°C for 5 h. Taking advantage from the high hardness of the alumina, the resulting BAPh/Al_2O_3 nanocomposites exhibited improved microhardness. The experimentally-obtained microhardness results are plotted in

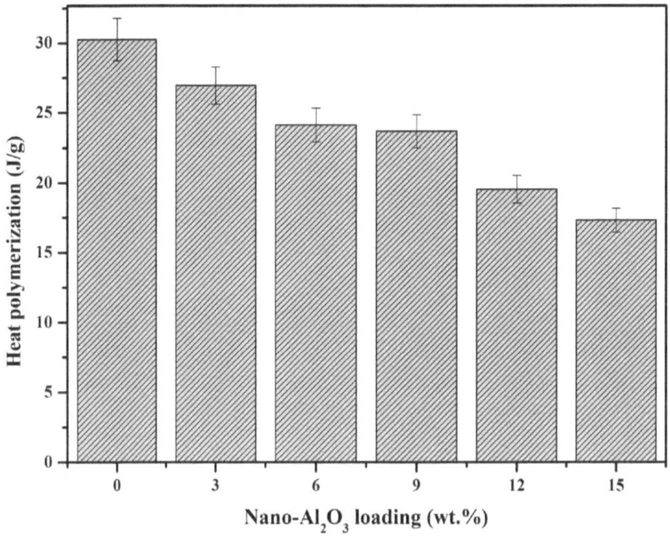

Figure 6.18 Evolution of the heat polymerization energy for the BAPh/Al_2O_3 nanocomposites. *BAPh*, bisphenol A-based PN; *Al_2O_3*, alumina or aluminum oxide. Reprinted from Derradji M, Ramdani N, Zhang T, Wang J, Gong L-d, Xu X-d, et al. Thermal and mechanical properties enhancements obtained by reinforcing a bisphenol-A based phthalonitrile resin with silane surface-modified alumina nanoparticles. Polym Compos 2017;38:1546–58.

Fig. 6.19. By analyzing these data, it appears that the neat, BAPh resin exhibited a microhardness of 395 MPa. This value systematically increased by adding more nanofillers contents. For example, at the maximum nano-Al_2O_3 loading of 15 wt.%, the nanocomposites exhibited the improved microhardness value of 505 MPa. The high hardness of the nanofillers with regard to that of the BAPh resin, along with their strong interactions and homogeneous dispersion into the matrix, played an important role in the changes observed during the microhardness test.

The tensile test was also performed and Young's modulus experiment results were compared with prediction obtained from several models such as Series, Halpin–Tsai, and Kerner. The Series model is used to describe simple and idealized systems; the composite modulus in this case depends only on the modulus of both filler and matrix as seen from the following equation [32]:

$$\frac{1}{E_c} = \frac{V_m}{E_m} + \frac{V_f}{E_f} \tag{6.1}$$

where E_c, E_f, and E_m are the modulus of the composite, filler, and matrix, respectively. V_f and V_m are the volume fraction of the filler and matrix, respectively.

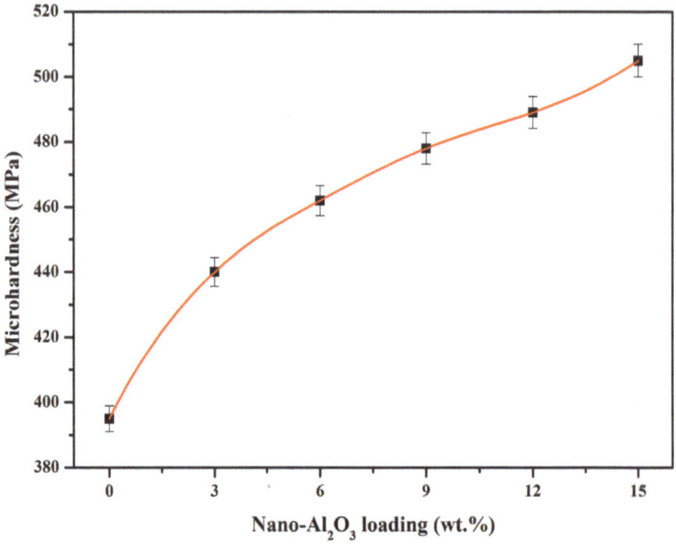

Figure 6.19 Evolution of the microhardness for the cured BAPh/Al_2O_3 nanocomposites. *BAPh*, bisphenol A-based PN; *Al_2O_3*, alumina or aluminum oxide. Reprinted from Derradji M, Ramdani N, Zhang T, Wang J, Gong L-d, Xu X-d, et al. Thermal and mechanical properties enhancements obtained by reinforcing a bisphenol-A based phthalonitrile resin with silane surface-modified alumina nanoparticles. Polym Compos 2017;38:1546–58.

Halpin–Tsai is a widely used model which can be described by the following equations [24,86]:

$$E_c = \frac{1 + \xi\eta V_f}{1 - \eta V_f} \tag{6.2}$$

$$\eta = \frac{E_f/E_m - 1}{E_f/E_m + \xi} \tag{6.3}$$

where ξ is a shape factor equal to 2 for spherical particles. In contrast with the Series model, the Halpin–Tsai equations include the shape factor which has a major influence on the modulus of the composite. Furthermore, the Halpin–Tsai model has been found to give accurate predictions for the epoxy/alumina and polybenzoxazine/silica systems. The Kerner model generalized by Lewis and Nielsen is given by [86]:

$$\frac{E_c}{E_m} = \frac{1 + ABV_f}{1 - B\varphi V_f} \tag{6.4}$$

$$A = \frac{7 - 5\vartheta}{8 - 10\vartheta} \tag{6.5}$$

$$B = \left(\frac{E_f}{E_m} - 1\right) \Big/ \left(\frac{E_f}{E_m} + A\right) \tag{6.6}$$

$$\varphi = 1 + \frac{1 - V_m}{{V_m}^2} V_f \tag{6.7}$$

where A is a constant that depends on both the shape of the filler and the Poisson's ratio of the resin. B is also a constant related to the modulus of the filler and that of the matrix. The maximum packing fraction of the reinforcing phase in the matrix is designated by φ.

The root-mean-square deviation for the tensile modulus is defined as follows:

$$\sigma_E = \left[\sum_{i=1}^{n} \frac{\left(E_c^{\mathrm{Exp}} - E_c^{\mathrm{Cal}}\right)^2}{n-m}\right]^{1/2} \tag{6.8}$$

where E_c^{Exp} is the experimentally measured tensile modulus and E_c^{Cal} is the calculated tensile modulus using the predictive models. Furthermore, n is the number of experimental points and m is the number of adjustable parameters. In this study, m was set to 1.

The Young's modulus experimental and calculated results are illustrated in Fig. 6.20, and Table 6.5 lists all the calculated parameters and the root-mean-square deviations. Before moving forward with the comparison between the models, it was clear that the addition of the nanofillers resulted in amelioration in the modulus of the BAPh matrix. Typically, the neat, BAPh resin modulus was found to be 2.9 GPa; this value was further increased to reach 4.3 GPa at 15 wt.% nanocontents. Meanwhile, the Halpin–Tsai and Kerner models reproduced the experimental data with a good accuracy and with a slight advantage for the Halpin–Tsai model. The Series model completely under estimated the experimental data and did not reflect the reality. The previous predictions were obtained by assuming that the nanoparticles are perfectly spherical and the shape factors were calculated accordingly. Therefore, to better estimate the modulus using this kind of nanofillers, the authors recalculated the shape factors by fitting the experimental data to those calculated by the models using the genetic algorithm of Matlab to minimize the root-mean-square deviations. As seen from Table 6.5, the Halpin–Tsai model gives a better accuracy with a shape factor (ξ) estimated to be 1.8272, and the Kerner model reproduces the experimental data more precisely with a shape factor (A) of 1.7831.

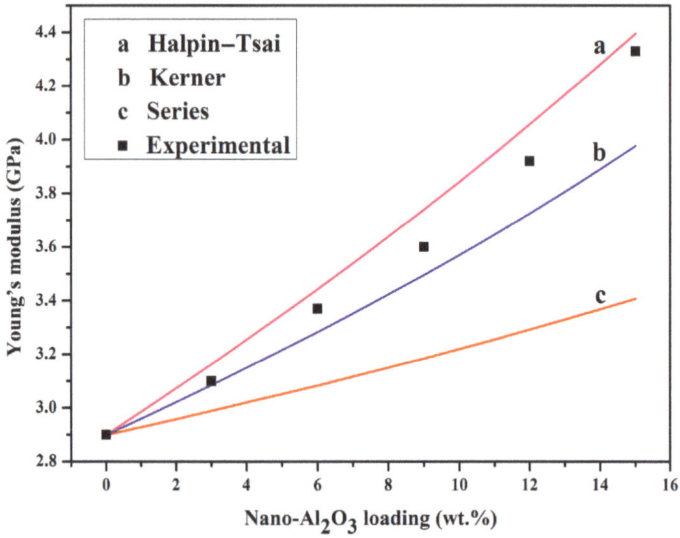

Figure 6.20 Calculations and experimental measurements of the cured BAPh/ Al_2O_3 nanocomposites Young's modulus. *BAPh*, bisphenol A-based PN; *Al_2O_3*, alumina or aluminum oxide. Reprinted from Derradji M, Ramdani N, Zhang T, Wang J, Gong L-d, Xu X-d, et al. Thermal and mechanical properties enhancements obtained by reinforcing a bisphenol-A based phthalonitrile resin with silane surface-modified alumina nanoparticles. Polym Compos 2017;38:1546–58.

Table 6.5 Values of the Models Parameters and Corresponding Root-Mean-Square Deviations

Models	Parameters	σ_E (GPa)
Series	–	0.5500
Halpin–Tsai	$\xi = 2.0000$, $\eta = 0.9782$	0.1019
	$\xi = 1.8272$, $\eta = 0.9794$	0.0602
Kerner	$A = 1.1122$, $B = 0.9845$	0.1867
	$A = 1.7831$, $B = 0.9845$	0.0544

Reprinted from Derradji M, Ramdani N, Zhang T, Wang J, Gong L-d, Xu X-d, et al. Thermal and mechanical properties enhancements obtained by reinforcing a bisphenol-A based phthalonitrile resin with silane surface-modified alumina nanoparticles. Polym Compos 2017;38:1546–58.

The impact of the alumina nanoparticles of the stress and strain is presented in Fig. 6.21. Clearly, there was a positive impact on the strength; it was highly increased, reaching 71 MPa at 15 wt.% nano-Al_2O_3 loading. On the other hand, the reinforcing phase reduced the

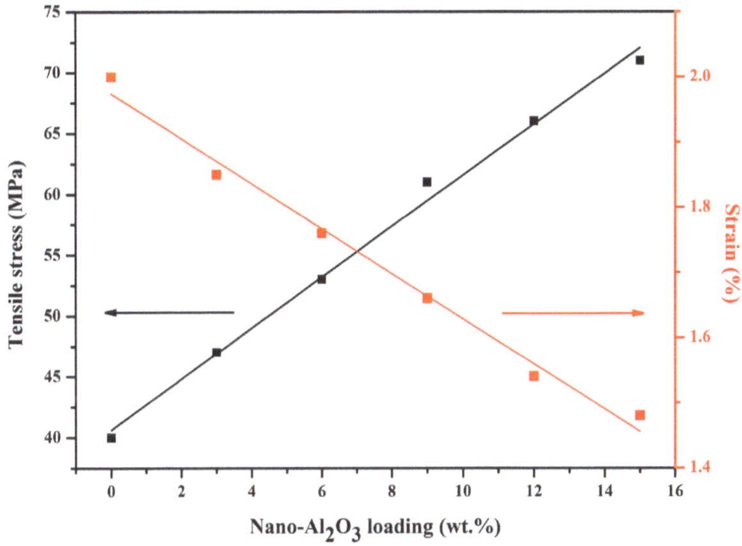

Figure 6.21 Tensile strength and strain of the cured BAPh/Al_2O_3 nanocomposites. *BAPh*, bisphenol A-based PN; *Al_2O_3*, alumina or aluminum oxide. Reprinted from Derradji M, Ramdani N, Zhang T, Wang J, Gong L-d, Xu X-d, et al. Thermal and mechanical properties enhancements obtained by reinforcing a bisphenol-A based phthalonitrile resin with silane surface-modified alumina nanoparticles. Polym Compos 2017;38:1546–58.

ductility of the neat BAPh matrix by 26%. Indeed, the alumina nanofillers reduced the BAPh chain mobility, resulting in a negative impact on the ductility.

The SEM observations were performed on the tensile fracture surfaces, the aim was to study the deformation mechanisms and understand the morphological changes brought by the silane-treated, alumina nanofillers. In Fig. 6.22 are reproduced the SEM images of the neat, BAPh resin and its nanocomposites of 3, 6, 9, 12, and 15 wt.% nanocontents. The unfilled resin showed smooth and featureless surface morphology confirming the brittle character of the BAPh matrix. Generally, a relatively low energy is necessary to fracture the neat PN sample resulting in a weak strength, as seen from the stress/strain curves presented in Fig. 6.21. Conversely, the alumina nanoparticles incorporation resulted in a rougher surface morphology and the amount of hackling was also increased as the amount of the nanofillers became more important. Moreover, the propagation of the fracture lines in the nanocomposites was hindered by the presence of the nanofillers. As a result, the crack path became more torturous and the long fracture lines were replaced by a multitude of microsized cracks. In addition, no signs of

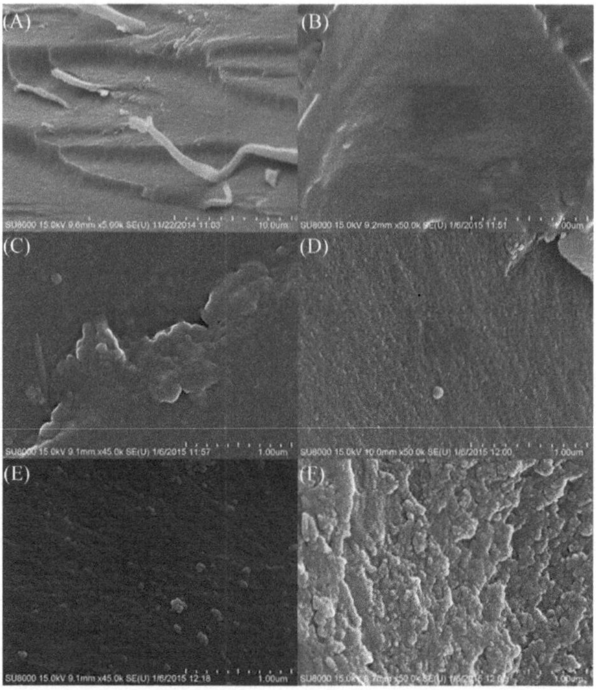

Figure 6.22 SEM images of the neat cured BAPh resin (A) and its nanocomposites of 3 (B), 6 (C), 9 (D), 12 (E), and 15 wt.% (F) nanocontents. *SEM*, scanning electron microscope; *BAPh*, bisphenol A-based PN. Reprinted from Derradji M, Ramdani N, Zhang T, Wang J, Gong L-d, Xu X-d, et al. Thermal and mechanical properties enhancements obtained by reinforcing a bisphenol-A based phthalonitrile resin with silane surface-modified alumina nanoparticles. Polym Compos 2017;38:1546–58.

nanoparticles debonding were observed which may be due to the good adhesion of the nanofillers with the BAPh matrix ensued by the silane surface modification.

To assess the nanoparticles dispersion inside the BAPh matrix, the authors used a transmission electron microscopy (TEM). The obtained TEM image of the BAPh-based nanocomposite at the maximum nanofillers amount of 15 wt.% is shown in Fig. 6.23. As seen from this figure, the treated alumina nanoparticles were homogeneously dispersed inside the PN matrix. Although some small agglomerations existed, their sized was inferior to 500 nm and they are also well dispersed.

The thermal and thermomechanical properties of the prepared materials were analyzed by TGA and DMA, and the obtained results are plotted in Figs. 6.24 and 6.25, respectively. Table 6.6 also resumes all the important thermal and thermomechanical properties. The incorporation

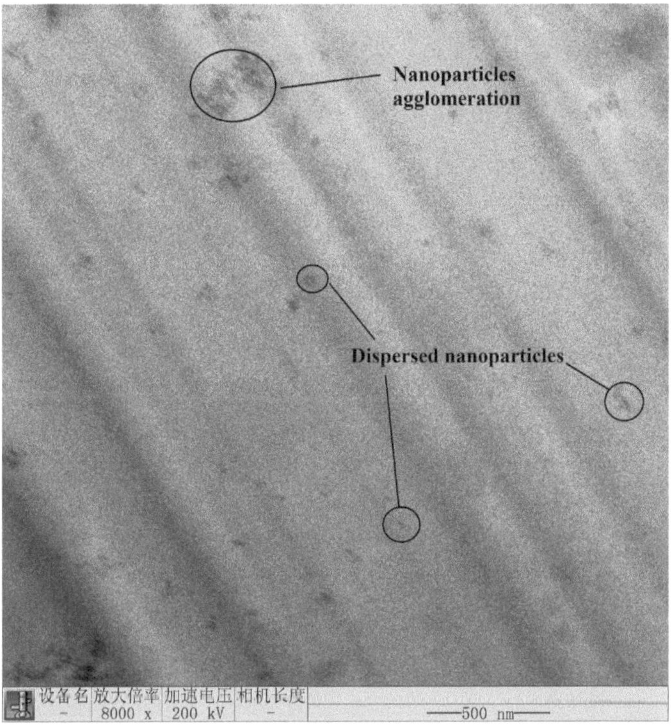

Figure 6.23 TEM image of the cured BAPh/Al_2O_3 nanocomposite at the maximum nanofillers' amount of 15 wt.%. *TEM*, transmission electron microscopy; *BAPh*, bisphenol A-based PN; *Al_2O_3*, alumina or aluminum oxide. Reprinted from Derradji M, Ramdani N, Zhang T, Wang J, Gong L-d, Xu X-d, et al. Thermal and mechanical properties enhancements obtained by reinforcing a bisphenol-A based phthalonitrile resin with silane surface-modified alumina nanoparticles. Polym Compos 2017;38:1546–58.

of 15 wt.% of the nanofillers resulted in improvements about 41 and 63°C in the $T_{5\%}$ and $T_{10\%}$, respectively and the char yield at 800°C reached 83.2%. The stiffness and glass transition temperature were also highly improved reaching 3.4 GPa and 346°C, respectively.

Shan et al., also reported about the use of alumina nanoparticles to reinforce the RPh resin [74]. The used nanofillers were first treated with a silane coupling agent and then the CPN was grafted on the surfaces of the nanoparticles. The surface modification of the nanofillers was confirmed by the FTIR observations. In contrast with the spectrum of the native alumina nanoparticles, the amino silane treated one showed new absorptions bands at 1549 and 1025 cm^{-1} attributed to the (N–H) in plane deformation and the (Si–O–Si) stretch, respectively.

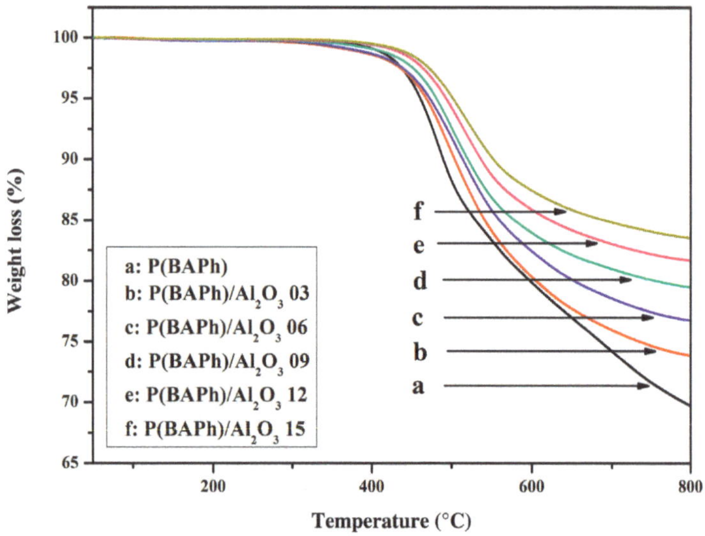

Figure 6.24 TGA curves of the cured BAPh/Al_2O_3 nanocomposites under nitrogen atmosphere. *TGA*, thermogravimetric analyses; *BAPh*, bisphenol A-based PN; *Al_2O_3*, alumina or aluminum oxide. Reprinted from Derradji M, Ramdani N, Zhang T, Wang J, Gong L-d, Xu X-d, et al. Thermal and mechanical properties enhancements obtained by reinforcing a bisphenol-A based phthalonitrile resin with silane surface-modified alumina nanoparticles. Polym Compos 2017;38:1546–58.

On the other hand, the CPN-grafted alumina spectrum displayed new bands at 1647, 1587, and 1245 cm^{-1} ascribed to the (C = O) stretching, (N–H) stretching, and (C–N) stretching, respectively. Moreover, the band at 2231 cm^{-1} confirmed the presence of the nitrile group on the nanofillers surfaces.

The TGA investigations of the raw alumina as well as the silane modified and CPN-grafted nanofillers revealed that the raw alumina only showed slight weight reductions even at 800°C. On the other hand, the silane and CPN-modified nanofillers showed char yields of about 94% and 91%, respectively. These reductions were attributed to the presence of organic groups less stable than the inorganic compounds. Thus, these TGA results corroborated those from the FTIR.

The nanocomposites were prepared according to the in situ polymerization process and cured in a muffle furnace following the curing schedule of 220°C for 2 h, 250°C for 4 h, 280°C for 6 h, and 320°C for 6 h. The bending properties of the chosen PN matrix and its related, alumina-based nanocomposites are summarized in Table 6.7. The neat RPh matrix strength was measured to be around 97.5 MPa and it was

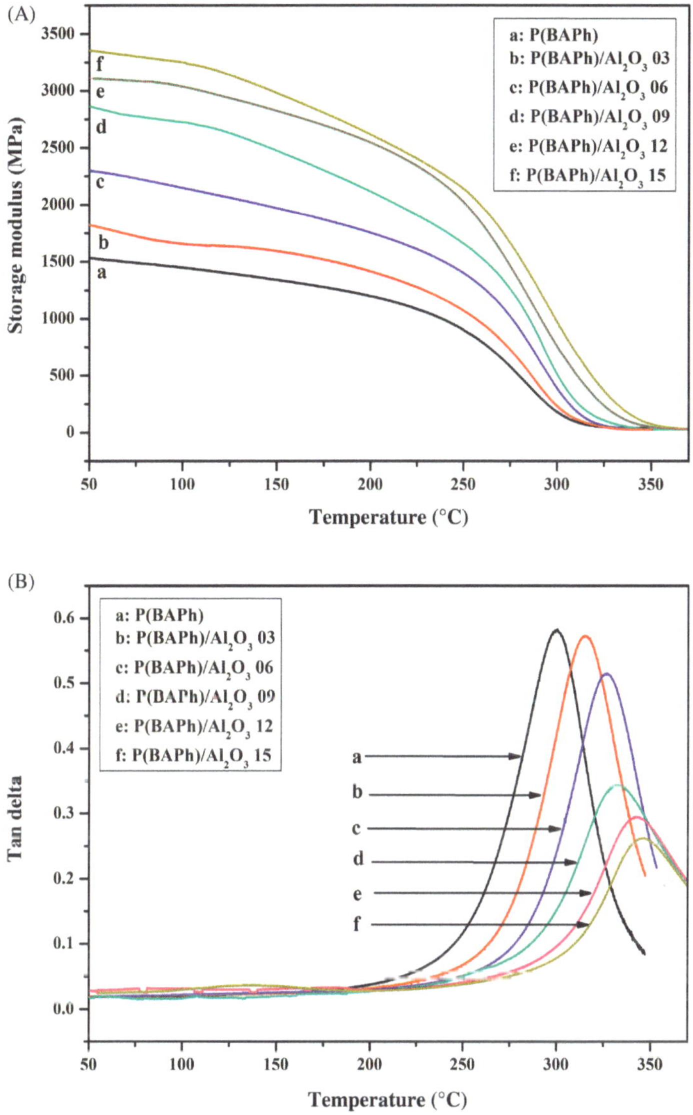

Figure 6.25 Evolution of storage modulus (A) and Tan delta (B) of the cured BAPh/Al_2O_3 nanocomposites. *BAPh*, bisphenol A-based PN; *Al_2O_3*, alumina or aluminum oxide. Reprinted from Derradji M, Ramdani N, Zhang T, Wang J, Gong L-d, Xu X-d, et al. Thermal and mechanical properties enhancements obtained by reinforcing a bisphenol-A based phthalonitrile resin with silane surface-modified alumina nanoparticles. Polym Compos 2017;38:1546–58.

Table 6.6 Thermal and Thermomechanical Properties of the P(BAPh)/Al_2O_3 Nanocomposites

Specimen Code	G' at 50°C (GPa)	T_g (°C)	$T_{5\%}$ (°C)	$T_{10\%}$ (°C)	Y_c (%,800°C)
P(BAPh)	1.5	300	460.9	489.7	69.5
P(BAPh)/Al_2O_3 03	1.8	316	468.4	503.9	73.8
P(BAPh)/Al_2O_3 06	2.3	326	472.9	514.1	76.8
P(BAPh)/Al_2O_3 09	2.8	333	481.2	520.4	79.5
P(BAPh)/Al_2O_3 12	3.2	342	493.3	537.0	81.7
P(BAPh)/Al_2O_3 15	3.4	346	501.8	552.2	83.2

Reprinted from Derradji M, Ramdani N, Zhang T, Wang J, Gong L-d, Xu X-d, et al. Thermal and mechanical properties enhancements obtained by reinforcing a bisphenol-A based phthalonitrile resin with silane surface-modified alumina nanoparticles. Polym Compos 2017;38:1546–58.

Table 6.7 Flexural Strength and Modulus of the PN/Al_2O_3 Nanocomposites by Shan et al. [74]

Amount of the Reinforcing Phase (wt.%)	Flexural Strength (MPa)		Flexural Modulus (GPa)	
	Native	Treated	Native	Treated
0	97.50		5.21	
1	109.01	118.79	5.32	6.03
3	120.19	150.97	5.62	6.33
5	85.44	97.68	3.67	4.37

Data extracted from Shan S, Chen X, Xi Z, Yu X, Qu X, Zhang Q. The effect of nitrile-functionalized nano-aluminum oxide on the thermomechanical properties and toughness of phthalonitrile resin. High Perform Polym 2016;29:113–23, Fig. 7.

higher than that obtained with the BAPh resin. On the other hand, both native and treated alumina showed improvements in the flexural strength to a certain extent. The bending strength reached its maximum value at the nanofillers loading of 3 wt.% for both types of alumina. For example, the values of 120 and 151 MPa at 3 wt.% loading of raw and surface modified alumina, respectively. Moreover, the further addition of the reinforcing phase to 5 wt.% ultimately resulted in a decrease in the bending strength. The flexural modulus variations exhibited the same trend. The increases in the bending properties before 3 wt.%

nanoloading were justified by the high degree of load transfer from the reinforcement to the PN matrix. Moreover, the presence of strong polar nitrile groups on the surface of the CPA-grafted alumina resulted in a good adhesion between the two phases. However, the nanoparticles agglomerations were more consequent at 5 wt.% and decreased the bending properties.

A comparative study between the alumina nanoparticles surface modification and the nanocomposites preparation techniques was opted for by Derradji et al. and Shan et al. The study revealed that the use of the amino silane coupling agent alone can allow a good state of dispersion and adhesion even at 15 wt.% of nanoloading, if a rigorous stirring and sonication are applied. Moreover, the use of a hot-press curing is preferable over the muffle furnace in enhancing the mechanical properties of the nanocomposites.

6.5.4 Properties of Phthalonitrile/TiO_2 Nanocomposites

Titania nanoparticles are well-known for their positive effect on the corrosion protective properties of high-performance, polymeric matrices [54]. For instance, excellent anticorrosion behavior was obtained with epoxy and polybenzoxazine reinforced with various amounts of titania nanoparticles [87,88]. Moreover, these ceramic nanofillers can also enhance the hardness, strength, thermal, and thermomechanical properties of these polymers. Derradji et al., prepared various, BAPh-based nanocomposites reinforced with silane surface modified, titania nanofillers (average nanoparticles diameter of 60 nm) and investigated their mechanical, thermal, and anticorrosion properties. The grafting of the silane on the nanofillers outer surfaces was proved by the FTIR technique. The spectra of both raw and treated titania are presented in Fig. 6.26. The silane grafting resulted in the appearance of new absorption bands such as those at 2940 and 2850 cm^{-1} were attributed to the asymmetric and symmetric ($-CH_2$) stretching vibrations, respectively. Moreover, the bands at 1455, 1305, and 998 cm^{-1} were ascribed to the vibration of (Si−O−C), the asymmetric, flexible vibration of (O−Si), and the stretching vibration of (Ti−O−Si), respectively.

The curing behavior of the BAPh resin reinforced with 6 wt.% of titania nanoparticles was also studied by FTIR and the results are plotted in Fig. 6.27. Clearly, as the curing reaction moved forward in time and temperature, the absorption at 2234 cm^{-1}, related to the nitrile groups, decreased in intensity. Meanwhile, the band at 1385 cm^{-1}, ascribed to the triazine structure, increased in intensity. These

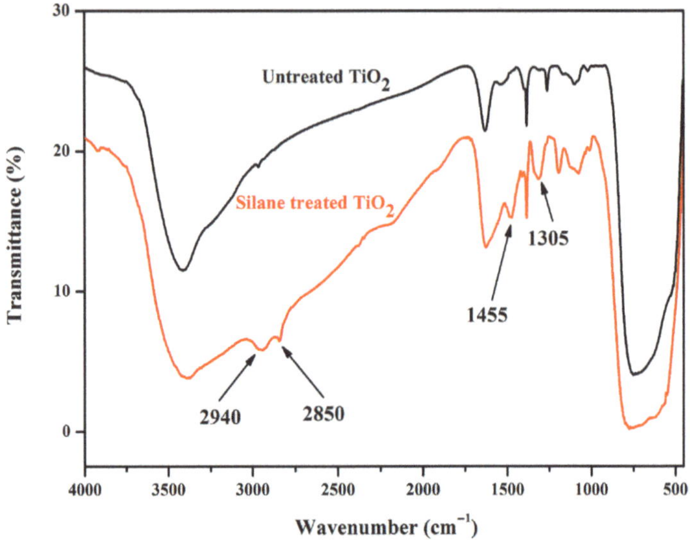

Figure 6.26 FTIR spectra of native and silane-treated TiO_2 nanoparticles. *FTIR*, Fourier infrared spectroscopy; *TiO_2*, titania or titanium oxide. Reprinted from Derradji M, Ramdani N, Zhang T, Wang J, Gong L-d, Xu X-d, et al. Effect of silane surface modified titania nanoparticles on the thermal, mechanical, and corrosion protective properties of a bisphenol-A based phthalonitrile resin. Prog Org Coat 2016;90:34–43.

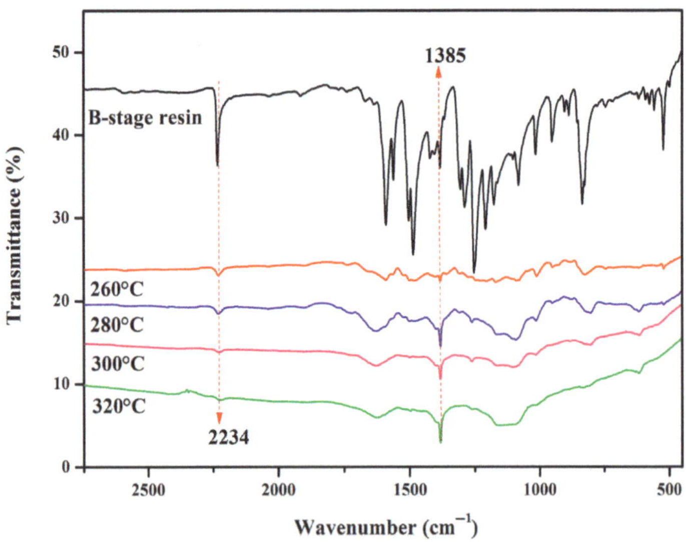

Figure 6.27 Spectra of the neat BAPh resin reinforced with 6 wt.% titania nanoparticles at different curing temperatures. *BAPh*, bisphenol A-based PN. Reprinted from Derradji M, Ramdani N, Zhang T, Wang J, Gong L-d, Xu X-d, et al. Effect of silane surface modified titania nanoparticles on the thermal, mechanical, and corrosion protective properties of a bisphenol-A based phthalonitrile resin. Prog Org Coat 2016;90:34–43.

observations confirmed that the addition reaction was not affected by the presence of the ceramic nanofillers. The triazine structure was also depicted at the early stages of the polymerization reaction, which indicated that the initiation of the cyano addition started before increasing the curing temperature. It is also worth noting that the remaining nitrile absorption indicated that the final polymeric network was essentially composed of polytriazine.

The BAPh/TiO_2 nanocomposites were prepared by a solution blending technique and the samples for the mechanical and thermal tests were cured by a hot press, according to the curing schedule of 240°C for 5 h, 260°C for 5 h, 280°C for 5 h, 300°C for 5 h, and 320°C for 5 h. In the same way to the other ceramic particles, the titania improved the hardness of the neat BAPh resin as it can be seen in Fig. 6.28. Indeed, the hardness of the neat BAPh resin which was about 400 MPa, was gradually enhanced to reach the value of 501 MPa at 6 wt.% nanoloading. The tensile investigations also revealed multiple ameliorations in the strength and modulus. As seen from Table 6.8, when moving from the neat BAPh resin to the nanocomposites of 6 wt.%, the improvements in the strength and modulus reached about 60% and 17%, respectively.

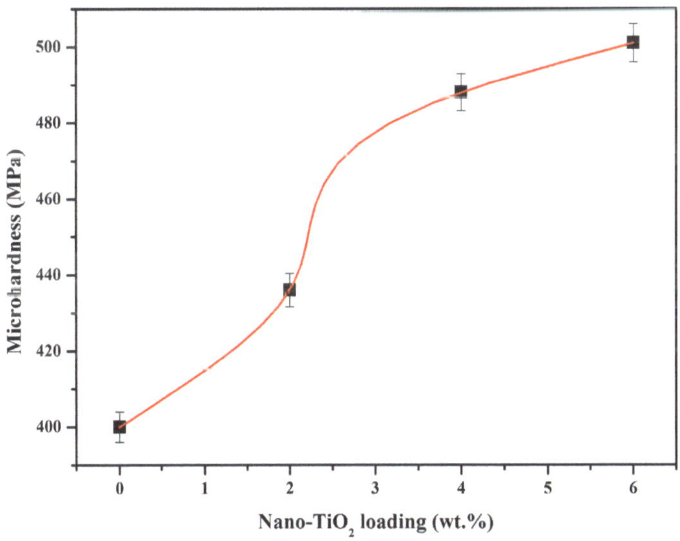

Figure 6.28 Microhardness of the cured BAPh/TiO_2 nanocomposites. *BAPh*, bisphenol A-based PN; *TiO_2*, titania or titanium oxide. Reprinted from Derradji M, Ramdani N, Zhang T, Wang J, Gong L-d, Xu X-d, et al. Effect of silane surface modified titania nanoparticles on the thermal, mechanical, and corrosion protective properties of a bisphenol-A based phthalonitrile resin. Prog Org Coat 2016;90:34–43.

Table 6.8 Strength and Modulus Values of the cured BAPh/TiO_2 Nanocomposites [54]

Specimen Code	Strength (MPa)	Modulus (GPa)
P(BAPh)	50	2.9
P(BAPh)TiO_2 02	62	3.0
P(BAPh)TiO_2 04	74	3.2
P(BAPh)TiO_2 6	80	3.4

Data extracted from Derradji M, Ramdani N, Zhang T, Wang J, Gong L-d, Xu X-d, et al. Effect of silane surface modified titania nanoparticles on the thermal, mechanical, and corrosion protective properties of a bisphenol-A based phthalonitrile resin. Prog Org Coat 2016;90:34–43, Fig. 4.

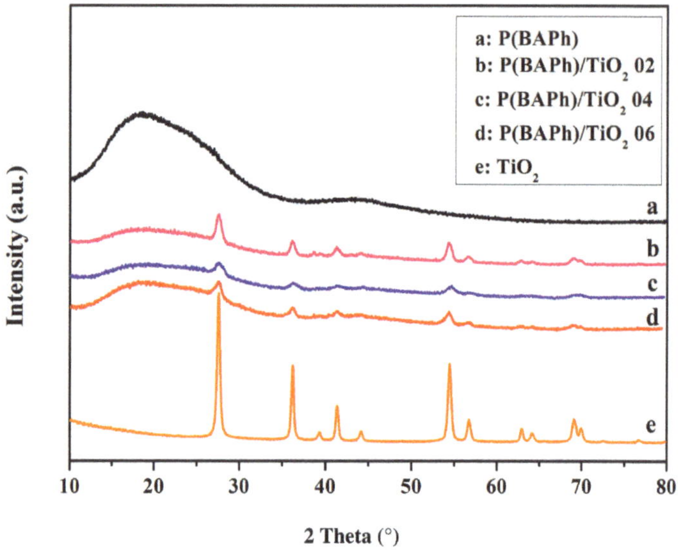

Figure 6.29 X-ray diffraction patterns of the TiO_2, unfilled BAPh matrix and its cured nanocomposites at 2, 4, and 6 wt.% nanoloading. *BAPh*, bisphenol A-based PN; *TiO_2*, titania or titanium oxide. Reprinted from Derradji M, Ramdani N, Zhang T, Wang J, Gong L-d, Xu X-d, et al. Effect of silane surface modified titania nanoparticles on the thermal, mechanical, and corrosion protective properties of a bisphenol-A based phthalonitrile resin. Prog Org Coat 2016;90:34–43.

Aiming to qualitatively and quantitatively confirm the presence of the titania nanoparticles in the PN matrix, the X-ray diffraction patterns of the TiO_2, BAPh resin and their cured nanocomposites at 2, 4, and 6 wt.% nanoloading are shown in Fig. 6.29. The amorphous nature of the BAPh matrix was confirmed by an X-ray diffractogram (XRD) containing a broad halo centered at $2\theta = 18.6$. On the other hand, the crystalline feature of the TiO_2 resulted in sharp and intense diffraction peaks. Interestingly, all the titania crystalline peaks were also depicted in the diffractogram of the nanocomposites; however, their intensity was proportional to the amount of the reinforcing phase.

The anticorrosion properties experimentations were carried out by an electrochemical impedance spectroscopy using an Autolab PGSTAT302 potentiostat in a conventional three-electrode cells. Carbon steel panels were used as the substrates and were first cleaned and degreased in acetone prior to apply the coatings. The neat BAPh resin and its TiO_2-based nanocomposites were deposed on the substrates by a dip coating process, and cured in a muffle furnace following the same curing schedule used for the samples designated for the mechanical and thermal tests. The carbon steel panel before and after the coating are shown in Fig. 6.30.

The protection of the metallic structures against marine corrosion using polymeric coatings is commonly used in the naval industry considering their feasibility and relatively low cost in contrast with other anticorrosion techniques. In order to prepare high-performance coatings, the chosen polymers need to exhibit a high degree of hydrophobicity as well as a good adhesion with the metallic structures. The PN resins are

Figure 6.30 Carbon steel plates before and after the coating. Reprinted from Derradji M, Ramdani N, Zhang T, Wang J, Gong L-d, Xu X-d, et al. Effect of silane surface modified titania nanoparticles on the thermal, mechanical, and corrosion protective properties of a bisphenol-A based phthalonitrile resin. Prog Org Coat 2016;90:34–43.

well-known for their outstanding water resistance (<1.5% by weight in most of the cases) which make them highly suitable to be used as a barrier coating to prevent against marine corrosion. Taking advantages of these interesting features, Derradji et al., investigated the corrosion protective properties of the BAPh resin and its related titania nanocomposites. The variation of the impedance modulus at the initial immersion stage for the carbon steel panel, the neat BAPh matrix, and all its related nanocomposites are presented in Fig. 6.31. The impedance modulus at low frequency values is commonly used to ascertain about the efficiency of the organic coatings. In other words, higher impedance modulus at low frequency values suggest ameliorated corrosion protective properties. As clearly seen from Fig. 6.31, the carbon steel substrate showed a low frequency modulus value of $8 \times 10^4\ \Omega\ cm^2$. This weak impedance modulus is logical since the carbon steel panels are highly affected by the marine corrosion. Meanwhile, the use of neat BAPh matrix as a coating material for this substrate ensued a consequent increase in the low frequency modulus value, which was measured to be $1.5 \times 10^9\ \Omega\ cm^2$. When compared with epoxy and polybenzoxazine,

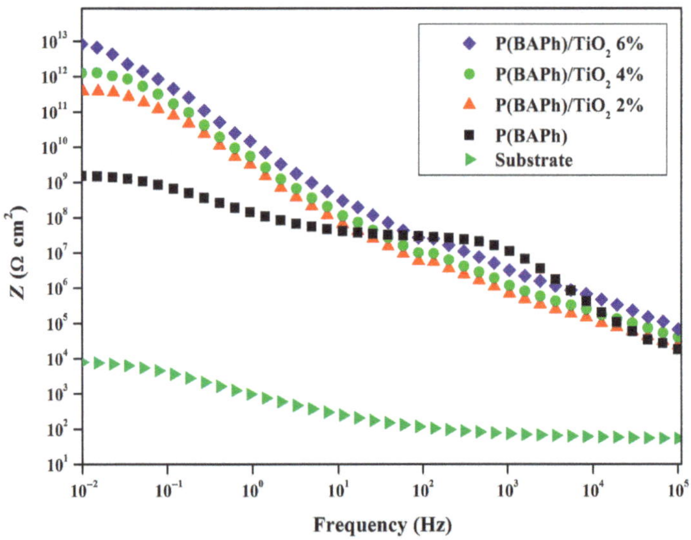

Figure 6.31 Evolution of the impedance modulus at the initial immersion stage for the substrate, the neat BAPh resin, and its related nanocomposites at 2, 4, and 6 wt.% nano-TiO_2 loading. *BAPh*, bisphenol A-based PN; *TiO_2*, titania or titanium oxide. Reprinted from Derradji M, Ramdani N, Zhang T, Wang J, Gong L-d, Xu X-d, et al. Effect of silane surface modified titania nanoparticles on the thermal, mechanical, and corrosion protective properties of a bisphenol-A based phthalonitrile resin. Prog Org Coat 2016;90:34–43.

this PN matrix was found to result in better anticorrosion performances. The introduction of the silane-treated, titania nanoparticles brought further enhancements in the low frequency modulus values. The nanocomposites showed a gradual increase in the impedance modulus values as the amount of the reinforcing phase increased. For instance, at 6 wt.% nano-TiO_2, the low frequency modulus reached the remarkable value of $9.6 \times 10^{12}\ \Omega\ cm^2$. Such a high value was never reached by any other thermoset reinforced with other ceramic fillers. The closest but still inferior value of $9.6 \times 10^{12}\ \Omega\ cm^2$ was reported by Ramdani et al where a BPA-based polybenzoxazine was reinforced with 20 wt.% of nano-Cr_2O_3 [68]. The superiority of the studied PN nanocomposites is due to the high hydrophobicity of the PN matrix and also to the positive role of the titania nanoparticles which reduced the water diffusion channels, and as a result, improved the performance of the coating.

The corrosion protective properties of the neat BAPh resin and its TiO_2-based nanocomposites were also investigated over an extended period of time to assess the long term performance of the coatings. The effect of different immersion times on the low frequency impedance modulus values is shown in Fig. 6.32. By analyzing the data, it was

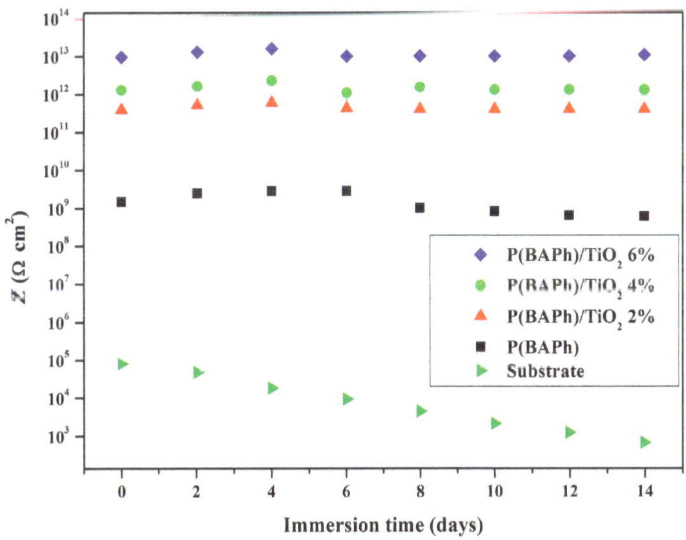

Figure 6.32 Evolution of the impedance modulus at low frequencies for all the samples at different immersion times. Reprinted from Derradji M, Ramdani N, Zhang T, Wang J, Gong L-d, Xu X-d, et al. Effect of silane surface modified titania nanoparticles on the thermal, mechanical, and corrosion protective properties of a bisphenol-A based phthalonitrile resin. Prog Org Coat 2016;90:34–43.

found that during the first 3 days of immersion, the impedances at low frequency values slightly increased due to the relaxation process that dissipated the stress generated by the high curing temperatures. Afterward, a slight decrease followed by a certain stability was noticed for all the samples after 5 days of immersion. Thus it is clear that the PN resin and its nanocomposites are good candidates to be used as advanced coating materials.

6.5.5 Properties of Phthalonitrile/ZnO Nanocomposites

The PN resins as any other polymeric materials can be affected by the UV lights which may have a negative impact on the polymeric network and result in catastrophic degradations in many important properties. Therefore it becomes necessary to assess the behavior of the PN resins when subjected to UV lights. Moreover, the incorporation of UV absorbing components within the PN resins can further enhance their UV-shielding performances. Ceramic particles, such as ZnO and TiO_2, can play that role if adequately introduced in the PN matrices. Derradji et al., reported on the use of ZnO nanoparticles (average nanoparticles diameter of 60 nm) to prepare high-performance, polymeric-based nanocomposites with improved thermal, mechanical, and UV-shielding properties [56]. The choice of the ZnO was mainly due to outstanding optical characteristics such as intensive UV and infrared absorptions. In addition, the nanostructured ZnO can bring multiple ameliorations in other properties because of its improved physicochemical properties including chemical stability, low dielectric constant, high luminous transmittance, high catalytic activity, and outstanding thermal stability.

Prior to the nanocomposites preparation, the nanoparticles were treated by the appropriate silane coupling agent to enhance their dispersion and adhesion with the PN matrix. The BAPh resin was chosen as the matrix. It should be noted the choice of the BAPh resin was mainly due to its improved properties and also for comparison purpose with other PN materials, since this kind of resin was extensively used to prepare high-performance, PN composites. The nanocomposites were prepared by a solution blending technique and cured by a hot-press following the curing procedure of 240°C for 5 h, 260°C for 5 h, 280°C for 5 h, 300°C for 5 h, and 320°C for 5 h.

The neat PN resin and all its related nanocomposites at 2, 4, 6 wt.% nano-ZnO were investigated by XRD to study the influence of the curing on both phases. Moreover, the XRD results can qualitatively assess the amount of the reinforcing phase. As seen in Fig. 6.33, where are

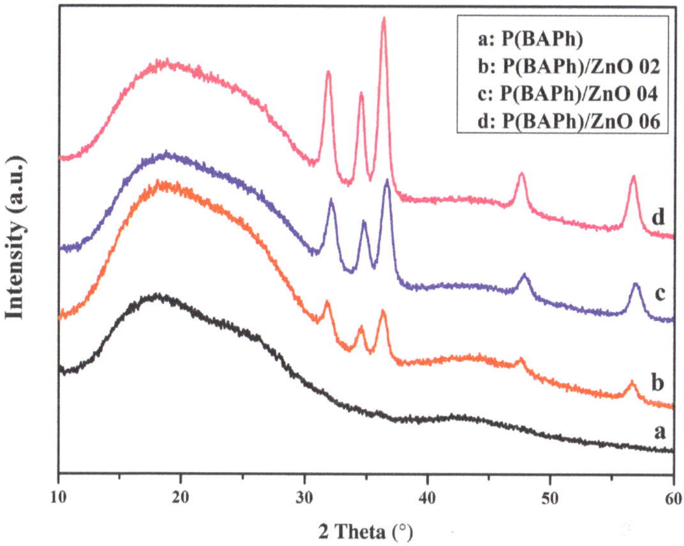

Figure 6.33 X-ray diffraction patterns of the ZnO, unfilled BAPh matrix and its cured nanocomposites at 2, 4, and 6 wt.% nanoloading. *ZnO*, zinc oxide; *BAPh*, bisphenol A-based PN. Reprinted from Derradji M, Ramdani N, Gong L-d, Wang J, Xu X-d, Lin Z-w, et al. Mechanical, thermal, and UV-shielding behavior of silane surface modified ZnO-reinforced phthalonitrile nanocomposites. Polym Adv Technol 2016;27:882–8.

presented the XRD results, the neat PN matrix showed an amorphous feature with a broad halo centered at $2\theta = 18.6$. Interestingly, the incorporation of the ZnO nanofillers resulted in the emergence of several crystalline peaks with intensities proportional to the amount of the nanoparticles. The previous observations implied that the nano-ZnO did not undergo any structural modifications upon curing.

The UV-shielding behavior was discussed in terms of percentage of absorbed lights. The UV–visible transmittance spectra for the studied materials are illustrated in Fig. 6.34. It can be clearly seen from this figure that the unfilled BAPh resin absorbed 78% of the UV lights. This value is close to that obtained with a BPA-based polybenzoxazine which is about 79% [89]. Furthermore, by adding more ZnO contents, the UV absorption markedly increased to reach 86.5% at the maximum nanofillers loading of 6 wt.%. These remarkable UV-shielding performances were attributed, first to the BAPh matrix which, even unfilled, still exhibited good results, and, also to the benefits of the treated ZnO nanoparticles that allowed more absorption of the UV lights.

Beside the improvements in the optical properties, the tensile investigations revealed consequent enhancements in the neat, BAPh resin

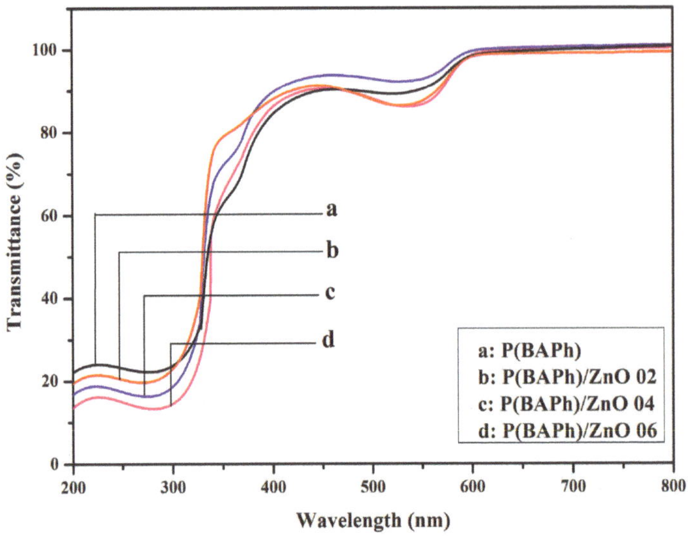

Figure 6.34 UV–Visible transmittance spectra of the neat resin and the cured BAPh/ZnO nanocomposites. *BAPh*, bisphenol A-based PN; *ZnO*, zinc oxide. Reprinted from Derradji M, Ramdani N, Gong L-d, Wang J, Xu X-d, Lin Z-w, et al. Mechanical, thermal, and UV-shielding behavior of silane surface modified ZnO-reinforced phthalonitrile nanocomposites. Polym Adv Technol 2016;27:882–8.

Table 6.9 Strength, Modulus, and Toughness Values of the cured BAPh/ZnO Nanocomposites [56]

Specimen Code	Strength (MPa)	Modulus (GPa)	Toughness (MPa)
P(BAPh)	40	2.9	0.4
P(BAPh)ZnO 02	46	3.2	0.4
P(BAPh)ZnO 04	56	3.8	0.5
P(BAPh)ZnO 6	61	4.3	0.5

strength, modulus, and toughness. All the tensile properties are listed in Table 6.9. By analyzing these data, it appeared that the strength was improved by 59% at the maximum nanofillers loading of 6 wt.%. The Young's modulus of the unfilled resin was measured to be 2.9 GPa and was largely enhanced as the amount of the reinforcing phase increased. Thus the nanocomposites of 2, 4, and 6 wt.% nano-ZnO exhibited the Young's modulus values of 3.2, 3.8, and 4.3 GPa, respectively. The toughness values, obtained by measuring the area under the stress–strain curves, also revealed that the neat, resin toughness was improved by 16%. Hence, it was confirmed that the ZnO nanoparticles

Table 6.10 Thermal and Thermomechanical Properties of the cured BAPh/ZnO Nanocomposites

Specimen Code	G' at 50°C (GPa)	T_g (°C)	$T_{5\%}$ (°C)	$T_{10\%}$ (°C)	Y_c (%, 800°C)
P(BAPh)	1.94	315	464.1	492.7	69.8
P(BAPh)ZnO 02	2.34	326	470.6	508.4	74.5
P(BAPh)ZnO 04	2.90	338	480.3	519.8	77.8
P(BAPh)ZnO 06	3.65	359	497.4	554.2	79.7

Reprinted from Derradji M, Ramdani N, Gong L-d, Wang J, Xu X-d, Lin Z-w, et al. Mechanical, thermal, and UV-shielding behavior of silane surface modified ZnO-reinforced phthalonitrile nanocomposites. Polym Adv Technol 2016;27:882–8.

not only improved the optical properties of the neat, PN resin but also enhanced its mechanical properties.

The thermal and thermomechanical investigations were carried out by TGA and DMA; respectively, and the results are listed in Table 6.10. The excellent thermal stability of the BAPh resin was positively affected by the introduction of ZnO nanofillers. Indeed, the thermal stability parameters such as the $T_{5\%}$, $T_{10\%}$, and Y_c reached 497, 554°C, and 79.7%, respectively. The ability of the nanofillers to restrict the motion inside the matrix and to insulate the polymeric material from the heat source is behind these thermal stability improvements. Moreover, the good dispersion of the nanofillers in the matrix afforded by the silane surface modification rendered the evacuation of the volatile products or small molecules difficult to achieve, and, as a result, the thermal stability was enhanced. The thermomechanical properties, namely the stiffness and T_g, showed the same positive trend. The stiffness of the neat BAPh resin increased from 1.9 to 3.7 GPa when moving from the unfilled resin to the nanocomposite of 6 wt.% nano-ZnO. The T_g also was enhanced by 44°C at the same nanofillers amount. The high modulus of the ZnO was behind the improvements in the neat BAPh stiffness and the increases in the T_g are attributed to the restriction in the motion inside the polymeric structures mainly caused by the rigid ZnO nanoparticles.

6.5.6 Properties of Phthalonitrile/BN Nanocomposites

As stated in the previously discussed sections, the choice of the ceramic fillers mainly depends on the targeted applications. For instance,

PN/TiO_2 nanocomposites were developed for high-performance, marine coatings and the PN/ZnO nanocomposites were designed for high UV-shielding efficiency. Following the same purpose, Derradji et al., intended to expand the applications of the PN material to the fabrication of electronic devices. It is well-know that such devices are made of materials able to effectively and quickly dissipate the heat generated during the operating time. Moreover, the PN resins as any other polymeric material, suffer from a poor thermal dissipation which may affect the performance, lifetime, and reliability of the PN-based, electronic devices. Therefore the authors proposed to upgrade the thermal conductivity of the neat, PN resins by incorporating high thermal conductive ceramic fillers such as BN and AlN. In order to achieve a high thermal conductivity, Derradji et al., proposed an experimental approach consisting on preparing highly-filled, PN nanocomposites based on silane surface-modified, hexagonal BN and cured by a hot compression molding technique [57]. Beside the choice of the BN nanoparticles as the reinforcing phase, which is clear since the BN presents the highest thermal conductivity among the ceramics, the aim of adding a high filler amount is to overcome the well-known scattering problem by the formation of a conductive network assured by an appropriate contact between the ceramic particles. However, adding a high filler amount may result in a decrease in the thermal conductivity and other important properties if the fillers are not well dispersed in the matrix. Hence, assuring a good dispersion of the fillers by an adequate treatment chemical treatment is necessary prior to the nanocomposites preparation. The hot compression molding technique is also favorable for the preparation of highly-filled, polymer composites considering the processability issues caused by high fillers loadings.

The typical BAPh resin was once more chosen as the matrix and reinforced with both native and silane treated, BN nanoparticles (average nanoparticles diameter of 60 nm). The BAPh/BN nanocomposites with various amounts of the reinforcing phase were prepared following the solution bending technique and cured by a hot-press according to the curing schedule of 240 to 320°C with increments of 20°C every 5 h. The thermal conductivity measurements were carried out by a thermal constants analyzer (TPS 2500S) equipped with a sensor type 5465 (Hot Disk, Sweden) and the tests were carried out on disk specimens with a diameter of 12.5 mm and thickness of 2 mm at 25.9°C. The thermal conductivity of the neat BAPh resin, its nanocomposites at different amounts, and state of surface of the nano-BN, and the results are depicted in Fig. 6.35. Obviously, the neat BAPh matrix exhibited the

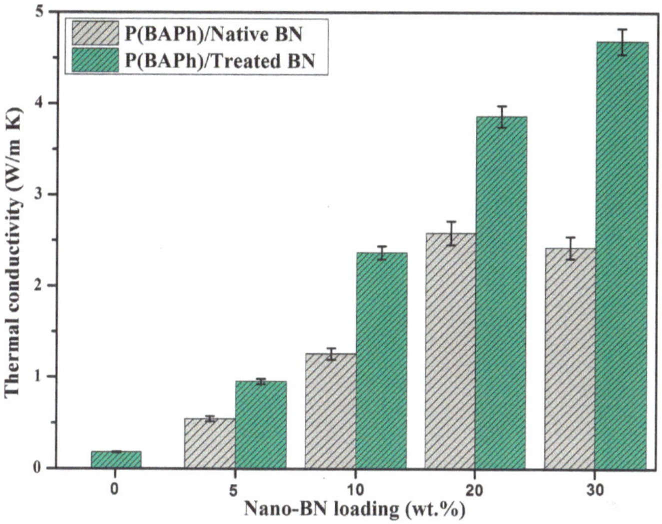

Figure 6.35 Thermal conductivity values of the neat BAPh resin its cured nanocomposites at different amounts and state of surface of the nano-BN. *BAPh*, bisphenol A-based PN; *BN*, boron nitride. Reprinted from Derradji M, Song X, Dayo AQ, Wang J, Liu W-b. Highly filled boron nitride-phthalonitrile nanocomposites for exigent thermally conductive applications. Appl Therm Eng 2017;115:630–6.

lowest thermal conductivity value of 0.18 W/m K. Meanwhile, the introduction of either native or silane-treated nano-BN resulted in nanocomposites with an improved thermal conductivity. For instance, at the lowest nanofillers amount of 5 wt.%, the native and treated BN-based nanocomposites showed the thermal conductivity values of 0.54 and 0.95 W/m K, respectively. At higher nano-BN contents, an exponential increase in the thermal conductivity was noticed for both types of nanocomposites. For example, the nanocomposites of 20 and 30 wt.% treated BN exhibited the exceptional thermal conductivity values of 3.86 and 4.69 W/m K, respectively. Although the native BN resulted in important enhancements, the thermal conductivity reached its maximum at 20 wt.% and further adding raw nano-BN resulted in a negative impact on the thermal conductivity. Thus it was clear that the nanoparticles treatment played a crucial role in avoiding the nanoparticle agglomerations and fostering the formation of a conductive network at high nanofillers loadings.

A TEM analysis was used to assess the formation of the conductive network at high nano-BN loadings. Fig. 6.36 illustrates the TEM images of the nanocomposites of 20 and 30 wt.% silane-treated nano-BN.

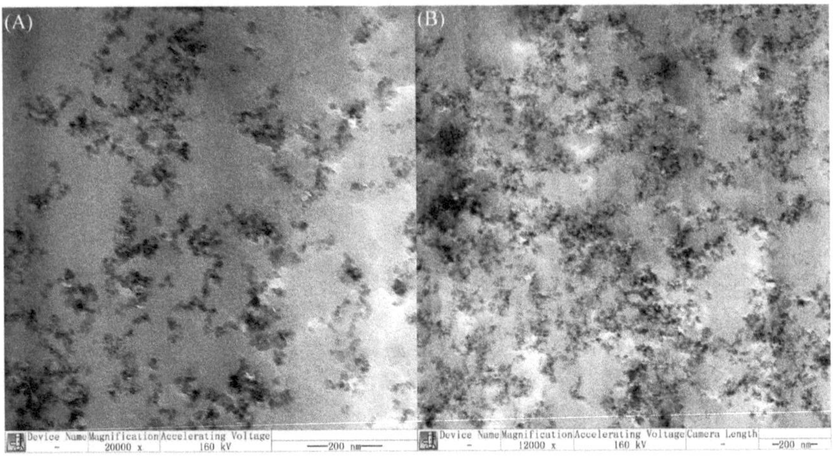

Figure 6.36 TEM images of the cured nanocomposites of 20 (A) and 30 wt.% (B) silane treated nano-BN. *TEM*, transmission electron microscopy; *BN*, boron nitride. Reprinted from Derradji M, Song X, Dayo AQ, Wang J, Liu W-b. Highly filled boron nitride-phthalonitrile nanocomposites for exigent thermally conductive applications. Appl Therm Eng 2017;115:630–6.

Clearly, the nanofillers tend to form a conductive network favorable for the heat transfer, and as the amount of the nanoparticles increased, the conductive paths were more interconnected leading to a more enhanced thermal conductivity. It is also important to point out that the formation of an effective conductive network at high fillers amounts is closely related to the dispersion of the fillers in the matrix. The surface modification of the fillers allowed an ameliorated nanoparticles dispersion. Therefore, at high fillers' amounts, the good dispersion resulted in effective conductive network. On the other hand, when native nanoparticles were used, the presence of large agglomerations prevented the formation of an effective conductive network and as a result the thermal conductivity is lower than that obtained with the silane surface-modified nano-BN.

The previous thermal conductivity measurements confirmed that the BAPh/BN nanocomposites can effectively be used in the electronic field as heat dissipating support materials. However, these nanocomposites should also possess good mechanical properties to withstand some of mechanical constraints that may rise during the work time. Therefore, the BAPh/BN nanocomposites were tested for their bending properties to see whether the BN nanoparticles can also enhance the mechanical properties of the neat PN matrix. The flexural strength and modulus of both native and treated, BN-based nanocomposites are reproduced in

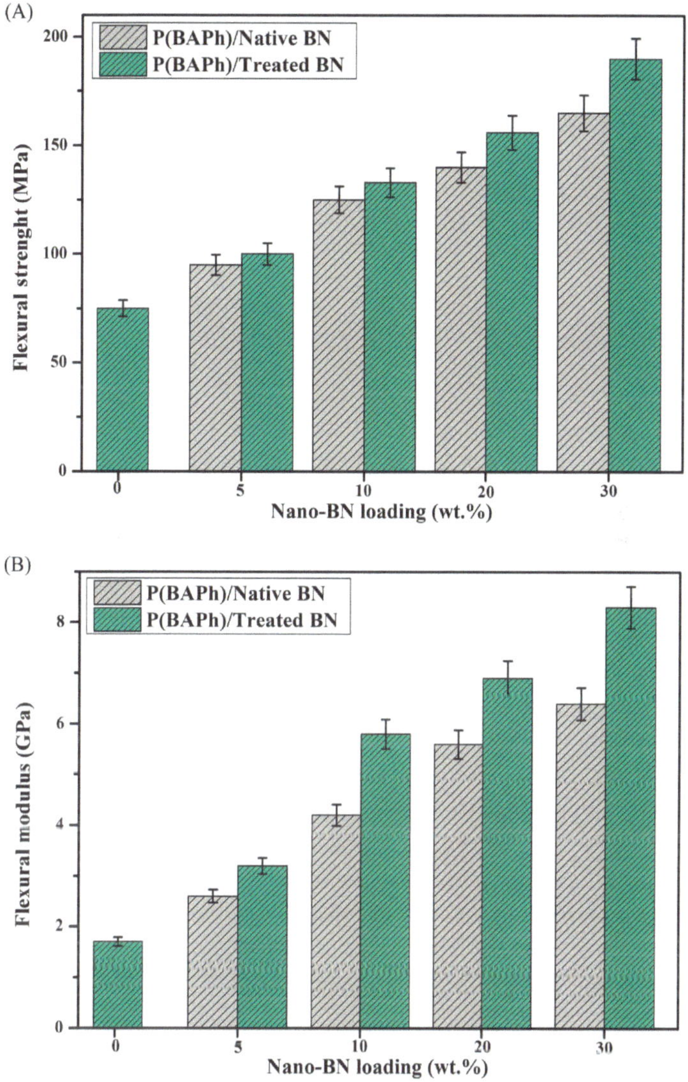

Figure 6.37 Flexural strength (A) and modulus (B) of the studied P(BAPh)/BN nanocomposites. *BAPh*, bisphenol A-based PN; *BN*, boron nitride. Reprinted from Derradji M, Song X, Dayo AQ, Wang J, Liu W-b. Highly filled boron nitride-phthalonitrile nanocomposites for exigent thermally conductive applications. Appl Therm Eng 2017;115:630–6.

Fig. 6.37. The unfilled BAPh resin exhibited a bending strength and modulus of 75 MPa and 1.7 GPa, respectively. Interestingly, all the nanocomposites exhibited ameliorated flexural properties compared to the neat PN matrix. Indeed, the bending strength and modulus reached

190 MPa and 8.3 GPa, respectively, at the maximum treated nano-BN loading of 30 wt.%. The native BN-based nanocomposites also resulted in improvements in the flexural properties but with a lower rate. These good findings were attributed to the high strength and modulus of the BN nanofillers and also to the good state of dispersion and adhesion of these nanoparticles with the matrix. In the light of these facts, it was also deducted that the BAPh/BN nanocomposites are promising materials to be used in electronic fields.

6.5.7 Properties of Phthalonitrile/ZrO_2 Nanocomposites

The preparation of the ceramic-based, PN micro- and nanocomposites was for a long time restricted to the use of the typical, PN monomers based on BPA or resorcinol precursors. Although their resulting polymers offered interesting features superior in many ways to other high-performance thermosets, some shortcomings related to their high melting points and brittleness were always pointed out as primary challenges to overcome. It should be also recalled that the PN resins can be grouped into either the first-generation resins or the oligomeric, second-generation, PN resins. The PN materials known as the first generation, including the typical BAPh and RPh resins, are obtained by reacting 4-nitrophthalonitrile with the diphenolate salts of bisphenols in a one-step reaction process. On the other hand, the second generation of the PN resins was designed to exhibit improved processability and toughness. The second-generation PN resins consist of oligomeric aromatic ether linkages between the two terminal PN units. These resins are obtained by a two-step reaction process starting with a nucleophilic displacement reaction involving a dihaloaromatic or dihaloaliphatic compound and an appropriate bisphenol, and finishing with an end-capping with 4-nitrophthalonitrile. The properties of the second-generation PN resins can be easily tailored by an adequate molecular design to meet specific needs. What is more important in the polymer composites field is that these materials can be used as advanced high-performance matrices for inorganic or fiber reinforced composites.

Derradji et al., reported on the preparation of two new, second-generation oligomeric containing aliphatic moieties, PN monomers/polymers and compared their mechanical and thermal properties with those of the first-generation PN resins [55]. Moreover, the newly-developed, PN resins were reinforced with ZrO_2 nanoparticles (average nanoparticles diameter of 50 nm) to investigate their possible use as advanced matrices for ceramic-based nanocomposites. The molecular structures of

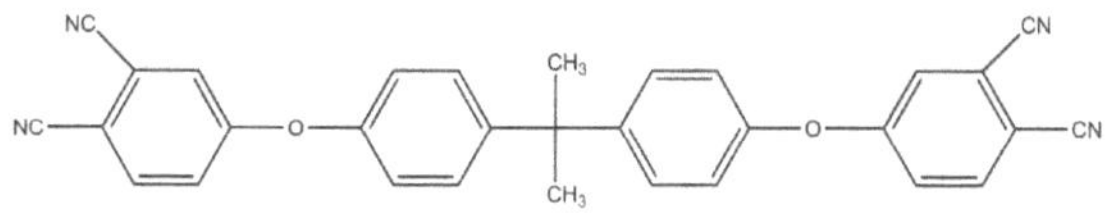

BAPh

First generation phthalonitrile monomer

C6-PN

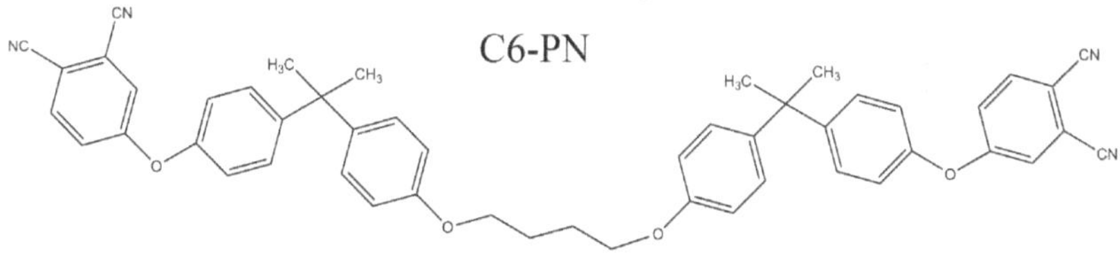

C4-PN

Second generation phthalonitrile monomers

Figure 6.38 Molecular structures of the first and second-generation of PN monomers and oligomers. *PN*, phthalonitrile. Reprinted from Derradji M, Feng T, Wang H, Ramdani N, Zhang T, Wang J, et al. New oligomeric containing aliphatic moiety phthalonitrile resins: their mechanical and thermal properties in presence of silane surface-modified zirconia nanoparticles. Iran Polym J. 2016;25:503–14.

the studied PN oligomers, noted C_4-PN and C_6-PN, are presented in Fig. 6.38. The curing behavior and processability were investigated by DSC and the obtained curves are plotted in Fig. 6.39. The curing of these PN oligomers were promoted by the introduction of the 10 wt.% of APN. As seen from Fig. 6.39, both PN oligomers showed one endothermic peak related to the melting and one exothermic peak ascribed to the curing process. It is also easy to notice that as the aliphatic spacer became longer, the processing window was enlarged. In fact, the C_4-PN and C_6-PN exhibited a processing window of 95 and 120°C, respectively. Both second-generation PN oligomers had a larger processing window than those of the first-generation PN monomers, for example, the BAPh monomer cured with the same type and amount of curing agent presented the narrow processing window of 50°C [51].

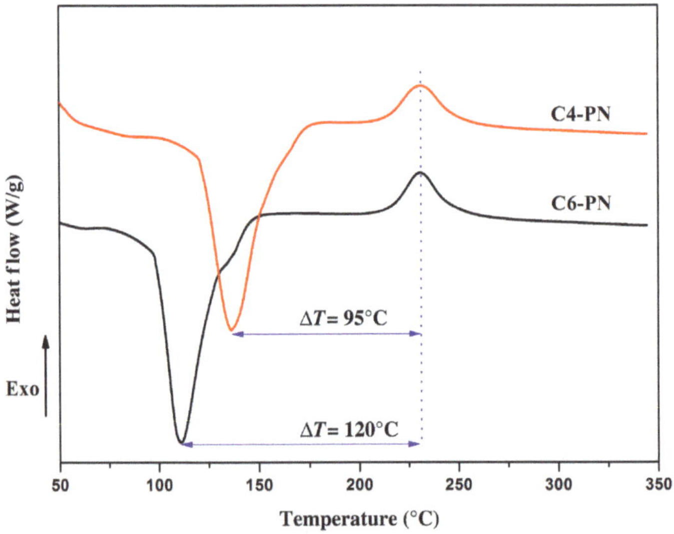

Figure 6.39 DSC thermograms of C_4-PN and C_6-PN oligomers cured with 10 wt.% APN. *DSC*, dynamic scanning calorimetry; *APN*, aminophenoxy PN. Reprinted from Derradji M, Feng T, Wang H, Ramdani N, Zhang T, Wang J, et al. New oligomeric containing aliphatic moiety phthalonitrile resins: their mechanical and thermal properties in presence of silane surface-modified zirconia nanoparticles. Iran Polym J. 2016;25:503–14.

The newly-developed, PN oligomers and their respective ZrO_2-based nanocomposites were prepared following the solution blending procedure and cured by a compression molding technique. The cured schedule adopted was 240°C for 5 h, 260°C for 5 h, 280°C for 5 h, 300°C for 5 h, and 320°C for 5 h. The prepared materials were investigated for their hardness and tensile properties; the aim was to study the impact of both the introduction of flexible moieties into the PN structure and the effect of the silane, surface-modified, zirconia nanoparticles on these important mechanical properties. The microhardness obtained results are presented in Fig. 6.40. The introduction of flexible moieties resulted in slight decreases in the microhardness. Indeed, the cured thermosets of C_4-PN and C_6-PN exhibited the microhardness values of 388 and 354 MPa. On the other hand, the BAPh resin has a microhardness value of 400 MPa. The results implied that the gain in processability slightly affected the hardness state of the PN resins. Furthermore, the introduction of different amounts of treated nano-ZrO_2 resulted in large ameliorations in the neat PN resins hardness. In fact, at the maximum nanofillers loading of 6 wt.% the increases in the microhardness reached 144 and 156 MPa for the nanocomposites prepared from C_4-PN and

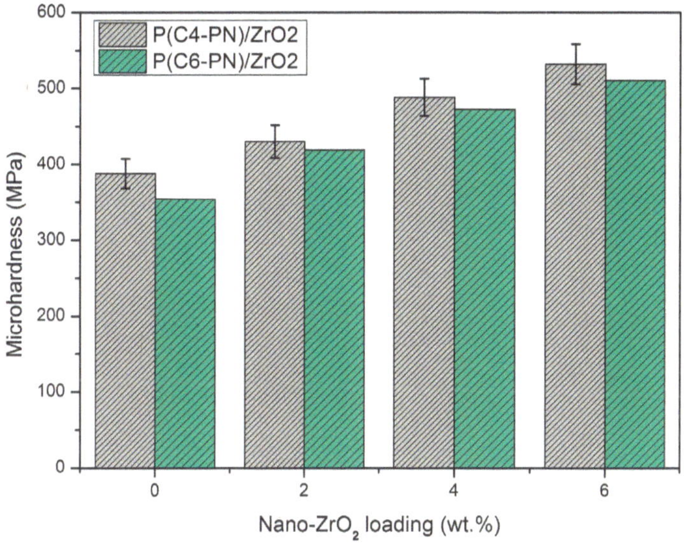

Figure 6.40 Microhardness of the cured C_4-PN/ZrO_2 and C_6-PN/ZrO_2 nanocomposites. Reprinted from Derradji M, Feng T, Wang H, Ramdani N, Zhang T, Wang J, et al. New oligomeric containing aliphatic moiety phthalonitrile resins: their mechanical and thermal properties in presence of silane surface-modified zirconia nanoparticles. Iran Polym J. 2016;25:503–14.

C_6-PN, respectively. Those findings confirmed that the slight reductions in the microhardness upon introducing flexible segments can be easily overcome by the introduction of fewer amounts of ceramic nanofillers.

The tensile properties of the studied materials were compared in terms of strength, ductility, and toughness. The strain/stress curves are shown in Fig. 6.41. It was noticed that as the aliphatic spacer became longer, the strength and ductility were enhanced. For example, the cured thermosets of C_4-PN and C_6-PN exhibited the strength value of 47 and 49 MPa, respectively. The ductility deduced from the strain value was measured to be 2.5% and 2.8% for the cured thermosets of C_4-PN and C_6-PN, respectively. In contrast, with the BAPh resin, the newly developed second-generation PN resins (C_4-PN and C_6-PN) exhibited the ameliorated toughness values of 0.67 and 0.76 MPa, respectively. Moreover, the zirconia nanoparticles brought positive changes in the strength and toughness of the unfilled, PN matrices. For instance, the strength and toughness of the C_6-PN based nanocomposites reached 100 and 0.96 MPa at the nano-ZrO_2 loading of 6 wt.%. The previous results confirmed the synergistic combination between the newly-developed, PN resins and the zirconia nanoparticles.

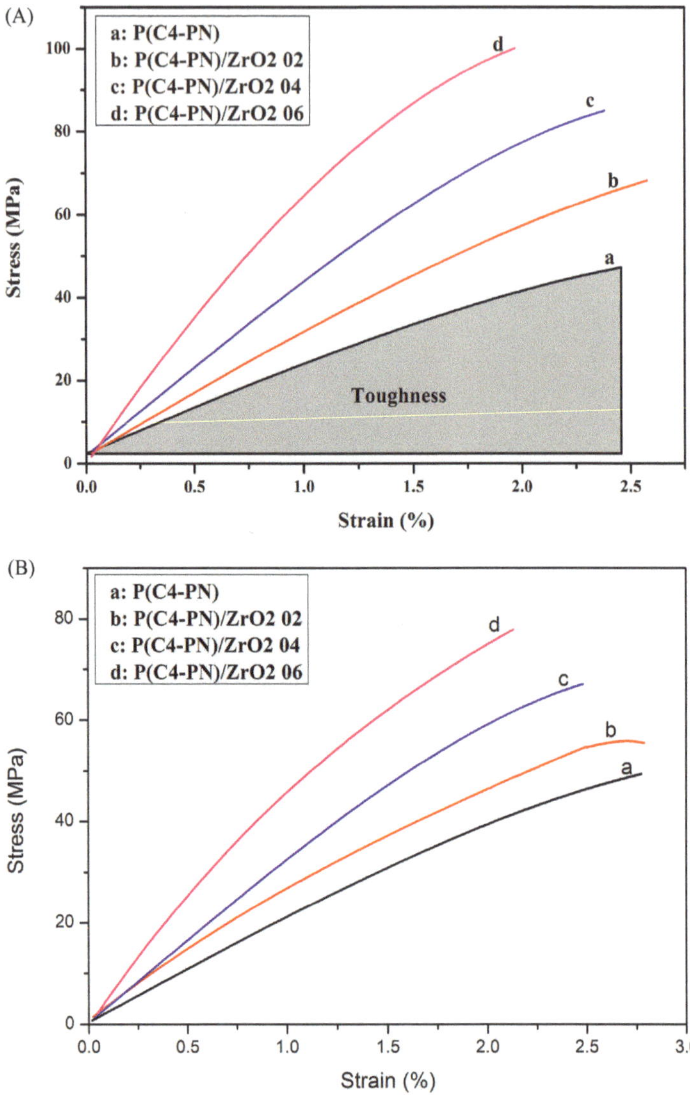

Figure 6.41 Tensile stress–strain curves of the cured nanocomposites C_4-PN/ZrO_2 (A) and C_6-PN/ZrO_2 (B). Reprinted from Derradji M, Feng T, Wang H, Ramdani N, Zhang T, Wang J, et al. New oligomeric containing aliphatic moiety phthalonitrile resins: their mechanical and thermal properties in presence of silane surface-modified zirconia nanoparticles. Iran Polym J. 2016;25:503–14.

The tensile fractured surface of the C_4-PN based nanocomposites were analyzed by SEM and the obtained micrographs are displayed in Fig. 6.42. The neat PN matrix showed smooth and featureless surface morphology confirming its brittle character. Moreover, the long fracture

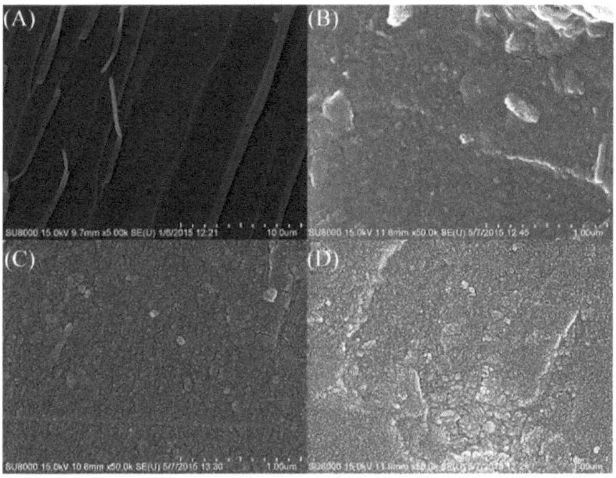

Figure 6.42 SEM Micrographs of the $P(C_4\text{-}PN)/ZrO_2$ fractured surface at various nano-ZrO_2 contents: neat resin (A), 2 wt.% (B), 4 wt.% (C), and 6 wt.% (D). *SEM*, scanning electron microscope. Reprinted from Derradji M, Feng T, Wang H, Ramdani N, Zhang T, Wang J, et al. New oligomeric containing aliphatic moiety phthalonitrile resins: their mechanical and thermal properties in presence of silane surface-modified zirconia nanoparticles. Iran Polym J. 2016;25:503–14.

lines indicated that a relatively small energy was required to break the neat PN resin sample. On the other hand, the fractured samples of the C_4-PN/ZrO_2 nanocomposites revealed multiple changes in the morphology. In fact, the long facture lines seen in the neat resin were replaced by a multitude of microcracks, indicating that the presence of the nanofillers hindered the propagation of the stress and resulted in tortuous crack paths. Moreover, the roughness and amount of hackling were proportional to the treated nano-ZrO_2 contents.

To fully understand the impact of the silane coupling agent on the dispersion state, a high-resolution TEM was used to compare the C_4-PN/ZrO_2 nanocomposites reinforced with both native and silane-treated ZrO_2. It is important to point out that the nanoparticles state of dispersion is the key parameter defining the properties of the final product. The TEM images obtained for both native and silane treated ZrO_2 based C_4-PN nanocomposites are presented in Fig. 6.43. As clearly seen this figure, the raw ZrO_2 resulted in large agglomerations, meanwhile, the silane coupling agent ensued a good dispersion of the nanofillers even at maximum reinforcing amount.

The impact of incorporating flexible linkage on the thermal stability was also studied by TGA and the results are displayed in Fig. 6.44. From the measurements of the thermal stability parameters, it appeared that the newly-developed, PN resins showed a reduced thermal

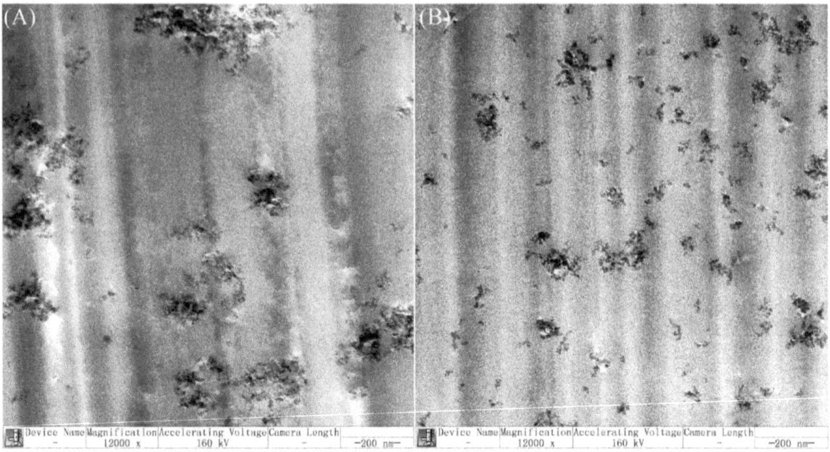

Figure 6.43 TEM images of both native (A) and silane treated (B) ZrO_2 based C_4-PN nanocomposites. *TEM*, transmission electron microscopy; *ZrO_2*, zirconia or zirconium oxide. Reprinted from Derradji M, Feng T, Wang H, Ramdani N, Zhang T, Wang J, et al. New oligomeric containing aliphatic moiety phthalonitrile resins: their mechanical and thermal properties in presence of silane surface-modified zirconia nanoparticles. Iran Polym J. 2016;25:503–14.

resistance in contrast with the first-generation PN resins. Nevertheless, these slight decreases are acceptable considering the numerous advantages obtained in terms of processability and mechanical properties. Furthermore, the addition of even few amounts of zirconia nanoparticles can recover the loss that came with the introduction of the aliphatic moieties. Thus it becomes evident that the second-generation PN resins and composites are also advanced materials for use in different applications. The main interesting feature of the second-generation PN is their good processability, which is highly favorable for the preparation of thick composites sections in the industry. This work also demonstrated that the mechanical and thermal properties of these second-generation PN resins can be easily enhanced by the addition of ceramic fillers. In a similar way to that of the first-generation resins, these new PN ones can be combined with different ceramic fillers to target specific properties in a close relationship with the intended applications.

6.5.8 Properties of Phthalonitrile/MAX Phase Nanocomposites

As stated in the introduction, the MAX phases are the second-generation of ceramics especially designed to be incorporated into polymeric, metallic, and ceramic matrices, aiming the preparation of

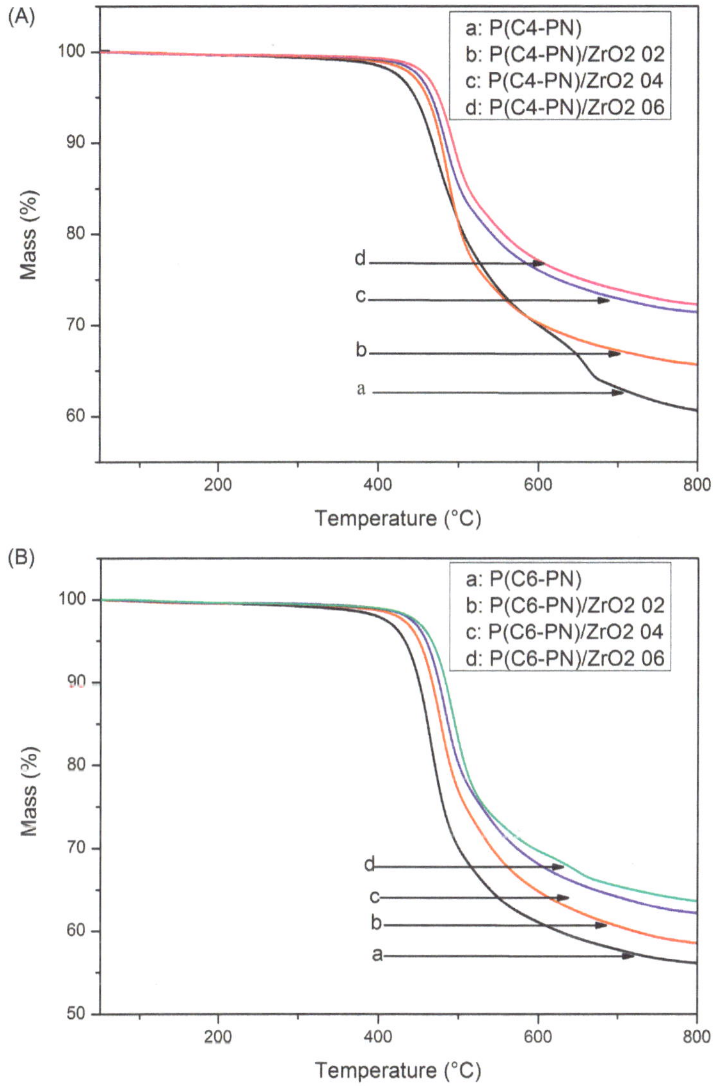

Figure 6.44 TGA curves of the P(C_4-PN)/ZrO_2 (A) and P(C_6-PN)/ZrO_2 (B) nanocomposites under nitrogen atmosphere. *TGA*, thermogravimetric analyses; *ZrO_2*, zirconia or zirconium oxide. Reprinted from Derradji M, Feng T, Wang H, Ramdani N, Zhang T, Wang J, et al. New oligomeric containing aliphatic moiety phthalonitrile resins: their mechanical and thermal properties in presence of silane surface-modified zirconia nanoparticles. Iran Polym J. 2016;25:503–14.

high-performance composites. The interesting features of the MAX phases include good thermal and electrical conductivities, easy machinability, low density, excellent wear resistance, high hardness and good oxidation resistance. The MAX phases are also known as "soft ceramics" because of their ability to enhance the ductility and eventually the toughness of different kind of matrices. For a long time, the development of hybrids from PN resins and ceramics was limited to changes in the molecular structures of the polymeric matrices or by finding new ways to enhance the state of adhesion and dispersion of the particles within the polymeric matrices. However, little attention was given to the reinforcing phase, although the recent advances in the ceramic field are now proposing a large panel of new advanced ceramic fillers. Taking advantage of the MAX phases interesting properties, Derradji et al. investigated the mechanical and thermal properties of the Ti_3SiC_2 MAX phase (average nanoparticles diameter of 50 nm) reinforced, PN nanocomposites [66].

The Ti_3SiC_2 MAX phase nanoparticles were previously treated with an amino silane coupling agent (aminopropyl trimethoxy silane) and used afterward as a new kind of reinforcement for the typical BAPh resin. The nanocomposites were prepared following the solution blending procedure and cured by a compression molding technique. The cured schedule adopted was 240°C for 5 h, 260°C for 5 h, 280°C for 5 h, 300°C for 5 h, and 320°C for 5 h. The resulting materials were investigated for thermal properties including the thermal conductivity and thermal stability. The thermal conductivity experimental measurements were compared with predictions from Series, Halpin–Tsai, and Kerner models and the results are shown in Fig. 6.45. The equations used are the same to those described in Section 5.3. The neat PN matrix exhibited a thermal conductivity value of 0.18 W/m K, this value was systematically enhanced as the amount of the reinforcing phase increased, reaching 0.6 W/m K at 15 vol.% nano-Ti_3SiC_2 which is about 233% gain in the thermal conductivity. All the used, theoretical models under estimated the nanocomposites thermal conductivities. At low nanofillers' contents (<5 vol.%), the predictions are acceptable; however, at higher nanofillers' contents, the nanoparticles start to form a conductive path. In other words, the nanofillers start to touch each other, facilitating the thermal transport and therefore increasing the thermal conductivity in a way that the models cannot predict. The formation of a conductive path at high nanofillers' amounts was confirmed by a high resolution TEM. As seen in Fig. 6.46, the nanofillers tend to form a conductive network favorable for the heat transfer. As the amount of the nanoparticles increases, conductive paths are more interconnected, leading to a more enhanced thermal conductivity. Moreover, the presented TEM image also confirmed that the formation of the conductive network was accompanied by a good state of

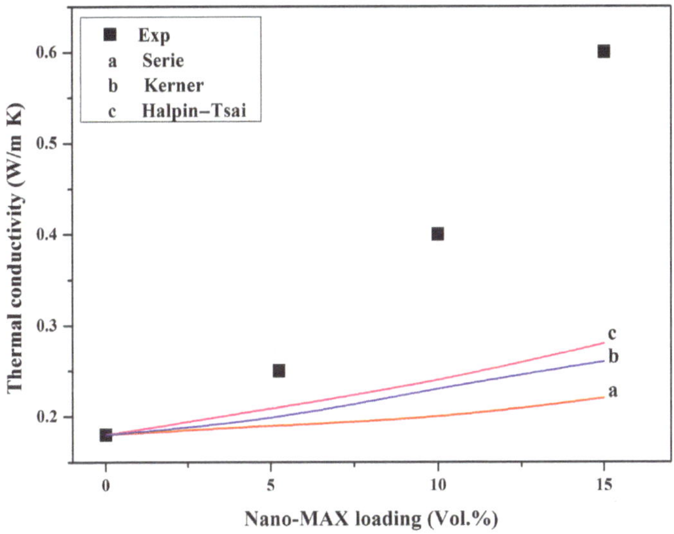

Figure 6.45 Thermal conductivity calculations and experimental measurements for the cured BAPh/MAX nanocomposites. *BAPh*, bisphenol A-based PN. Reprinted from Derradji M, Henniche A, Wang J, Dayo AQ, Ouyang Jh, Liu Wb, et al. High performance nanocomposites from Ti_3SiC_2 MAX phase and phthalonitrile resin. Polym Compos 2017. doi:10.1002/pc.24401.

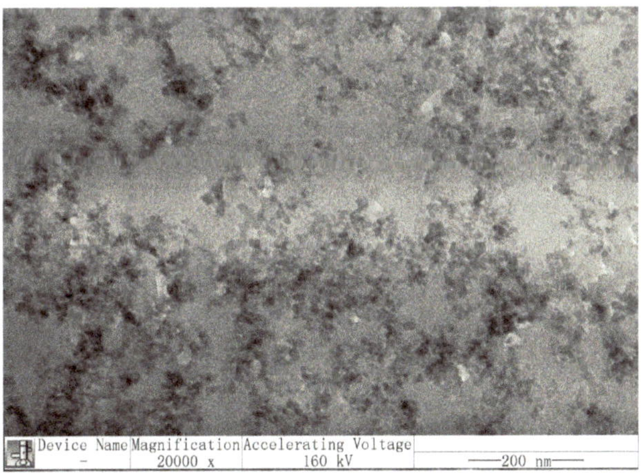

Figure 6.46 High resolution TEM image of the P(BAPh)/MAX nanocomposite at 15 vol.% nano-MAX. *TEM*, transmission electron microscopy; *BAPh*, bisphenol A-based PN. Reprinted from Derradji M, Henniche A, Wang J, Dayo AQ, Ouyang Jh, Liu Wb, et al. High performance nanocomposites from Ti_3SiC_2 MAX phase and phthalonitrile resin. Polym Compos 2017. doi:10.1002/pc.24401.

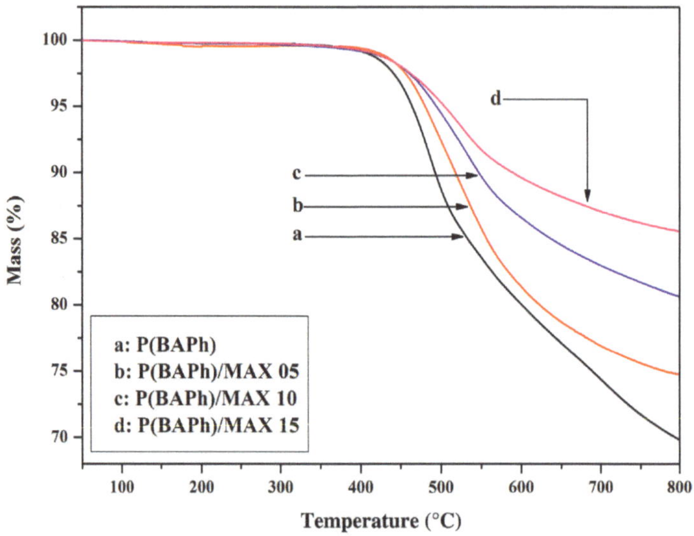

Figure 6.47 TGA curves of the cured BAPh/MAX nanocomposites at various Ti_3SiC_2 contents. *BAPh*, bisphenol A-based PN. Reprinted from Derradji M, Henniche A, Wang J, Dayo AQ, Ouyang Jh, Liu Wb, et al. High performance nanocomposites from Ti_3SiC_2 MAX phase and phthalonitrile resin. Polym Compos 2017. doi:10.1002/pc.24401.

dispersion and no microsized agglomerations were seen, even at the maximum nanofillers loading.

The excellent thermal stability of the PN resins is what gives the PN based materials their unique properties. Therefore, the preparation of any PN based composite should either maintain or preferably extend the thermal resistance. The MAX phase thermal stability is a way above that of any polymeric matrices, and as a result, their incorporation into the PN resin was expected to positively affect the overall thermal resistance. The TGA curves obtained under nitrogen atmosphere and the resulting thermal stability parameters are displayed in Fig. 6.47 and Table 6.11, respectively. The excellent thermal stability of the neat PN resin was confirmed by a $T_{5\%}$ and $T_{10\%}$ of 464 and 493°C, respectively, and a char yield at 800°C of 69.8%. The addition of the reinforcing phase further enhanced the neat resin thermal stability. In fact, a significant increase in all the thermal stability parameters was noticed upon adding the nano-MAX phase. At the maximum nanofillers loading, $T_{5\%}$ reached the values of 504°C, $T_{10\%}$ was 590°C, and the char yield at 800°C remained at 85.5%. Indeed, the outstanding thermal stability of the MAX phase nanoparticles was behind the previous results.

Table 6.11 Thermal Stability Parameters of the P(BAPh)/Ti_3SiC_2 Nanocomposites

Specimen Code	$T_{5\%}$ (°C)	$T_{10\%}$ (°C)	Y_c (%, 800°C)
P(BAPh)	464	493	69.8
P(BAPh)/MAX 05	480	518	74.6
P(BAPh)/MAX 10	494	547	80.6
P(BAPh)/MAX 15	504	590	85.5

Reprinted from Derradji M, Henniche A, Wang J, Dayo AQ, Ouyang Jh, Liu Wb, et al. High performance nanocomposites from Ti_3SiC_2 MAX phase and phthalonitrile resin. Polym Compos 2017. doi:10.1002/pc.24401.

Moreover, the good state of dispersion of the nanofillers also highly affected the motions inside the polymeric matrix and acted as a shielding material that rendered the evacuations of small molecules or volatile products difficult to achieve. As a result, the starting decomposition temperatures were enhanced.

The high ductility and strength of the MAX phase was also expected to result in PN nanocomposites with enhanced toughness. Therefore, the tensile test was performed and the obtained results such as the stress/strain curves and toughness are presented in Figs. 6.48 and 6.49, respectively. The addition of thc MAX phase affected the stress and strain in a different way to the traditional ceramics. For instance, as the nanofillers amount increased, the strength, which is the value of stress before breaking occurs, also increased reaching the value of 84 MPa at 15 vol.% nano-MAX. However, the nanocomposites strains remained at values close to that of the neat resin (about 1.6%). The MAX phases are ductile or soft ceramics and therefore their addition in brittle matrices such as the PN resins do not affect the polymer ductility. These phenomena were not seen with the traditional ceramics such as Al_2O_3, TiO_2 where the ductility of the PN was negatively affected [53,54]. The toughness of all the prepared materials was calculated by integrating the area under the stress/strain curves. As expected, the soft ceramics increased the toughness of the neat resin by 79.4%, 91.2%, and 123.5% for the MAX phase amounts of 5, 10, and 15 vol.%, respectively. These improvements in the PN matrix toughness are one of the best ever reached in the field of ceramics based PN composites. For instance, the SiC and ZnO particles brought slight ameliorations in the toughness about 19% and 16%, respectively [51,56].

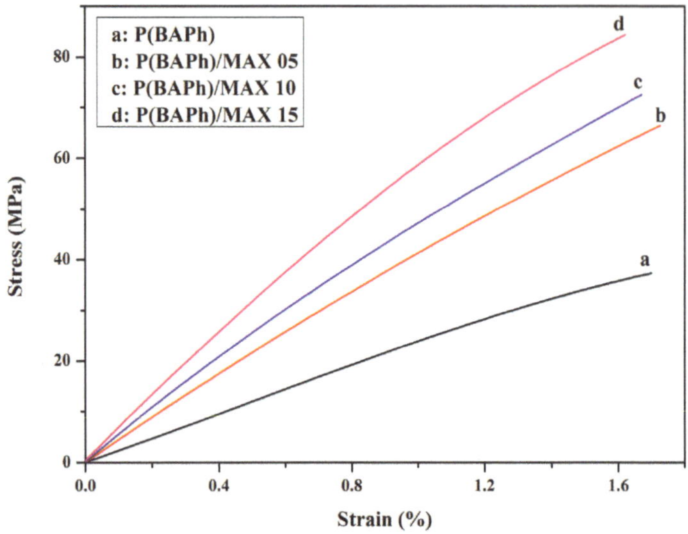

Figure 6.48 Stress/strain curves of the cured BAPh/MAX nanocomposites. *BAPh*, bisphenol A-based PN. Reprinted from Derradji M, Henniche A, Wang J, Dayo AQ, Ouyang Jh, Liu Wb, et al. High performance nanocomposites from Ti_3SiC_2 MAX phase and phthalonitrile resin. Polym Compos 2017. doi:10.1002/pc.24401.

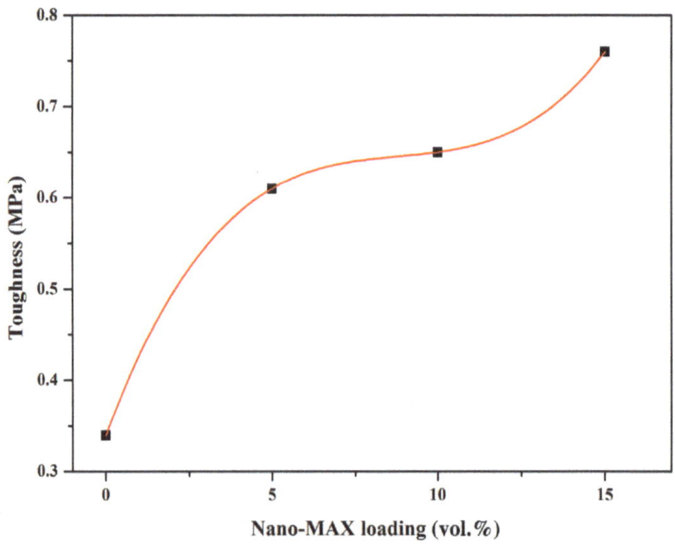

Figure 6.49 Toughness of the P(BAPh)/MAX nanocomposites at different amounts of the nano-Ti_3SiC_2. *BAPh*, bisphenol A-based PN. Reprinted from Derradji M, Henniche A, Wang J, Dayo AQ, Ouyang Jh, Liu Wb, et al. High performance nanocomposites from Ti_3SiC_2 MAX phase and phthalonitrile resin. Polym Compos 2017. doi:10.1002/pc.24401.

6.6 Conclusions

The present chapter reviewed the recent advances in the field of ceramic-based micro- and nanocomposites. It was deduced that the neat PN resins properties can be easily tailored to meet the need of different industries. Globally, all of the studied ceramics brought various ameliorations in the thermal and mechanical properties. Moreover, ceramics like TiO_2, ZnO, and BN resulted in materials with improved anticorrosion, UV-shielding performances, and thermal conductivity, respectively. The PN/ceramic composites properties were step-by-step compared with other high-performance polymers reinforced with similar fillers, and the results showed a clear advantage for the PN materials. The recent developments in the PN monomers molecular design were also highlighted and their resulting, ceramic-based composites are an important area to be addressed.

References

[1] Jeon I-Y, Baek J-B. Nanocomposites derived from polymers and inorganic nanoparticles. Materials 2010;3:3654–74.
[2] Kumar B, Fellner J. Polymer–ceramic composite protonic conductors. J Power Sources 2003;123:132–6.
[3] Durand J, Vardavoulias M, Jeandin M. Role of reinforcing ceramic particles in the wear behaviour of polymer-based model composites. Wear 1995;181:833–9.
[4] Kumar B, Scanlon LG. Polymer–ceramic composite electrolytes. J Power Sources 1994;52:261–8.
[5] Jordan J, Jacob KI, Tannenbaum R, Sharaf MA, Jasiuk I. Experimental trends in polymer nanocomposites—a review. Mater Sci Eng A 2005;393:1–11.
[6] Pomogailo AD. Hybrid polymer–inorganic nanocomposites. Russ Chem Rev 2000;69:53–80.
[7] Móczó J, Pukánszky B. Polymer micro and nanocomposites: structure, interactions, properties. J Ind Eng Chem 2008;14:535–63.
[8] Vasile C. Handbook of polyolefins. New York: CRC Press; 2000.
[9] Kango S, Kalia S, Celli A, Njuguna J, Habibi Y, Kumar R. Surface modification of inorganic nanoparticles for development of organic–inorganic nanocomposites—a review. Prog Polym Sci 2013;38:1232–61.
[10] Ting RY, Keller TM, Price TR, Poranski Jr CF. Characterization of the cure of diether-linked phthalonitrile resins. ACS Symp Ser 1982;195:337–50.

[11] Achar BN, Fohlen GM, Parker JA. Synthesis and characterization of the novel type of polymerizable bisphthalonitrile monomers. J Polym Sci A: Polym Chem 1986;24:1997–2010.

[12] Keller TM. Phthalonitrile-based high temperature resin. J Polym Sci A: Polym Chem 1988;26:3199–212.

[13] Warzel ML, Keller TM. Tensile and fracture properties of a phthalonitrile polymer. Polymer 1993;34:663–6.

[14] Sastri SB, Keller TM. Phthalonitrile polymers: cure behavior and properties. J Polym Sci A: Polym Chem 1998;37:2105–11.

[15] Sastri SB, Armistead JP, Keller TM. Phthalonitrile–carbon fiber composites. Polym Compos 1996;17:816–22.

[16] Sastri SB, Armistead JP, Keller TM. Phthalonitrile–glass fabric composites. Polym Compos 1997;18:48–54.

[17] Keller TM, Dominguez DD. High temperature resorcinol-based phthalonitrile polymer. Polymer 2005;46:4614–18.

[18] Zhang T, Wang J, Derradji M, Ramdani N, Wang H, Lin ZW, et al. Synthesis, curing kinetics and thermal properties of a novel self-promoted fluorene-based bisphthalonitrile monomer. Thermochim Acta 2015; 602:22–9.

[19] Laskoski M, Neal A, Schear MB, Keller TM, Ricks-Laskoski HL, Saab AP. Oligomeric aliphatic–aromatic ether containing phthalonitrile resins. J Polym Sci A: Polym Chem 2015;53:2186–91.

[20] Laskoski M, Schear MB, Neal A, Dominguez DD, Ricks-Laskoski HL, Hervey J, et al. Improved synthesis and properties of aryl ether-based oligomeric phthalonitrile resins and polymers. Polymer 2015;67:185–91.

[21] Derradji M, Wang J, Liu W-b. High performance ceramic-based phthalonitrile micro and nanocomposites. Mater Lett 2016;182:380–5.

[22] Yan H, Ning R, Liang G, Huang Y, Lu T. The effect of silane coupling agent on the sliding wear behavior of nanometer ZrO_2/bismaleimide composites. J Mater Sci 2007;42:958–65.

[23] Goyal R, Tiwari A, Mulik U, Negi Y. Novel high performance Al_2O_3/poly (ether ether ketone) nanocomposites for electronics applications. Compos Sci Technol 2007;67:1802–12.

[24] Goyal RK, Tiwari AN, Negi YS. Microhardness of PEEK/ceramic micro- and nanocomposites: correlation with Halpin–Tsai model. Mater Sci Eng A 2008;491:230–6.

[25] Hon K, Gill T. Selective laser sintering of SiC/polyamide composites. CIRP Ann—Manuf Technol 2003;52:173–6.

[26] Zhao H, Li RK. A study on the photo-degradation of zinc oxide (ZnO) filled polypropylene nanocomposites. Polymer 2006;47:3207–17.

[27] Zhao H, Li RK. Crystallization, mechanical, and fracture behaviors of spherical alumina-filled polypropylene nanocomposites. J Polym Sci B: Polym Phys 2005;43:3652–64.

[28] Ash B, Schadler L, Siegel R. Glass transition behavior of alumina/polymethylmethacrylate nanocomposites. Mater Lett 2002;55:83–7.
[29] Khaled S, Sui R, Charpentier PA, Rizkalla AS. Synthesis of TiO_2–PMMA nanocomposite: using methacrylic acid as a coupling agent. Langmuir 2007;23:3988–95.
[30] Rhee SH, Choi JY. Preparation of a bioactive poly (methyl methacrylate)/silica nanocomposite. J Am Ceram Soc 2002;85:1318–20.
[31] Ng C, Schadler L, Siegel R. Synthesis and mechanical properties of TiO_2–epoxy nanocomposites. Nanostruct Mater 1999;12:507–10.
[32] Naous W, Yu X-Y, Zhang Q-X, Naito K, Kagawa Y. Morphology, tensile properties, and fracture toughness of epoxy/Al_2O_3 nanocomposites. J Polym Sci B: Polym Phys 2006;44:1466–73.
[33] Li Y-Q, Fu S-Y, Mai Y-W. Preparation and characterization of transparent ZnO/epoxy nanocomposites with high-UV shielding efficiency. Polymer 2006;47:2127–32.
[34] Carballeira P, Haupert F. Toughening effects of titanium dioxide nanoparticles on TiO_2/epoxy resin nanocomposites. Polym Compos 2009;31:1241–6.
[35] Gupta S, Ramamurthy PC, Madras G. Synthesis and characterization of flexible epoxy nanocomposites reinforced with amine functionalized alumina nanoparticles: a potential encapsulant for organic devices. Polym Chem 2011;2:221–8.
[36] Sheng X, Akinc M, Kessler MR. Rheology and dynamic mechanical analysis of bisphenol E cyanate ester/alumina nanocomposites. Polym Eng Sci 2010;50:302–11.
[37] Sheng X, Akinc M, Kessler MR. The effects of alumina and silica nanoparticles on the cure kinetics of bisphenol E cyanate ester. Polym Eng Sci 2010;50:1075–84.
[38] Qin W, Peng D, Wu X, Liao J. Study on the resistance performance of TiO_2/cyanate ester nano-composites exposed to electron radiation. Nucl Instrum Methods Phys Res, Sect B: Beam Interact Mater At 2014;325:115–19.
[39] Qin D, Gu A, Liang G, Yuan L. A facile method to prepare zirconia electrospun fibers with different morphologies and their novel composites based on cyanate ester resin. RSC Adv 2012;2:1364–72.
[40] Cai H, Yan F, Xue Q. Investigation of tribological properties of polyimide/carbon nanotube nanocomposites. Mater Sci Eng A 2004;364:94–100.
[41] Xie S-H, Zhu B-K, Li J-B, Wei X-Z, Xu Z-K. Preparation and properties of polyimide/aluminum nitride composites. Polym Test 2004;23:797–801.
[42] Chiang P-C, Whang W-T, Tsai M-H, Wu S-C. Physical and mechanical properties of polyimide/titania hybrid films. Thin Solid Films 2004;447:359–64.

[43] Tsai M-H, Liu S-J, Chiang P-C. Synthesis and characteristics of polyimide/titania nano hybrid films. Thin Solid Films 2006;515:1126–31.
[44] Ramdani N, Wang J, Wang H, Feng Tt, Derradji M, Liu Wb. Mechanical and thermal properties of silicon nitride reinforced polybenzoxazine nanocomposites. Compos Sci Technol 2014;105:73–9.
[45] Ramdani N, Derradji M, Wang J, Liu W-b, Mokhnache E-O. Effects of aluminium nitride silane-treatment on the mechanical and thermal properties of polybenzoxazine matrix. Plast Rubber Compos 2016;45:72–80.
[46] Ramdani N, Derradji M, Feng T-t, Tong Z, Wang J, Mokhnache E-O, et al. Preparation and characterization of thermally-conductive silane-treated silicon nitride filled polybenzoxazine nanocomposites. Mater Lett 2015;155:34–7.
[47] Wu Y, Zeng M, Jin H, Xu Q, Lu L, Wang J, et al. Effects of glass-to-rubber transition on the friction properties of ZrO_2 reinforced polybenzoxazine nanocomposites. Tribol Lett 2012;47:389–98.
[48] Wang X, Chen D, Ma W, Yang X, Lu L. Polymerization of bismaleimide and maleimide catalyzed by nanocrystalline titania. J Appl Polym Sci 1999;71:665–9.
[49] Kurahatti R, Surendranathan A, Srivastava S, Singh N, Kumar AR, Suresha B. Role of zirconia filler on friction and dry sliding wear behaviour of bismaleimide nanocomposites. Mater Des 2011;32:2644–9.
[50] Vadivel M, Kumar MSC, Selvam V, Shobana S, Narayan SA. Thermal and morphological studies on bismaleimide modified ZnO nanoparticles. Asian J Chem 2014;26:7160.
[51] Derradji M, Ramdani N, Zhang T, Wang J, Feng TT, Wang H, et al. Mechanical and thermal properties of phthalonitrile resin reinforced with silicon carbide particles. Mater Des 2015;71:48–55.
[52] Derradji M, Ramdani N, Zhang T, Wang J, Lin Z-w, Yang M, et al. High thermal and thermomechanical properties obtained by reinforcing a bisphenol-A based phthalonitrile resin with silicon nitride nanoparticles. Mater Lett 2015;149:81–4.
[53] Derradji M, Ramdani N, Zhang T, Wang J, Gong L-d, Xu X-d, et al. Thermal and mechanical properties enhancements obtained by reinforcing a bisphenol-A based phthalonitrile resin with silane surface-modified alumina nanoparticles. Polym Compos 2017;38:1546–58.
[54] Derradji M, Ramdani N, Zhang T, Wang J, Gong L-d, Xu X-d, et al. Effect of silane surface modified titania nanoparticles on the thermal, mechanical, and corrosion protective properties of a bisphenol-A based phthalonitrile resin. Prog Org Coat 2016;90:34–43.
[55] Derradji M, Feng T, Wang H, Ramdani N, Zhang T, Wang J, et al. New oligomeric containing aliphatic moiety phthalonitrile resins: their mechanical and thermal properties in presence of silane surface-modified zirconia nanoparticles. Iran Polym J 2016;25:503–14.

[56] Derradji M, Ramdani N, Gong L-d, Wang J, Xu X-d, Lin Z-w, et al. Mechanical, thermal, and UV-shielding behavior of silane surface modified ZnO-reinforced phthalonitrile nanocomposites. Polym Adv Technol 2016;27:882–8.

[57] Derradji M, Song X, Dayo AQ, Wang J, Liu W-b. Highly filled boron nitride-phthalonitrile nanocomposites for exigent thermally conductive applications. Appl Therm Eng 2017;115:630–6.

[58] Barsoum MW, El-Raghy T. Synthesis and characterization of a remarkable ceramic: Ti_3SiC_2. J Am Ceram Soc 1996;79:1953–6.

[59] Tzenov NV, Barsoum MW. Synthesis and characterization of Ti_3AlC_2. J Am Ceram Soc 2000;83:825–32.

[60] Wang X, Zhou Y. Layered machinable and electrically conductive Ti_2AlC and Ti_3AlC_2 ceramics: a review. J Mater Sci Technol 2010;26:385–416.

[61] Wang S, Zhu S, Cheng J, Qiao Z, Yang J, Liu W. Microstructural, mechanical and tribological properties of Al matrix composites reinforced with Cu coated Ti_3AlC_2. J Alloys Compd 2017;690:612–20.

[62] Mahesh K, Balanand S, Raimond R, Mohamed AP, Ananthakumar S. Polyaryletherketone polymer nanocomposite engineered with nanolaminated Ti_3AlC_2 ceramic fillers. Mater Des 2014;63:360–7.

[63] Zhang H, Wang L, Chen Q, Li P, Zhou A, Cao X, et al. Preparation, mechanical and anti-friction performance of MXene/polymer composites. Mater Des 2016;92:682–9.

[64] Wang W-J, Li C-W, Chen K-P. Electrical, dielectric and mechanical properties of a novel Ti_3AlC_2/epoxy resin conductive composites. Mater Lett 2013;110:61–4.

[65] Henniche A, Derradji M, Wang J, Liu W-b, Ouyang J-h, Medjahed A. High-performance polymeric nanocomposites from phthalonitrile resin and silane surface–modified Ti_3AlC_2 MAX phase. High Perform Polym 2017. Available from: https://doi.org/10.1177/0954008317699678.

[66] Derradji M, Henniche A, Wang J, Dayo AQ, Ouyang Jh, Liu Wb, et al. High performance nanocomposites from Ti_3SiC_2 MAX phase and phthalonitrile resin. Polym Compos 2017. Available from: https://doi.org/10.1002/pc.24401.

[67] Ihom PA, Nor JI, Usman W. Recent trends and developments in surface hardening technology a focus on new ceramics for surface modification in industry. Int J Sci Eng Res 2012;3:1–22.

[68] Ramdani NLW, Wang J, Derradji M. Ceramic based polybenzoxazine micro- and nanocomposites. In: Ishida H, Froimowicz P, editors. Advanced and emerging polybenzoxazine science and technology. 1st ed. Amsterdam: Elsevier; 2017.

[69] Corriu RJ, Leclercq D. Recent developments of molecular chemistry for sol–gel processes. Angew Chem Int Ed Engl 1996;35:1420–36.

[70] Ahmad Z, Mark J. Polyimide–ceramic hybrid composites by the sol–gel route. Chem Mater 2001;13:3320–30.
[71] Brinker CJ, Scherer GW. Sol–gel science: the physics and chemistry of sol–gel processing. New York: Academic Press; 2013.
[72] Ou Y, Yang F, Yu ZZ. A new conception on the toughness of nylon 6/silica nanocomposite prepared via in situ polymerization. J Polym Sci B: Polym Phys 1998;36:789–95.
[73] Yuchun O, Feng Y, Jin C. Interfacial interaction and mechanical properties of nylon 6-potassium titanate composites prepared by in-situ polymerization. J Appl Polym Sci 1997;64:2317–22.
[74] Shan S, Chen X, Xi Z, Yu X, Qu X, Zhang Q. The effect of nitrile-functionalized nano-aluminum oxide on the thermomechanical properties and toughness of phthalonitrile resin. High Perform Polym 2016;29:113–23.
[75] Liu M, Jia K, Liu X. Effective thermal conductivity and thermal properties of phthalonitrile-terminated poly (arylene ether nitriles) composites with hybrid functionalized alumina. J Appl Polym Sci 2015;132:41595.
[76] Keller T.M. Epoxy phthalonitrile polymers. United States Patent US 4619986. 1986 Oct 28.
[77] Dominguez DD, Keller TM. Phthalonitrile–epoxy blends: cure behavior and copolymer properties. J Appl Polym Sci 2008;110:2504–15.
[78] Rykowski J.J. Silane coupling agents. United States Patent US 4179537. 1979 Dec 18.
[79] Plueddemann EP. Chemistry of silane coupling agents. Silane coupling agents. New York: Springer; 1991. p. 31–54.
[80] Rong MZ, Zhang MQ, Zheng YX, Zeng HM, Walter R, Friedrich K. Structure–property relationships of irradiation grafted nano-inorganic particle filled polypropylene composites. Polymer 2001;42:167–83.
[81] Kickelbick G. Concepts for the incorporation of inorganic building blocks into organic polymers on a nanoscale. Prog Polym Sci 2003;28:83–114.
[82] Shi G, Zhang MQ, Rong MZ, Wetzel B, Friedrich K. Friction and wear of low nanometer Si_3N_4 filled epoxy composites. Wear 2003;254:784–96.
[83] Sato M, Yamamoto T, Takeuchi M, Yamanouchi K. Humidity sensitivity of Lamb waves on composite polyimide/ZnO/Si_3N_4 structure. Jpn J Appl Phys 1993;32:2380.
[84] Hongxia Y, Pengbo L, Junping Z, Rongchang N. Mechanical properties of novel bismaleimide nanocomposites with Si_3N_4 nanoparticles. J Reinf Plast Compos 2010;29:1515–22.
[85] Yan H, Zhang M, Liu C, Zhang J. Nano-Si_3N_4/epoxidized silane/cyanate ester composites for electronic packaging. Polym bull 2013;70:2923–33.
[86] Nielsen LE, Landel RF. Mechanical properties of polymers and composites. New York: CRC Press; 1994.

[87] Shi H, Liu F, Yang L, Han E. Characterization of protective performance of epoxy reinforced with nanometer-sized TiO_2 and SiO_2. Prog Org Coat 2008;62:359–68.

[88] Escobar J, Poorteman M, Dumas L, Bonnaud L, Dubois P, Olivier M-G. Thermal curing study of bisphenol A benzoxazine for barrier coating applications on 1050 aluminum alloy. Prog Org Coat 2015;79:53–61.

[89] Rajamanikam R, Pichaimani P, Kumar M, Muthukaruppan A. Optical and thermomechanical behavior of benzoxazine functionalized ZnO reinforced polybenzoxazine nanocomposites. Polym Compos 2017;38:1881–9.

Index

Note: Page numbers followed by "*f*" and "*t*" refer to figures and tables, respectively.

A

Acid-curing agents, 27
AFPN. *See* 9,9-Bis(4-aminophenyl)-2,7-(3,4-dicyanophenoxy)-fluorene (AFPN)
Al_2O_3. *See* Aluminum oxide (Al_2O_3)
Aliphatic alcohols, 9
Aliphatic spacers, 6–7
  synthetic route of PN monomers with, 5*f*
Aliphatic-aromatic ether, 93
Alkene, 108
  alkene-functional PN, 109–112
Alkoxy or phenoxy linkages, 4
Alkyne, 114
Alkyne-functional phthalonitrile, 112–118, 113*f*
  synthetic route of propargyl-functional, PN oligomers, 117*f*
Allyl-functional, PN polymer, 111–112
Allyl-functional groups, 149
Allyl-functional phthalonitrile/bismaleimide, 219–222, 220*f*
Allyl-functional PN/BZPN, 206–209
AlN. *See* Aluminum nitride (AlN)
Alumina. *See* Aluminum oxide (Al_2O_3)
Alumina nanoparticles, 303–304, 334–338
Aluminum iso-propoxide, 301–302
Aluminum nitride (AlN), 296–297
Aluminum oxide (Al_2O_3), 296–297, 328–329
  properties of phthalonitrile/Al_2O_3 nanocomposites, 328–341
Ameliorated processability, 295–296
Amine
  activated cyano addition reaction, 16*f*
  impact on curing behavior, 23, 24*f*
  salt curing agents, 26–28
Amine-cured BAPh polymers, 62–63
Amine-curing agent, 68–69, 124, 253–254, 258*t*
Amino functionalities, 109
4-Amino-*N*-{(3,4-dicyanophenoxy) phenyl} benzamides PN monomers, 39, 39*f*
Aminophenoxyphthalonitrile (APN), 69–70, 303–304, 329–330
  3-APN, 279–280
  4-APN, 23, 24*f*, 39–41, 46–49, 47*f*, 123, 135, 193
    curing agent, 77
  monomers, 278–279
4-Aminophenylsulfone, 17, 62–63
4-Aminophthalonitrile, 17
Aminopropyl trimethoxy silane, 364–366
Anticorrosion properties, 304, 341, 345
p-APB, 183
APN. *See* Aminophenoxy-phthalonitrile (APN)
Aromatic ether, 112
Aromatic spacers, 6–7

B

BAPh. *See* Bisphenol A-based PN (BAPh)
BDB. *See* 2,6-Bis (4-diaminobenzoxy) benzonitrile (BDB)

Bending test, 285–286
Benzimidazole, 108, 175–176
Benzimidazole-functional phthalonitrile (BIPN), 151–154, 152*f*
Benzophenone, 88–90
Benzoxazine, 108, 175–176, 275–276
  chemical structure of PN monomer bearing, 14, 14*f*
  PN/benzoxazine, 197–209
Benzoxazine-functional phthalonitrile (BZPN), 140–151, 141*f*, 146*f*, 198–200, 199*f*, 217–219, 282–283
  allyl-functional phthalonitrile/ benzoxazine, 206–209
  BZPN/bismaleimide, 217–219
  BZPN/GF composites, 285
  BZPN/phenylnitrile-functional benzoxazine, 198–201
  complex viscosity as function
    of temperature, 149*f*
    of time, 150*f*
  DSC results and curing conditions, 142*t*
  flammability properties, 143*t*
  monomer, 14, 15*f*
  reaction mechanism, 147*f*
  SEM images of BZPN/BPA, 148*f*
  thermal properties, 143*t*
  typical PN/BZPN, 201–206
Bifunctional PN monomers, 4–5
Biphenyl PN (BPh), 61*f*, 64, 177, 177*f*, 249–250
  BPh/GF composite
    panels, 274
    properties comparison, 277*t*
  BPh/IM-7 prepegs, 250, 251*t*
    flammability results, 254*t*
BIPN. *See* Benzimidazole-functional phthalonitrile (BIPN)
2,6-Bis (4-diaminobenzoxy) benzonitrile (BDB), 23, 24*f*
Bis(3-(3,4-dicyanophenoxy)phenyl) phenyl phosphate (RPPhPN), 268
1,3-Bis(3-aminophenoxy)-benzene (*m*-APB), 17–18, 18*f*, 20, 23, 64, 67, 74–75, 120–122, 249–250, 253–254
3,5-Bis(3,4-dicyanophenoxy)aniline (CPA), 29
  synthetic route and chemical structure of, 29*f*
4,4'-Bis(3,4-dicyanophenoxy) biphenyl-based PN (BPh), 18
9,9-Bis(4-aminophenyl)-2,7-(3,4-dicyanophenoxy)-fluorene (AFPN), 43–44, 44*f*, 45*f*
2,6-Bis(4-hydroxybenzoyl)pyridine, 162
9,9-Bis{(3,4-dicyanophenoxy)phenyl} fluorene (BPF), 43–44, 44*f*
Bis{4-(4-aminophenoxy) phenyl} sulfone (*p*-BAPS), 20, 20*f*, 23, 67, 183, 253–254, 256–257
Bismaleimide (BMI), 108, 175–176, 290
  BMI-functional PN, 135–136, 135*f*
  monomers, 217
  PN/bismaleimide, 217–222
    BZPN/bismaleimide, 217–219
  resins, 1, 135
4,4′-Bismaleimidodiphenylmethane, 135
Bisphenol A (BPA), 10, 10*f*, 59–60, 222–223, 275–276
  polymer, 77
  polysulfone, 9, 10*f*
Bisphenol A-based PN (BAPh), 10, 61*f*, 64
  BAPh-based copolymers, 190
  BAPh/Al_2O_3 nanocomposites, 330–331, 331*f*
  BAPh/BN nanocomposites, 352–353
  BAPh/BZPN/GF composites, 286

flexural strength and modulus, 287*t*
BAPh/GF composites, 280–281
flexural strength and modulus, 282*t*
thermal and thermo-oxidative properties, 281*t*
BAPh/propargyl blends, 222–223
BAPh/Si_3N_4 nanocomposites, 325–326, 326*t*
BAPh/SiC composites, 312–313, 313*f*
monomers, 187–189, 279–280, 303–304
nanocomposites, 159–161
oligomers, 204–205
polymer, 63*t*, 77
resin, 352–353
modulus, 333
Bisphenol A6F-based PN (6FPh), 61*f*, 64
Bisphenol compounds, 80–81
Bisphenol S, 9, 10*f*
Bisphenol S-based PN (BSPh), 60–61, 61*f*
Bisphenol-A6F (BPF), 59–60
Bisphenol-based resins. *See* Phthalonitrile resins (PN resins)
Bisphenols, 59–60
Bisphenols-based PN polymers, 139
Bispropargyl ether of BPA, 222–223, 222*f*
Blends formation, 175–176
BMI. *See* Bismaleimide (BMI)
Boron carbide, 296–297
Boron nitride (BN), 296–297
properties of phthalonitrile/BN nanocomposites, 351–356
BPA. *See* Bisphenol A (BPA)
BPF. *See* Bisphenol-A6F (BPF)
BPh. *See* 4,4'-Bis(3,4-dicyanophenoxy)biphenyl-based PN (BPh)
BSPh. *See* Bisphenol S-based PN (BSPh)
B-stage, 6–7, 9, 263, 303–304
BZPN. *See* Benzoxazine-functional phthalonitrile (BZPN)

C

C10-PN/T-300 prepegs, 248
Carbon fibers (CF), 241, 243–270. *See also* Glass fibers (GF)
processes of making CF from pitch and PAN precursors, 244*f*
properties, 243–247
reinforced phthalonitrile composites, 247–270
complex viscosity variations of BPh prepolymers, 255*f*
effects of hydroxyl groups contents, 260*f*
flexural and short beam tests results, 264*t*
LOI, 262*t*
mechanical properties of CF-reinforced composites, 266*t*
Phthalonitrile Blends/Carbon Fibers Composites, 265–270
Phthalonitrile/Carbon Fibers Composites, 247–258
room temperature mechanical properties, 249*t*, 251*t*
SEM images of fractured surfaces of CF composites, 261*f*, 267*f*
storage modulus and Tan Delta variations, 257*f*
X-functional phthalonitrile/carbon fibers composites, 259–265
tensile and modulus properties of commercial, 246*t*
Carbon nanotubes from PN precursors, 101–103, 103*f*
Carbon steel panels, 345, 345*f*
Carbonization process, 101–103

Carboxyl-terminated butadiene acrylonitrile rubber (CTBN rubber), 233–234
PN/CTBN rubber, 233–235
4-(4-Carboxylphenoxy) PN monomers (CPN monomers), 309–310
CPN-grafted alumina spectrum, 337–338
CPN-modified nanofillers, 338
Catastrophic decomposition temperatures, 82
Ceramic fillers, 296–297
properties, 299–301
sizes, 295
Ceramic nanofillers, 341
Ceramic nanoparticles, 298
Ceramic particles, 295–296, 348
agglomerations, 298–299
Ceramic reinforcements, first generation of, 296–297
Ceramic-based phthalonitrile micro- and nanocomposites, 356. *See also* Fiber-reinforced phthalonitrile composites
PN/Al_2O_3 nanocomposite properties, 328–341
PN/BN nanocomposite properties, 351–356
PN/MAX phase nanocomposite properties, 362–368
PN/Si_3N_4 nanocomposite properties, 322–327, 324*f*
PN/SiC microcomposite properties, 310–322, 311*f*
PN/TiO_2 nanocomposite properties, 341–348, 343*f*
PN/ZnO nanocomposite properties, 348–351, 350*t*
PN/ZrO_2 nanocomposite properties, 356–362
preparation methods of hybrid materials, 301–305
in situ polymerization, 303–304
sol–gel processing, 301–302, 302*f*
solution blending technique, 304–305
studied ceramic filler properties, 299–301
surface modification of ceramic particles, 305–310
chemical treatments, 305–308
grafting of synthetic polymers, 308–310
thermal and mechanical properties of various ceramic fillers, 300*t*
CF. *See* Carbon fibers (CF)
C-glass GF, 271
Chromene, 116, 119*f*
Chromium oxide (Cr_2O_3), 296–297
Claisen rearrangement, 223
C*n*-PN polymers, 247–248
Conductive path, 298, 353–354, 364–366
Cone calorimeter, 212
Conventional fabrication techniques, 80
Conventional impregnation process, 283
Conventional industrial processes, 150–151
Copolymers, 107–108, 269
formation, 175–176
Cost-effective method, 175–176
CPA. *See* 3,5-Bis(3,4-dicyanophenoxy)aniline (CPA)
CPN monomers. *See* 4-(4-Carboxylphenoxy) PN monomers (CPN monomers)
Cr_2O_3. *See* Chromium oxide (Cr_2O_3)
Critical stress intensity factors (K_{IC}), 210–211, 211*f*
Cross-linking degrees, 62–63, 99–100
C-stage, 6–7, 9, 17

CTBN rubber. *See* Carboxyl-terminated butadiene acrylonitrile rubber (CTBN rubber)
Cure conditions, effects of, 73–74, 74*t*
Cured PN/novolac copolymers, 210–211
Curing agent, 178, 183
Curing agents/cure behavior relationship
  metals and metallic salts curing agents, 7–8
  organic amine-curing agents, 16–25
  phenols curing agents, 8–15, 10*f*
  strong organic acid/amine salt curing agents, 26–28
Curing behavior, 68–69, 82, 107, 138–139, 176, 178, 182–183, 185–187, 249–250
  of BAPh resin, 341–343
  BAPh/novolac blends, 214
  of cyanate ester-functional, 156–157
  double-stage, 217–218
  DSC analyzes, 219–220
  and mechanism
    curing agents/cure behavior relationship, 7–28
    curing behavior/monomer structure relation, 5–7
    self-catalyzed phthalonitrile monomers and oligomers, 29–50, 35*f*
  of neat BAPh resin, 312
  neat BPh and epoxy, 183–184
  PEEKOH, 226
Curing schedule, 71–72
Cyanate esters, 1, 108
  cyanate ester-functional PN, 155–158, 156*f*
  resins, 1
Cyclodehydration in acetic acid, 152–153

D

DAPN. *See* 4-(4-(3,5-Diaminobenzoyl) phenoxy) PN (DAPN)
DBPA. *See* Diallylbisphenol-A BPA (DBPA)
DDA. *See* Dynamic dielectric analysis (DDA)
DDS. *See* 4,4'-Diaminediphenyl sulfone (DDS)
Dehydration catalysis, 152–153
Diallylbisphenol-A BPA (DBPA), 14–15, 109, 285
  DBPA-Ph, 289–290
4,4'-Diaminediphenyl sulfone (DDS), 23, 24*f*, 123, 187–189, 265–266
4,4'-Diamino-diphenyl ether (ODA), 303–304
4-(4-(3,5-Diaminobenzoyl) phenoxy) PN (DAPN), 33–36
4,4′-Diaminodiphenyl ether, 159
Differential scanning calorimetry (DSC). *See* Dynamic scanning calorimetry (DSC)
4,4′-Difluorobenzophenone, 165
Diglycidyl ether of bisphenol A, 183
Dihaloaliphatic compounds, 80–81, 93
Dihaloaromatic compounds, 80–81
3,5-Dihydroxybenzoic acid, 152–153
Dilution process, 320–321
Dimethyl formamide (DMF), 38
Dip coating process, 304
Dispersion behavior of inorganic particles, 308, 309*f*
DMA. *See* Dynamic mechanical analysis (DMA)
DMF. *See* Dimethyl formamide (DMF)
DSC. *See* Dynamic scanning calorimetry (DSC)
Dual behavior, 297

Dynamic dielectric analysis (DDA), 248
Dynamic mechanical analysis (DMA), 72–73, 178–182, 187, 248, 303–304, 320–321
Dynamic scanning calorimetry (DSC), 10, 13, 15*f*, 93, 109, 110*f*, 120*f*, 153*f*, 157*f*, 178, 190, 204, 250, 312
  analysis, 219–220
  curve
    of BPh monomers, 21, 22*f*
    of CPA monomers, 30*f*
    of DAPN monomers, 36*f*
    of PN prepolymers, 25, 25*f*
  thermograms, 144–145, 145*f*
    of unfilled BAPh matrix, 330*f*

E

E-glass GF, 271, 274, 283
Electro acceptors, 26
Endothermic transition, 109, 124
Epoxy, 1
  epoxy-based composites, 1
  PN/epoxy, 182–197
    blends, 183–192
    flame retardancy properties of BAPh/epoxy copolymers, 191*t*
    flexural strength variations of epoxy, 192*f*
    functional-phthalonitrile/epoxy blends, 195–197
    self-catalyzed, 192–195
    storage modulus, 188*f*, 189*f*
    thermal properties of BPh/epoxy copolymers, 186*t*
  resin, 182–183, 248
Exigent applications, 1, 55, 58–59, 243–244
Exotherms, 116–118

F

F-178/T-300 composites, 248–249
Fiber-reinforced composites (FRC), 241
Fiber-reinforced phthalonitrile composites, 242–243. *See also* Ceramic-based phthalonitrile micro-and nanocomposites
  carbon fibers/phthalonitrile composites, 243–270
  glass fibers/phthalonitrile composites, 270–291
Fiber-reinforced polymers (FRP), 241–242
Filament winding, 19–20, 66–67, 151–152, 243
First-generation of phthalonitrile
  materials, 356
  monomer, 357*f*
  resins, 56, 59–80
    mechanical and thermomechanical properties, 71–77, 72*t*
    thermal aging resistance and flammability properties, 66*t*
    thermal properties of BAPh polymer, 63*t*
    thermal resistance properties, 60–71
    water absorption properties, 78–80
Flame resistance, 136–137, 210, 252–253
Flammability
  measurements, 212
  properties, 66*t*
  tests, 143–144, 277–278
Fourier transform infrared (FTIR) spectroscopy, 10, 18–19, 109–110, 111*f*, 116, 118*f*, 190, 218–219, 310–311
  of AFPN, 45*f*
  analyses, 311–312
  of native and silane-treated Al_2O_3 nanoparticles, 329*f*
  of native and silane-treated Si_3N_4nanoparticles, 323*f*

of native and silane-treated TiO_2 nanoparticles, 342*f*
of prepolymers and polymers, 49*f*
spectra of neat BAPh resin, 312*f*
6FPh. *See* Bisphenol A6F-based PN (6FPh)
FRC. *See* Fiber-reinforced composites (FRC)
Free hydroxyl groups, 10, 285–286
Friedel–Crafts acylation, 162
FRP. *See* Fiber-reinforced polymers (FRP)
FTIR spectroscopy. *See* Fourier transform infrared (FTIR) spectroscopy
"Fully cured" PN composites, 256–257
Functional-phthalonitrile/epoxy blends, 195–197

G

General purpose (GP), 245
Glass fibers (GF), 270–291. *See also* Carbon fibers (CF)
oxide designations and respective weight ranges, 272*t*
physical properties, 273*t*
properties, 270–274
reinforced phthalonitrile composites, 274–291
phthalonitrile blends/glass fiber composites, 286–291, 288*t*
phthalonitrile/glass fiber composites, 274–282
X-functional phthalonitrile/glass fiber composites, 282–286
Glass transition temperature, 1, 58–59, 107, 175, 242–243, 295–296
GP. *See* General purpose (GP)
Grafting
monomer, 298–299
of synthetic polymers, 308–310
"Grafting from" technique, 309, 310*f*
"Grafting to" method, 309, 310*f*

H

Halogen-free flame retardant, 190–191
Halpin–Tsai model, 331–333
Heat polymerization method, 9, 17
Hexcel F-178 composites, 248–249
High modulus (HM), 243–244, 247
High strength (HT), 241, 271
High-performance (HP), 241–243
polymers, 1, 66, 80, 275–276, 295–296
bismaleimide resins, 135
PN resins, 175
polyimides, 118–119
thermosets, 1, 55, 243, 295–296, 322
epoxy resins, 182–183
HM. *See* High modulus (HM)
HP. *See* High-performance (HP)
HPNM monomers, 32
HPNM-a monomers, 31*f*, 32
HPNM-b monomers, 32
HPNM-c monomers, 32–33
HPNM-d monomers, 32–33
HT. *See* High strength (HT)
HT GF. *See* S-glass GF
Hydrolysable group, 305–306
Hydroxyl(s), 116, 119*f*
groups, 146–147
hydroxyl activated cyano addition reaction, 8, 9*f*

I

IM. *See* Intermediate modulus (IM)
Imide, 108, 175–176
imide-functional PN, 118–135, 121*f*, 123*f*
melt viscosity as function, 133*f*
synthetic route and designation, 132*f*
thermal properties, 125*t*
monomers, 118–119
Impedance modulus, 345–347, 346*f*, 347*f*

Impregnation process, 280
In situ polymerization, 301, 303–304
Initial decomposition temperature, 145–146
Inorganic composites, 304
Interaromatic aliphatic chain length, 58–59, 59*t*
Intermediate modulus (IM), 247
Ionization constant (Ki), 26
Ionizing radiations, 103–104
Isoindoline, 10–12, 12*f*, 68–69, 137–138, 157
Isothermal DSC scans, 204–205

K
Kerner model, 331, 333, 364–366

L
Laminates, 241
Lewis acids, 26–27
Limiting Oxygen Index (LOI), 190–191, 262*t*
Low-modulus type CF, 247

M
Magnesium phthalocyanines preparation, 2
Mannich bridge (–CH2–NR–CH2–), 198
Mannich-based networks, 142–143
m-APB. *See* 1,3-Bis(3-aminophenoxy)-benzene (*m*-APB)
MAX phase, 299–301, 305, 362–364
  properties of phthalonitrile/MAX phase nanocomposites, 362–368
MAX phase nanoparticles, 364–366
"MAX phases", 297, 305
m-BAPS, 20, 20*f*, 25, 66–67
MDA. *See* 4'-Methylene dianiline (MDA)
Mechanical properties
  of BAPh resin, 310–311
  of BAPh/epoxy blends, 191–192
  of first-generation PN resins, 71–77, 72*t*, 93–98
  PN matrices, 107–108
  polymers, 341
Mechanical test shape requirements, 303–304
Melt viscosity of PN prepolymers, 21–23, 22*f*
MeSiPN, 41–42, 42*f*
Mesophage, 245
Meta-amine-based polymers, 127
Metallic salts curing agents, 7–8
Metallic structures protection, 345–347
Metals salts curing agents, 7–8
4'-Methylene dianiline (MDA), 63–64
Micro-composites, 295
MIL-STD-2031 flame test, 100–101
Molecular
  compositions, 99–100
  design, 107–108
  weight analyses, 2
Monofuctional PN monomers, 4–5, 36–37, 37*f*
Monomer(s), 69–70, 308
  curing behavior/monomer structure relation, 5–7
  structure, 71–72
Multifunctional PN monomers, 4–5

N
N-Methyl-2-pyrrolidone. *See N*-Methylpyrrolidinone (NMP)
N-Methylpyrrolidinone (NMP), 10–11, 250
Namco 5208 composites, 248–249
Nanocomposites, 295, 338–341
  preparation, 348
Nanofillers, 353–354

1,6 Naphthyl-functional, PN monomers, 159
Naphthyl-functional PN, 158–161, 159*f*, 160*f*
Native nanoparticles, 353–354
Native treated SiC particles, 314–315
Naval research laboratory (NRL), 2–4, 59–60
Neat BAPh resin, 345. *See also* Bisphenol A-based PN (BAPh)
  corrosion protective properties of, 347–348
  curing behavior of, 312
  matrix, 352–353
  SEM images of, 336*f*
  spectra, 312*f*
  toughness and strength of, 315–317
Neat PN resin, 348–349
4-Nitrophthalonitrile, 109
NMP *N*-Methylpyrrolidinone (NMP)
Non-PN monomers, 278–279
Nonalkyne, 114
Noncatalyzed alkyne-functional, 113–114
Novolac, 108, 175–176
  novolac-containing PN, 116–118
  novolac-functional curing, 265–266
  novolac-functional PN, 136–140, 137*f*
    TGA curves under nitrogen atmosphere, 140*f*
  PN/novolac, 210–217
  prepolymers, 155–156
NRL. *See* Naval research laboratory (NRL)
Nucleophilic displacement reaction, 56

O

OCN groups, 158, 158*f*
ODA. *See* 4,4'-Diamino-diphenyl ether (ODA)
ODPA. *See* 4,4′-Oxydiphthalic dianhydride (ODPA)
Oligomeric molecules, 298–299
Oligomers, 82, 183, 301–302
Organic amine-curing agents, 16–25
Organic monomers, 301–302
Organofunctional group (Y), 305–306
Ortho-amine-based polymer, 125–127
Owens Corning Textile Products, 270–271
Oxazine ring-opening, 142
Oxidative aging data, 84*t*, 90*t*, 92*t*
4,4′-Oxydiphthalic dianhydride (ODPA), 131

P

PAN. *See* Polyacrylonitrile (PAN)
Para-amine-based polymers, 127
Paraformaldehyde, 144
p-BAPS. *See* Bis{4-(4-aminophenoxy) phenyl} sulfone (*p*-BAPS)
Peak heat release (PHR), 66, 143–144, 252–253
PEEK. *See* Poly(ether ether ketone) (PEEK)
PEEKOH groups. *See* Poly(ether ether ketone)hydroxyl groups (PEEKOH groups)
PEEKPN. *See* PN-terminated poly (ether ether ketone) (PEEKPN)
PEN. *See* Poly(arylene ether nitrile) (PEN)
2PEN-BPh, 24*f*
Phenolic curing agent, 11, 77, 80–81
Phenolic resin, 108
Phenols curing agents, 8–15, 10*f*
Phenoxy substituted imine, 210
1,2-Phenylenediamine, 152–153
Phenylethynyl
  groups, 131

Phenylethynyl (*Continued*)
reactivity, 131–133
Phenylnitrile-functional benzoxazine, 198–200, 199*f*
BZPN/phenylnitrile-functional benzoxazine, 198–201
DSC thermograms of, 200*f*
Phenylpolyhedral oligomeric silsesquioxane (POSS), 195
Phosphorous containing acids, 26–27
PHR. *See* Peak heat release (PHR)
Phthalocyanine, 11–12, 12*f*, 157
Phthalonitrile resins (PN resins), 1, 56–59, 107, 175, 242–243, 295–296, 348
actual and potential applications, 100–104
applications, 100–101, 101*f*
blends, 287–289
C_4-PN, 359
C_6-PN, 359
and curing behavior, 60, 66–67, 125, 305–306
curing behavior and mechanism, 5–50
diazonium salt, 137–138
early, 56–59
first-generation PN resins, 59–80
general structure of PN monomers, 3*f*
matrix, 322
monomers, 55–56, 60, 69–70, 175, 183, 278–279, 301–302
preparation by nitro displacement, 6*f*
monomers/oligomers, 120
nanocomposites, 304
oligomers, 80–81, 88, 113–114
overview and historical development, 1–5
phthalocyanine structure, 2*f*
PN blends/carbon fibers composites, 265–270
PN/bismaleimide, 217–222
PN/carbon fibers composites, 247–258
PN/CF composites, 269, 270*t*
PN/novolac, 210–217
PN/propargyl, 222–226
PN-based alloys, 176
PN-based materials, 55, 100–101, 175, 299–301
polymers, 55, 97
second-generation PN resins, 80–100
shear modulus and T_g variations, 59*t*
synthetic route of PN monomers with aliphatic spacer, 5*f*
Phthalonitrile-based blends and copolymers, 176–235
phthalonitrile/benzoxazine, 197–209
phthalonitrile/bismaleimide, 217–222
phthalonitrile/CTBN rubber, 233–235
phthalonitrile/epoxy, 182–197
phthalonitrile/novolac, 210–217
phthalonitrile/PEEK, 231–233
phthalonitrile/PEN, 226–231
phthalonitrile/phthalonitrile, 176–182
phthalonitrile/propargyl, 222–226
PN monomers, 175
polymeric
materials, 175
systems, 176
Phthalonitrile/Al_2O_3 nanocomposites properties, 328–341
Phthalonitrile/BN nanocomposites properties, 351–356
Phthalonitrile/epoxy, 182–197
blends, 183–192
functional-phthalonitrile/epoxy blends, 195–197
self-catalyzed, 192–195

Phthalonitrile/glass fibers composites, 274–282
  effect of postcure conditions, 275*t*
  properties comparison of BPh/GF composites, 277*t*
  smoke and combustion data of several GF-reinforced composites, 278*t*
  storage modulus and Tan Delta variations, 279*f*
Phthalonitrile/MAX phase nanocomposites properties, 362–368
Phthalonitrile/PEN, 226–231
Phthalonitrile/phthalonitrile, 176–182
  chemical structures of BPh, 177*f*
  complex viscosity variations over temperature, 179*f*
  storage modulus and loss factor variations, 181*f*
Phthalonitrile/Si_3N_4 nanocomposite properties, 322–327, 324*f*
Phthalonitrile/SiC microcomposite properties, 310–322, 311*f*
Phthalonitrile/TiO_2 nanocomposite properties, 341–348, 343*f*
Phthalonitrile/ZnO nanocomposite properties, 348–351, 350*t*
Phthalonitrile/ZrO_2 nanocomposite properties, 356–362
Pitch precursors, 244–245
Pitch-based CF remain, 245–246
PMR. *See* Polyimide resin (PMR)
PMR-15. *See* Polyimide resin-15 (PMR-15)
PN resins. *See* Phthalonitrile resins (PN resins)
PN-terminated poly(ether ether ketone) (PEEKPN), 232, 265–267
3PN3PEODPA oligomers, 133–134
3PNODPA monomers, 133–134
Polarization optical microscope (POM), 317, 317*f*
Poly(arylene ether nitrile) (PEN), 226–228
  PEN-OH, 291
  PN/PEN, 226–231
    thermal properties of BPh/PEN copolymers, 229*t*
Poly(arylene ether nitrile)-functional PN polymers, 166–170, 168*f*
  DSC curves, 169*f*
  TGA curves, 169*f*
Poly(ether ether ketone) (PEEK), 108, 164, 231–232, 265–266
  PEEK-like PN resin applications, 100–101, 102*f*
  PN/PEEK, 231–233
Poly(ether ether ketone)-functional PN, 164–166
Poly(ether ether ketone)hydroxyl groups (PEEKOH groups), 232, 265–266
Polyacrylonitrile (PAN), 244–245
Polybenzoxazine, 108, 140–141, 197–198
Polyimide resin (PMR), 1, 242–243
Polyimide resin-15 (PMR-15), 63–64
  PMR-15/IM-7 composites, 251–252, 251*t*
Polyimide(s), 1, 108, 118–119, 127–129, 129*f*
Polymer(s), 1, 32, 66, 304
  chemistry, 108
  composites, 305–306
  grafting, 309–310
  matrix, 242, 296–297
  nanocomposites, 298
  polymeric-based structures, 103–104
  structure, 70–71
  systems, 176
Polymerization, 56–57, 66–67, 110–111
  reaction, 10–11, 120–122, 138, 285–286

POM. *See* Polarization optical microscope (POM)
POSS. *See* Phenylpolyhedral oligomeric silsesquioxane (POSS)
Postcure conditions, effects of, 73–74, 74*t*
Postcure effect, 74–75, 252, 253*t*
Potassium carbonate (K_2CO_3), 38, 88, 109
Potassium hydroxide, 60
PPN. *See* 4-{3-(Prop-2-yn-1-yloxy) phenoxy} benzene-1,2-dicarbonitrile (PPN)
Prepegs consolidation technique, 274–275
Processability, 21–23, 41–42, 68–69, 82, 88, 127–129, 175–176, 178, 182–183, 185–187, 193, 197–198, 278–279
  bispropargyl ether of BPA, 222–223
  of blends, 194, 212
  blends, 286
  of epoxy, 190
  of monomers, 248
  of neat BPh resin and prepegs, 250
  PN monomers, 178, 231
4-{3-(Prop-2-yn-1-yloxy)phenoxy} benzene-1,2-dicarbonitrile (PPN), 268
Propargyl, 226
  PN/propargyl, 222–226
    DSC curves of neat BAPh and bispropargyl monomers, 225*f*
    PN cure reaction pathways in propargyl functions, 227*f*
    polymerization mechanism of propargyl ether functions, 224*f*
  propargyl-functional, 118
Pyridine, 161
  pyridine-functional PN, 161–163, 162*f*, 164*f*
2,6-Pyridinedicarbonyl chloride, 162
Pyrolysis, 64–65
Pyromellitic dianhydride, 123

R

Radiant heat fluxes, 212
Reactive carbon–cobalt nanocompound, 101–103
Reinforcing phase, 334–335, 351–352
Relaxation process, 320–321
Researched curing agents, 55–56
Resin infusion molding (RIM), 243, 276–277
Resin transfer molding (RTM), 19, 66–67, 214, 323–325
Resorcinol, 23, 23*f*
Resorcinol-based PN (RPh), 60–61, 61*f*
  monomers, 303–304
  polymers, 23, 154
  resin, 67
Rheological measurements, 93, 124, 131–133, 138–139, 154, 190
Rheometric analysis, 178
Rheometric studies, 19, 37
RIM. *See* Resin infusion molding (RIM)
Root-mean-square deviation, 333, 334*t*
Roving, 271
RPh. *See* Resorcinol-based PN (RPh)
RPPhPN. *See* Bis(3-(3,4-dicyanophenoxy)phenyl) phenyl phosphate (RPPhPN)
RTM. *See* Resin transfer molding (RTM)

S

Scanning electron microscopy (SEM), 260–261, 317–318, 318*f*. *See also* Transmission electron microscopy (TEM)
  images of neat BAPh resin, 336*f*

investigations, 326–327
observations, 335–336
Second-generation phthalonitrile resins, 56, 80–100, 356
mechanical and thermomechanical properties, 93–98
thermal resistance properties, 82–93
water absorption properties, 99–100
Self-catalyzed, PN monomers, 45–46
Self-catalyzed phthalonitrile monomers and oligomers, 29–50, 35*f*
Self-catalyzed phthalonitrile/epoxy blends, 192–195
Self-polymerization of allyl groups, 285–286
SEM. *See* Scanning electron microscopy (SEM)
Series model, 331, 333
S-glass GF, 271
Short carbon fibers, 245–246
Si_3N_4. *See* Silicon nitride (Si_3N_4)
SiC. *See* Silicon carbide (SiC)
Silane, 338
coupling agents, 305–306, 308*f*
general formula, 305–306, 306*f*
process of hydrolytic depositions, 307*f*
silane treated SiC particles, 314–315
Silica (SiO_2), 271
Silicon carbide (SiC), 296–297
microparticles, 322
properties of phthalonitrile/SiC microcomposites, 310–322, 311*f*
Silicon nitride (Si_3N_4), 296–297
properties of phthalonitrile/Si_3N_4 nanocomposites, 322–327, 324*f*
SiPNs, 39–41, 41*f*, 43*f*
Sluggish curing behavior, 55–56
Sodium hydroxide, 60
"Soft ceramics". *See* MAX phase
Sol–gel process(ing), 301–302, 302*f*
Solution blending technique, 301, 304–305
Solvent polymerization, 9, 17
Solvent resistance, 130–131
Stannous chloride, 7–8
Storage modulus, 127, 128*f*, 155*f*, 165*f*
Strain/stress curves, 359, 360*f*
of BAPh/MAX nanocomposites, 368*f*
Strong organic acid curing agents, 26–28
Superior flame resistance, 175, 182–183, 190–191, 217, 295–296
Surface modification of ceramic particles, 305–310
chemical treatments, 305–308
grafting of synthetic polymers, 308–310

T

T-300 8H CF, 265–266
T-300 CF. *See* Thornel 300 CF (T-300 CF)
T-700 CF, 263
Tan Delta
peak, 320–321
variations, 114, 115*f*, 256
TEM. *See* Transmission electron microscopy (TEM)
Tensile fractured surface of C_4-PN based nanocomposites, 360–361
Tensile test, 303–304, 331, 367
Tetrabutyl titanate, 301–302
Tetraethyloxysilane, 301–302
Tetrahydrofuran (THF), 265–266
TGA. *See* Thermogravimetric analysis (TGA)
Thermal behavior, 84–85, 87*t*

Thermal conductivity value, 352–353, 353*f*
Thermal decomposition process, 57
Thermal investigations, 351
Thermal oxidative stability, 124–125
Thermal properties, 71–72, 92, 114, 125*t*, 127–129, 304, 336–337, 340*t*, 341
  of BAPh/ZnO nanocomposites, 351*t*
  RPh Polymers, 70*t*
Thermal relaxation behavior, 151
Thermal resistance, 62–63
  properties, 60–71, 82–93
Thermal stability, 73–74, 108, 111–112, 125–127, 130*t*, 134–135, 145–146, 182–183
  investigations, 318–320, 319*f*
  MAX phase, 366–367
  parameters of BAPh/SiC composites, 320*t*
  of PN resins, 366–367
  by TGA, 118
Thermal stability parameters, 70–71, 70*t*, 71*t*, 86*t*
Thermo-oxidative behavior, 84–85, 87*t*
Thermo-oxidative stability tests, 60–61, 145–146
Thermogravimetric analysis (TGA), 10, 60–61, 125–127, 126*f*, 182, 219, 318–320. *See also* Dynamic mechanical analysis (DMA)
  curves, 62*f*, 65*f*, 83*f*, 85*f*, 89*f*, 134–135, 134*f*
  investigations of raw alumina, 338
Thermomechanical investigations, 351
Thermomechanical properties, 71–77, 72*t*, 93–98, 110–111, 163, 336–337, 340*t*
  of BAPh/ZnO nanocomposites, 351*t*
  first-generation resins, 71–77, 72*t*
Thermomechanical test shape requirements, 303–304
Thermosets, 1, 59
  cured, 93–95
  high-performance, 1, 217, 229
  matrices, 242–243
  PN, 28
  void-free, 303–304
THF. *See* Tetrahydrofuran (THF)
Thornel 300 CF (T-300 CF), 247–248
THR. *See* Total heat release (THR)
Titania, 296–297, 343
  crystalline peaks, 344–345
  nanoparticles, 341
Titanium oxide (TiO_2), 296–297
  BAPh/TiO_2 nanocomposites, 343, 344*t*
  properties of phthalonitrile/TiO_2 nanocomposites, 341–348, 343*f*
  TiO_2-based nanocomposites, 347–348
  X-ray diffraction patterns, 344*f*
p-Toluenesulfonic acid (TsOH), 27–28, 28*f*
  TsOH-based 11-PN polymer, 82
Total heat release (THR), 66, 143–144, 213*t*
TPnPh-POSS, 195
Transmission electron microscopy (TEM), 336. *See also* Scanning electron microscopy (SEM)
  analysis, 353–354
  images of nanocomposites, 354*f*
Triazine, 11–12, 12*f*, 157
Triazine-linked, PN/CF composites, 263, 264*f*, 264*t*
Trisilanol POSS, 195
TsOH. *See* *p*-Toluenesulfonic acid (TsOH)

U
Ultrahigh modulus, 247
US Navy flammability MIL-STD-2031 requirements, 277–278
UV shielding behavior, 296–297, 349
UV-Vis analysis, 10–11
UV–visible transmittance spectra, 349, 350*f*

V
Vacuum assisted resin transfer molding (VARTM), 19, 151
Vacuum infusion process (VIP), 243, 268
Vacuum-assisted RTM, 66–67, 323–325
VARTM. *See* Vacuum assisted resin transfer molding (VARTM)
Vertical burning test (UL-94), 190–191
Vinyl ester, 175–176
VIP. *See* Vacuum infusion process (VIP)
ViSiPN, 41–43, 42*f*

W
Water
  absorption properties, 78–80, 99–100
  repellent behavior, 78–79
  uptake
    properties, 56
    second-generation PN polymers, 99*t*
    test, 277–278

X
X-functional phthalonitrile/carbon fiber composites, 259–265
X-functional phthalonitrile/glass fiber composites, 282–286
X-functional PN materials, 49–50, 108
X-functional PN monomers and polymers
  alkene-functional PN, 109–112
  alkyne-functional PN, 112–118
  BIPN, 151–154
  bismaleimide-functional PN, 135–136
  BZPN, 140–151
  cyanate ester-functional PN, 155–158
  imide-functional PN, 118–135
  naphthyl-functional PN, 158–161
  novolac-functional PN, 136–140
  poly(arylene ether nitrile)-functional PN, 166–170
  poly(ether ether ketone)-functional PN, 164–166
  pyridine-functional PN, 161–163
X-ray diffraction patterns, 344–345
  of TiO_2, 344*f*
  of ZnO, 349*f*

Y
Young's modulus, 333
  experiment results, 331
  unfilled resin, 349–351

Z
Zinc oxide (ZnO), 296–297
  properties of PN/ZnO nanocomposites, 348–351, 350*t*
Zirconia, 296–297
  dispersion, 305
  nanoparticles, 296–297, 361–362
Zirconium oxide (ZrO_2), 296–297
  properties of PN/ZrO_2 nanocomposites, 356–362

Kolibri 360 GmbH
Ferdinand-Jühlke-Straße 7
99095 Erfurt, Deutschland
produktsicherheit@kolibri360.de